Sir Kenneth Green Library
All Saints

Please return or renew by the date
stamped below, or on recall

Telephone renewals: 0161 247 6130

Do not mark this book in any way. A charge will be made for any damage.

Natural Computing Series

Series Editors: G. Rozenberg
Th. Bäck A.E. Eiben J.N. Kok H.P. Spaink

Leiden Center for Natural Computing

Springer

Berlin
Heidelberg
New York
Hong Kong
London
Milan
Paris
Tokyo

Laura F. Landweber · Erik Winfree (Eds.)

Evolution as Computation

DIMACS Workshop, Princeton,
January 1999

With 127 Figures and 13 Tables

 Springer

Editors

Prof. Dr. Laura F. Landweber
Princeton University
Department of Ecology
and Evolutionary Biology
Princeton, NJ 08544
USA

lfl@princeton.edu

Prof. Dr. Erik Winfree
Caltech
Computer Science and
Computation and Neural Systems
Pasadena, CA 91125
USA

winfree@caltech.edu

Series Editors

G. Rozenberg (Managing Editor)
rozenber@liacs.nl
Th. Bäck, A.E. Eiben, J.N. Kok, H.P. Spaink

Leiden Center for Natural Computing
Leiden University
Niels Bohrweg 1
2333 CA Leiden, The Netherlands

Library of Congress Cataloging-in-Publication Data

Evolution as computation: DIMACS workshop, Princeton, January 1999/Laura
Landweber, Erik Winfree (eds.).
     p. cm. – (Natural computing series)
  Includes bibliographical references.
  ISBN 3540667091 (alk. paper)
   1. Evolutionary programming (Computer science) 2. Genetic algorithms. I.
Landweber, Laura F. (Laura Faye), 1967- II. Winfree, Erik, 1969- III. DIMACS
(Group) IV. Series.
  QA76.618.E84 2000
  005.1–dc21   00-039471

ACM Computing Classification (1998): F.1.1, F.2.2, I.2.8, J.3

ISBN 3-540-66709-1 Springer-Verlag Berlin Heidelberg New York

Springer-Verlag Berlin Heidelberg New York,
a member of BertelsmannSpringer Science+Business Media GmbH
http://www.springer.de

© Springer-Verlag Berlin Heidelberg 2002
Printed in Germany

Cover Design: KünkelLopka, Werbeagentur, Heidelberg
Typesetting: Camera ready by the author
SPIN 10723333   45/3142SR – 5 4 3 2 1 0 – Printed on acid-free paper

Introduction:
Evolution and Computation:
Where Do They Meet?

Erik Winfree and Laura F. Landweber

Evolutionary algorithms originated in the 1970s from the realization that computer programs can use the biologically inspired processes of genetic mutation, recombination, and selection to solve optimization problems that are analytically intractable. Successful applications to a wider gamut of problems became possible with the addition of new mechanisms such as generalized recombination, gene duplication, "developmental" evaluation of genomes, and unbounded genome size. Has this experience distilled concepts of importance for understanding biological evolution?

Meanwhile, biological research has gone far beyond the central results known in the 1970s. Many types of genetic modifications influence genome evolution, including variable-rate mutations, rearrangement by transposition, retroviral insertion, and editing. Our understanding of how cellular and developmental processes are programmed by genetic circuitry is yielding insights into the evolution of evolvability. Has this wealth of experience uncovered processes of importance to computer science and in turn shed light on the computational properties of cells?

These were the questions in our minds as we were organizing the DIMACS Workshop on Evolution as Computation. Our goal was to draw together computer scientists and molecular biologists to explore the combinatorial processes that drive evolution.

The lead article by **Jim Shapiro** raises one of the central issues this workshop explored: biological genetic systems are more complex than standard dogmas presume. The subtext – and sometimes overt text – of much modern work on evolution is that independent random mutations in the genome are the source of genetic variation upon which selection acts: there is no "genetic intelligence" that picks just the right mutation to remedy today's challenge. However, as Shapiro's chapter outlines, the truth lies somewhere in between: a host of enzymes actively control the rates and mechanisms of mutations and genome rearrangements. Shapiro concludes, "it is now plausible at the molecular level to conceive of concerted, non-random changes in the genome guided by cellular computing networks during episodes of evolutionary change. Thus, just as the genome has come to be seen as a highly sophisticated information storage system, its evolution has become a matter of highly sophisticated information processing." How can we understand the evolutionary consequences of these cellular algorithms for genetic change?

A similar question is a pressing challenge for computer science researchers studying evolutionary algorithms – a class of algorithms designed to exploit the principles of Darwinian evolution to solve mathematical optimization problems (reviewed by **Thomas Bäck, Joost Kok, and Grzegorz Rozenberg**). Evolutionary algorithms maintain a population of digital "genomes," each representing a potential solution to the problem. Each genome's "fitness" is a mathematical function to be optimized. Improved solutions are found by subjecting the population of genomes to repeated rounds of selection, in which only the most fit survive, followed by reproduction, which introduces genomic variation. In the simplest version, which reflects the most common hypothesis of biological evolution, mutations arise independently at each genomic locus, and each locus affects the fitness independently.

Concrete algorithms for evolutionary optimization must make each of these somewhat fuzzy notions precise – making possible a rigorous understanding of how each detail of the algorithm affects the evolutionary process. For example, how do mutation rates, population size, and selection pressure affect the speed of evolutionary optimization? An understanding of these issues is important not only for practitioners of evolutionary algorithms, who want fast results for practical problems, but also for those who study biological evolution, who may be interested in why viruses often have increased mutation rates and "hot spots" or in how the immune system can most rapidly identify and select the proper response to an antigen. This latter system is studied in this volume by **Patricia and Theodore Theodosopoulos**, who consider a trade-off between the safety of small mutational steps vs. large steps that may lead more quickly to a solution. **Thomas Bäck, Joost Kok, and Grzegorz Rozenberg** review more formal theoretical results for the simplest versions of evolutionary algorithms, and they go on to study empirically whether the predicted advantages of large populations hold true for more complex optimization problems. Extending the theoretical analysis to moderate forms of cooperativity among genetic loci, **James Crutchfield and Erik van Nimwegen** develop an analytic framework, called statistical dynamics, for studying epochal evolution driven by point mutations.

Unfortunately, few analytic results are available for more complex and realistic situations, because real-world problems – both in biology and in engineering – often can't be phrased in terms of independent or mildly linked factors. A change that makes one thing better may make something else worse. Consequently, the application of evolutionary algorithms to real-world problems has been largely an empirical art. Many extensions to the simple evolutionary algorithm discussed above have been tried – and some have proved to be essential. Consider the analog electronic circuit design problem tackled by **John Koza, Forrest Bennett, David Andre, and Martin Keane**. Their genetic programming algorithms have produced impressive results – creatively coming up with the same designs that were, in fact, patented in the last century. Two choices were key to their success: (1) they used a highly "evolvable" representation of digital circuits, where each digital genome specified a developmental program for growing the circuit, and (2) they used forms of variation, including mutation and recombination, uniquely suited to their representation of programs as LISP expressions.

If representation and mechanisms of variation are so critical for applying digital evolution to difficult optimization problems, so too must they be critical for the even more difficult problems surmounted by natural evolution. Is this what we are seeing in the complex genome system architecture and natural genetic engineering described by Shapiro? The relationship of representation and variation will be an exciting area for future exploration, both *in vivo* and *in silico*.

Several speakers at this conference took a deeper look at the representations used in biological organisms and at how they evolve. The genetic code – the relation of DNA sequence to protein sequence – is perhaps the most fundamental of all biological codes, and its origin still remains an enigma after roughly 35 years of research. **Stephen Freeland** and **Guy Sella and David Ardell** give complementary views of how the code's present structure was shaped by natural selection. Their premise is that, to be an "evolvable" representation, the genetic code must be arranged to minimize the negative effects of genetic errors (due to either mutation or mistranslation). Both groups found that the "universal" genetic code – the one found in nearly every organism on earth – falls in the best 0.0001% of all possible triplet codes, according to their criteria. It is also clear, as reviewed by **Robin Knight**, that biochemical, historical, and metabolic factors have together influenced many aspects of the genetic code.

Moving up a level, **Mark Ptashne and Alexander Gann** ask what architectures for genetic regulatory networks have the flexibility to evolve efficiently? They propose that the many weak interactions among regulatory proteins upstream from the gene are particularly evolvable. For example, rearranging the localization of binding domains – a form of variation that crossover events during recombination are particularly suited to producing – smoothly changes the regulatory effect, resulting in new regulatory functions.

The ability to generate new – but not disastrously fatal – variations by rearranging the order of genetic elements is demonstrated dramatically by work on the bacteriophage T7 by **Drew Endy** and colleagues. Using computer simulation based on detailed knowledge of the phage's genome and all the proteins involved in the phage's life cycle, they examined how the order of the genetic elements (genes, promoters, terminators, etc.) in the genome affects the rate of viable virus production. The simulations indicated that the genetic program is not brittle – most variations worked, to some degree – and that the wild-type genome is in at least the best 2% of all possible permutations of its 122 genetic elements. They were then able to go back into the laboratory to generate, by artificial genetic engineering, several of the positional mutants studied in the simulations, thus providing a real-world test of the simulation's results. This research, coupling detailed simulations with systematic experimental studies of organisms with synthetic or modified genomes, may develop into "a predictive, system-level biology, grounded at the genetic level." To this end, **Roger Brent** describes new technology to dissect genetic regulatory networks using molecules that they term "peptide aptamers," which disrupt protein–protein interactions in a cell by competitive binding. Such tools also can be used to test the functional significance of protein polymorphisms by interacting with and inactivating specific allelic forms.

Genome rearrangement occurs in all organisms by a variety of mechanisms and at many different scales. In this regard, some organisms are more equal than others. A pair of papers by **Andrzej Ehrenfeucht, David Prescott, and Grzegorz Rozenberg** and by **Laura Landweber and Lila Kari** juxtapose two different computational views of the spectacular gene rearrangements that take place in ciliates, a genetically sophisticated group of unicellular eukaryotes. These organisms posess two nuclei: the transcriptionally active *macronucleus* and the functionally inert germline *micronucleus*. During the reproductive cycle, the macronucleus is generated by extensive splicing and rearrangements of the micronuclear source code; in some cases a single gene must be assembled (unscrambled) from over 50 pieces in the micronucleus. This is an example of the kinds of complex algorithms that can be involved in directing genetic change. We have yet to understand how, or whether, these complex genetic manipulations help ciliates evolve – but we now have some hints. Ptashne and Gann's and Endy's studies show how recombination and transposition may be ideally suited to evolve genetic regulatory circuits; Koza and colleagues' novel digital recombination is fruitfully matched to developmental LISP programs; might ciliates' genetic engineering similarly produce a form of variation uniquely adept at evolving the cell's genetic programs?

Is the complexity we see in natural genetic engineering just due to baroque physical phenomena, or must we concede that purposeful computation – information processing – is going on? **Ron Weiss, George Homsy, and Tom Knight** address the converse question: could natural genetic mechanisms be used to build *in vivo* digital circuits for computations of our own choice? These authors propose using the splicing and dicing of elements from natural genetic regulatory networks – promoter and operator sequences, genes for DNA binding proteins – to create synthetic circuits where signals are represented by protein concentrations instead of electrical voltages. They analyze a simple logical element, the inverter, out of which arbitrary digital circuits can be built, and argue that the tools are now available for experimental implementations in bacteria. It follows that evolution is exploring a space of possible genomes equivalent to, or at least containing, all possible computer programs.

Can we attribute aspects of higher-level system organization to evolutionary pressures on how the cell computes? **Charles Ofria and Chris Adami** take as an example the remarkable fact that some phage genomes contain overlapping reading frames – where a single gene locus codes as many as three distinct proteins encoded by the same DNA sequence. Creating an artificial world populated by digital "organisms" (programs) that reproduce themselves and compete for resources, they observe a similar phenomenon: parts of an organism's code may evolve to be used for multiple purposes. In their digital world, they can easily track statistics, and thus evaluate the advantages (multiple-purpose code results in smaller, faster-reproducing organisms) and disadvantages (multiple-purpose code makes beneficial mutations less likely) as they play out during the course of evolution.

Eric Baum and Igor Durdanovic – in contrast to other authors in this volume – argue that biological evolution is a poor choice for computer science to emulate: runaway biological evolution can lead to non-adaptive traits, such as the peacock's tail, and thus cannot be relied on for global optimization. Instead, economic models, where one can impose rules designed to ensure that competition results in added value, motivate Baum's algorithms for machine learning.

What will it take to understand evolution acting on programs? It is certainly very different from the picture of evolution acting on independent traits that we considered earlier. Correlations, from the very direct and immediate to the very subtle and delayed, are the very stuff that computations are made of. The experience of computer scientists working with evolutionary algorithms suggests that the classical assumption of variation restricted to point mutations and crossover during recombination is not sufficient, or at least is not efficient, for evolving the hierarchical architectures that computer scientists and nature itself must both struggle with. More complex forms of variation are needed, specifically suited to the structures being evolved. This may be one role of natural genetic engineering.

The papers found in this book were submitted by speakers at the Workshop for Evolution as Computation shortly after the conference; they appear here essentially unchanged in content. Many thanks go to Dr. Hans Wössner and Ingeborg Mayer at Springer-Verlag for careful editorial comments and persistent encouragement, and especially to Christine Ortega and Karolyn Knoll for their help bringing this project to fruition.

Contents

Authors' Affiliations

Christoph Adami
Digital Life Laboratory
Jet Propulsion Laboratory
4800 Oak Grove Drive
Pasadena, CA 91109
USA
adami@caltech.edu

David Andre
Division of Computer Science
University of California
Berkeley, CA 94720
USA
dandre@cs.berkeley.edu

David H. Ardell
Department of Molecular Evolution
Uppsala University
Norbyvägen 18C
751 24 Uppsala
Sweden
dave.ardell@ebs.uu.se

Thomas Bäck
CTO and Vice President
NuTech Solutions, Inc.
NuTech Solutions GmbH
Martin-Schmeisser-Weg 15
44227 Dortmund
Germany
Thomas.Baeck@nutechsolutions.com

Eric B. Baum
NEC Research Institute
4 Independence Way
Princeton, NJ 08540
USA
eric@research.nj.nec.com

Forrest H. Bennett III
Chief Scientist
Genetic Programming Inc.
Box 1669
Los Altos, CA 94023
USA
forrest@evolute.com

Roger Brent
The Molecular Sciences Institute
2168 Shattuck Avenue
Berkeley, CA 94704
USA
brent@molsci.org

James P. Crutchfield
Santa Fe Institute
1399 Hyde Park Road
Santa Fe, NM 87501
USA
chaos@santafe.edu

Igor Durdanovic
NEC Research Institute
4 Independence Way
Princeton, NJ 08540
USA
igor@uni-paderborne.de

Andrzej Ehrenfeucht
Department of Computer Science
Engineering Center
ECOT 717, 430 UCB
University of Colorado at Boulder
Boulder, CO 80309
USA
andrzej@cs.colorado.edu

Drew Endy
Department of Biology
Massachusetts Institute of Tech.
31 Ames Street, Room 68-132
Cambridge, MA 02139
USA
endy@MIT.edu

Stephen J. Freeland
Department of Biological Sciences
University of Maryland
1000 Hilltop Circle
Catonsville, MD 21250
USA
freeland@umbc.edu

Alexander Gann
Cold Spring Harbor Laboratory
1 Bungtown Road
Cold Spring Harbor, NY 11724
USA
gann@cshl.edu

George E. Homsy
Artificial Intelligence Laboratory
Massachusetts Institute of Tech.
Boston, MA 02139
USA
ghomsy@ai.mit.edu

Lila Kari
Department of Computer Science
University of Western Ontario
London, ON, N6A 5B7
Canada
lila@csd.uwo.ca

Martin A. Keane
Chief Scientist
Econometrics Inc.
111 E. Wacker Dr.
Chicago, IL 60601
USA
makeane@ix.netcom.com

Robin D. Knight
Department of MCD Biology
University of Colorado
Boulder, CO 80309
USA
rob@spot.colorado.edu

Thomas F. Knight, Jr.
Artificial Intelligence Laboratory
Massachusetts Institute of Tech.
Boston, MA 02139
USA
tk@ai.mit.edu

Joost N. Kok
LIACS, Leiden University
for Natural Computing
Niels Bohrweg 1, 2333 CA Leiden
The Netherlands
joost@liacs.nl

John R. Koza
Consulting Professor
Section on Medical Informatics
Department of Medicine M.C.5479
Medical School Office Building
251 Campus Drive
Stanford University
Stanford, CA 94305
USA
koza@stanford.edu

Laura F. Landweber
Department of Ecology
and Evolutionary Biology
Princeton University
Princeton, NJ 08544
USA
lfl@princeton.edu

Erik van Nimwegen
Center for Studies in Physics
and Biology
The Rockefeller University
1230 York Ave. Box 75
New York, NY 10021
USA
erik@golem.rockefeller.edu

Charles Ofria
Department of Computer Science
and Engineering
Michigan State University
3115 Engineering Building
East Lansing, MI 48824
USA
ofria@cse.msu.edu

David M. Prescott
Department of MCD Biology
University of Colorado
Boulder, CO 80309
USA
prescotd@spot.colorado.edu

Mark Ptashne
Molecular Biology Program
Sloan-Kettering Cancer Institute
New York, NY 10021
USA
M-Ptashne@ski.mskcc.org

Grzegorz Rozenberg
LIACS, Leiden University
Advanced Computer Science
Niels Bohrweg 1, 2333 CA Leiden
The Netherlands
rozenber@liacs.nl

Guy Sella
Department of Applied Mathematics
and Computer Science
The Weizmann Institute
Rehovot 76100
Israel
sella@wisdom.weizmann.ac.il

James A. Shapiro
Department of Biochemistry
and Molecular Biology
University of Chicago
920 E. 58th Street
Chicago, IL 60637
USA
jsha@midway.uchicago.edu

Patricia K. Theodosopoulos
185 Freeman Street, 743
Cambridge, MA 02446
USA
ptaetheo@rcn.com

Theodore V. Theodosopoulos
185 Freeman Street, 743
Cambridge, MA 02446
USA
ted_theodosopoulos@msn.com

Ron Weiss
Department of Electrical Engineering
Princeton University
Princeton, NJ 08544
USA
rweiss@princeton.edu

Erik Winfree
Departments of Computer Science and
Computation and Neural Systems
California Institute of Tech.
Pasadena, CA 91125
USA
winfree@caltech.edu

Genome System Architecture and
Natural Genetic Engineering

James A. Shapiro

Abstract. Molecular genetics reveals three aspects of genome organization and reorganization that provide opportunities for formulating new views of the evolutionary process:

1. Organization of the genome as a hierarchy of systems (not units) determining many aspects of genetic function (only some of which are specifying protein and RNA sequences);
2. The presence of many repetitive DNA elements in the genome which do not encode protein or RNA structure but serve as the physical basis for functional integration; and
3. The operation of regulated cellular natural genetic engineering systems capable of rearranging basic genomic components throughout the genome in a single cell generation.

Thus, concerted, non-random changes in the genome guided by cellular computing networks are plausible at the molecular level.

1   Introduction

As we begin to apply information science approaches to evolutionary questions, it is essential to ask a foundational question: Can we use the standard repertoire of basic concepts in genetics, based largely on Mendelian studies in the first half of this century, or do we need to formulate new fundamental principles based on the last half century of molecular genetics? The argument of this paper is that new concepts are necessary because molecular genetics has revealed phenomena completely unanticipated by classical genetics and has provided a number of computation-like insights into questions of genome organization, genome expression, and genome reorganization (Shapiro, 1999).

It is useful to think of the genome as an information storage organelle integrated into the life of each cell. A primary genomic function is to encode proteins and RNA molecules that do many diverse jobs in the cell. This coding information must be dynamically accessible at the right times and in the right amounts so that complex cellular programs are executed properly. Each genome contains information for many programs: cellular housekeeping functions, emergency responses such as damage repair, formation of differentiated cell types during development, and even formation of distinct organisms as part of a complex life cycle (e.g., the same genome gives rise to a caterpillar and a butterfly). The need to access, coordinate, and distinguish between these various classes of coding information means that cells have systems for addressing different regions of the genome. The addressing system depends upon a distinct class of genetic information that must be repetitive in nature because it marks multiple regions of the genome to be expressed or to be silent at the same times. The genome must also have a high degree of physical organization, and that too depends on repetitive information indicating how the DNA is to be compacted.

Functionally, the genome needs to do more than express protein and RNA coding information. It has to be replicated, proofread so that errors and damage can be corrected, distributed equitably to daughter cells at cell division, and reprogrammed when necessary (i.e., the sequence information has to be altered or rearranged). Sometimes reprogramming is part of the normal organismal life cycle and sometimes it happens episodically under emergency conditions (McClintock, 1984).

2   The Conceptual Legacy of Molecular Genetics

A series of historical benchmarks provides the context for understanding why molecular genetics leads to new ways of thinking about genomes and evolution. Darwin's *On the Origin of Species by Means of Natural Selection* appeared in the middle of the 19th Century, and Mendelism was rediscovered in 1900. The next four decades saw the development of classical genetics and the formulation of the neo-Darwinian Modern Synthesis to explain evolution based on random mutation to generate new forms of independent Mendelian units (genes). About midcentury, we learned that DNA carries hereditary information (Avery *et al.*, 1944) and that its structure facilitates both replication and coding (Watson and Crick, 1953). The operon theory (Jacob and Monod, 1961) taught us that "genes" are actually divisible systems composed of protein-coding information ("structural genes") and cis-acting control sites in the DNA. These sites are recognized by regulatory proteins to ensure proper expression of the cognate structural genes. The importance of recognizing control sites as a new kind of genetic determinant cannot be overemphasized. The sites were soon discovered to be shared between multiple genetic loci, creating multilocus genomic systems for coordinately regulated expression, and they served as the prototype for other repetitive cis-acting sites that play important roles in replication, repair, DNA compaction, and chromosome segregation, as well as gene expression. A few years later, the importance of repetitive DNA elements as major quantitative components of genomes was recognized by physical methods (Britten and Kohne, 1968). Another midcentury discovery of the highest significance for thinking about evolution was the demonstration that cells contain built-in agents of genome restructuring which can alter patterns of gene expression (McClintock, 1951, 1987). Molecular studies amply confirmed McClintock's initially controversial findings and showed that all cells possess biochemical systems for genome restructuring (Bukhari *et al.*, 1976; Shapiro, 1983; Berg and Howe, 1989; Sherratt, 1995). Collectively, these discoveries set the stage for thinking of genomes as hierarchically integrated systems capable of biologically controlled change rather than as collections of autonomous genetic units subject to individual evolution by random variation.

3   Systems Rather Than Units

The molecular genetic dissection of individual genetic loci is exemplified by the history of the *E. coli lac* operon, whose analysis was the basis of the operon theory. The locus that started out as a point on the genetic map of *E. coli* in the late 1940s has today become a composite of interacting regulatory sites and coding sequences (Reznikoff, 1992; Shapiro, 1997). Without the proper arrangement of these diverse elements, the proteins involved in lactose metabolism would not be expressed under the appropriate conditions (i.e., in the absence of glucose and the presence of lactose or some other inducer). By virtue of its transcriptional promoter region, which contains a binding site for the cyclic AMP (cAMP) receptor protein, the *lac* operon is integrated into a distributed system of catabolic systems, all of which contain a similar binding site and are thus regulated by the abundance of cAMP in the cell (Saier *et al.*, 1996). In *E. coli* cells, elevated cAMP levels serve as a chemical representation of the absence of glucose in the growth medium.

It is worth mentioning that not only is the *lac* operon required for *E. coli* to grow on lactose as a growth substrate, but so are the *cya* locus, which encodes adenylate cyclase (the enzyme that produces cAMP from ATP), and *crp*, which encodes the cAMP receptor protein in addition to the genetic loci encoding the proteins of glycolysis. Thus, a simple phenotype, lactose utilization, results from expression of multiple genetic loci, each of which is itself a system of transcriptional signals and coding sequences. When we consider more complex phenotypes, such as synthesis of a cellular organelle or development of a particular multicellular structure, then the sequentially coordinated involvement of many genetic loci becomes an obvious necessity (Monod and Jacob, 1961).

Even individual protein-coding sequences are no longer viewed as units but as systems of distinct domains, each encoding a separate structural feature or functional region of the protein (Doolittle, 1995). For example, the LacI repressor protein has distinct domains for DNA binding, inducer binding, and multimerization of the subunits (Miller, 1996). Such domains can be separated genetically and combined with new domains to encode hybrid proteins with novel functional characteristics. An example of such genetic engineering of protein domains is the widely used "two-hybrid system" designed to search for physical interactions between proteins. This system works by stitching on the appropriate domains to convert one interacting protein into a DNA-binding molecule and its partner into a transcriptional activator so that, attached to each other, they reconstitute a fully functional positive transcription factor (Fields and Song, 1989). Computer searches of newly discovered proteins generally do not seek matches of the entire sequence but rather look for signatures of specific domains which give clues to the various functionalities linked together in the molecule.

4   Repetitive DNA and Genome System Architecture

The change from thinking of the genome as a collection of autonomous units to conceptualizing it as an interactive, hierarchically organized system of systems, much like the software needed to run a computer, implies that genomes contain a multiplicity of codes for the various aspects of genomic function (Trifonov and Brendel, 1986):

- Protein synthesis and processing
- Transcription and RNA processing
- DNA replication and chromosome segregation
- Chromatin structure
- DNA rearrangement
- Methylation and other epigenetic modification (imprinting)

These codes are carried in large part by cis-acting DNA sites (repetitive sequence motifs) which serve as recognition elements to integrate the activities of transcription, replication, imprinting, and rearrangement complexes at distinct genomic regions.

Distributed recognition sites constitute one class of repetitive DNA elements within the genome, but there are many other classes as well. These include larger repeated structures dispersed throughout the genome, such as mobile genetic elements (Shapiro, 1983; Berg and Howe, 1989; Sherratt, 1995) and SINES (small interspersed nucleotide elements) that are reverse transcripts of various nuclear RNA sequences (Deininger, 1989). In mammals, the SINE elements are major genomic constituents, and human genomes contain over 500,000 *Alu* elements derived from 7S RNA.

Other classes of repetitive DNA sequences are clustered at a few locations in the genome and often comprise hundreds or thousands of tandem repetitions of DNA motifs that range from a very few base pairs up to several hundred base pairs (satellite DNA, Sagot and Myers, 1998). Cytologically, these clustered tandem arrays can be seen to have different compaction and replication properties from the main sections of the chromosomes ("euchromatin") and have been termed "heterochromatin" by cytogeneticists (Weiler and Wakimoto, 1995).

Although there is a widespread belief that repetitive DNA has no function, since it does not code for protein structure, there is an extensive literature documenting many roles for repetitive DNA elements and heterochromatin.

Clustered repeats have major effects on gene expression and genome distribution to progeny cells. Three examples illustrate this. The phenomenon known as "position effect" in *Drosophila* reflects the effects that proximity to heterochromatic regions has on the expression of genetic loci normally located in euchromatin. Heterochromatin suppresses their expression during development and sometimes creates quite special phenotypic patterns of expression, such as eyes with pigment only in certain sectors (Weiler and Wakimoto, 1995; Henikoff, 1992; Karpen, 1994). Many organisms have large tandem arrays surrounding their centromeres (Csink and Henikoff, 1998), and when placed at new locations, these arrays can help organize the formation of new centromeres, as in human artificial chromosomes (Harrington *et al.*, 1997). During meiosis in the *Drosophila* germ line, chromosome pairing is necessary for proper chromosome segregation; when pairing by homology and crossing over does not occur, proper distribution is assured by blocks of repetitive DNA (McKee *et al.*, 1992; Karpen *et al.*, 1994; Demburg *et al.*, 1996).

Given the important roles that have been documented for both dispersed and clustered repetitive DNA elements, it is likely that changes in the distribution of repetitive DNA are critical events in evolutionary change.

Amplifying on the argument that repetitive elements provide hierarchical addressing for the genome, we can hypothesize that the distribution of repeats organizes the functional "system architecture" of each genome (Shapiro, 1991, 1999). Alteration of system architecture governing how the various regions of the genome function during cellular and organismal life cycles may well be more important for the speciation and evolution of new forms than changes in individual proteins, which are largely shared between distinct organisms (Duboule and Wilkins, 1998). This hypothesis predicts that repetitive DNA will be more highly specific taxonomically than protein-coding sequences, and that seems to be the case (Dover, 1982; Epplen *et al.*, 1994; Elder and Turner, 1995). Sibling species have indistinguishable phenotypes and often hybridize to produce viable but sterile progeny; thus, their protein complements are isofunctional. However, such species often differ sharply in their complements of repetitive DNA (Beerman, 1977; Dowsett, 1983). Mammals all share the same complements of proteins, but each mammalian order has its own complement of highly repetitive SINE elements (Deininger, 1989) and centromeric repeats (Willard, 1990).

5 Natural Genetic Engineering

A second requirement of the idea that evolution involves changes in genome system architecture is that cells possess the capacity to carry out the kind of cut-and-splice natural genetic engineering needed to reorganize repetitive and coding sequences so that they can construct novel functional genomic systems. This capacity could be inferred from McClintock's discovery of "controlling elements," which can move from one site to another in the genome and alter chromosome structure and patterns of gene expression (McClintock, 1951, 1965, 1978, 1987). She emphasized the ability of these elements to build up systems that connected genetic loci at more than one site in the genome (McClintock, 1956). Molecular genetic analysis of mobile elements in all organisms has amply confirmed the generality of McClintock's observations and revealed a variety of biochemical mechanisms by which DNA sequences can spread throughout the genome (Bukhari *et al.*, 1977; Shapiro, 1983; Berg and Howe, 1989; Sherratt, 1995). It is particularly important to note that certain mechanisms, such as gene conversion, replicative transposition at the DNA level, and retrotransposition via RNA intermediates inherently amplify the number of copies of a particular mobile DNA element and can thus lead to rapid accumulation of repeated sequences in the genome. All mobile elements that have been characterized contain regulatory motifs, such as promoters, enhancers, terminators, and splice sites, so that they alter the control of adjacent DNA sequences when they insert at new locations (e.g., Errede *et al.*, 1981). The evidence that insertions of mobile elements have created new regulatory configurations during evolution is accumulating (e.g., Brosius, 1991; Britten, 1997). In thinking about how evolution can bias the potential for useful outcomes to the action of mobile elements, we should keep in mind the fact that the movement of defined segments of DNA containing specific regulatory sequences is in itself a highly nonrandom process.

In many organisms, natural genetic engineering has been incorporated into the normal life cycle. In the mammalian immune system, the problem of using a finite amount of DNA to encode a virtually infinite repertoire of antigen-binding molecules (immunoglobulins and T cell receptors) is solved in the appropriate lymphocyte lineages by activating the construction of functional protein-coding loci from a family of smaller genomic segments (Blackwell and Alt, 1989). In addition to multiplying the number of final coding sequences by joining different combinations of segments, the cutting and splicing events are flexible and include the ability to incorporate newly synthesized untemplated DNA sequences. Thus, any two segments can be joined to encode many distinct amino acid sequences, tremendously amplifying immunoglobulin and T cell receptor diversity. In the case of newly constructed determinants that encode antigen-binding immunoglobulins, two further genetic engineering events occur as the lymphocytes develop: (1) a distinct class of DNA rearrangement event connects the "variable region," determining antigen specificity, to different "constant regions" that determine the localization and function of different immunoglobulin classes (IgM, IgD, IgA, IgG), and (2) a poorly understood process called somatic hypermutation specifically induces multiple nucleotide substitutions into the variable region (without modifying nearby flanking sequences) to generate mutant antibodies, which can then be selected for higher antigen affinity. The immune system thus illustrates a developmentally regulated cascade of natural genetic engineering processes to solve a specific biological problem.

A second case of natural genetic engineering as part of the normal life cycle involves the ciliated protozoa, the organisms in which catalytic RNA (Cech, 1983) and telomerase (Blackburn, 1991) were discovered. These protozoa are single cells with two distinct genomes: a transcriptionally silent diploid germline genome containing large chromosomes enclosed in a small "micronucleus" and a transcriptionally active polyploid somatic genome containing many short minichromosomes enclosed in a large "macronucleus" (Prescott, 1992, 1997, this volume). Following each mating, a new zygote micronucleus is formed in the exconjugant cell. Before the newly mated cell undergoes mitosis, the zygote micronucleus divides, the old micro- and macronuclei in the cell degenerate, and one of the daughter micronuclei develops into the new zygote macronucleus. Macronuclear development comprises multiple DNA processing events that include endoreplication of the germ line chromosomes, fragmentation of the chromosomes into thousands of segments, removal of most fragments, reassembly of the remaining fragments into functional coding units, and capping of the reassembled minichromosomes with telomeres. Reassembly of fragments into functional coding units is a complex process of genetic rearrangement. Individual protein determinants can have many discrete segments in the micronuclear chromosomes that have to be reordered to produce a functional coding sequence in a macronuclear minichromosome (Prescott, 1992, 1997). The ciliates thus demonstrate that living cells are capable of genome-wide reorganization in a single cell generation, massively fragmenting the germ-line genome and correctly stitching together the appropriate DNA segments to construct a functional somatic genome.

6   Regulatory Circuits and Control of Natural Genetic Engineering Activities

DNA rearrangements that have become part of the normal life cycle must be subject to a very high degree of biological regulation. Otherwise, they would not occur at the proper time, in the proper cells, and at the proper sites in the genome to fulfill their functions. In contrast, the natural genetic engineering events in evolution that create novel genome architectures and genomic determinants for new phenotypes are not programmed but occur episodically. This does not mean, however, that they are unresponsive to biological control circuits. In fact, the study of the biochemical mechanisms underlying genetic change reveals multiple connections to regulatory systems, as is true of all cellular activity. The ability to respond to biological inputs and activate natural genetic engineering functions means that genetic variability can change. This solves the superficial paradox between organisms displaying genomic stability during normal reproduction and organisms with a fluid genome at moments of crisis, containing internal agents of change that may create useful novelty. We increasingly view cellular regulatory circuits (signal transduction networks) as computational systems capable of evaluating multiple cellular inputs and making decisions about which biochemical or biomechanical functions to activate and inactivate (e.g., Bray, 1990). Thus, it is central to the theme of *evolution as computation* to examine some of the ways that natural genetic engineering functions can be controlled.

At the first level, the timing and degree of activity of natural genetic engineering systems respond to environmental and physiological inputs. In bacteria, these inputs include starvation (Foster, 1993; Shapiro, 1997), irradiation (Walker, 1996), and temperature (van der Lelie *et al.*, 1992). A number of DNA mobilization functions are activated (at least in part) by intercellular signalling molecules, such as DNA uptake in bacteria (Grossman, 1995; Gwinn *et al.*, 1996) and Ty3 transposition in yeast (Kinsey and Sandmeyer, 1995). In mammals, Mouse Mammary Tumor Virus (MMTV) and related human endogenous retroviruses respond to steroids and other growth factors, both positively and negatively (Cato *et al.*, 1989).

Another important activating factor for several classes of transposable and retrotransposable elements is mating between distinct stably interbreeding populations. It makes sense to assume that inability to find a mate from the parental population is the kind of crisis situation that can lead to an episode of evolutionary change. Interpopulational mating frequently derepresses the activity of mobile elements in the chromosomes of one population and leads to the phenomenon known as hybrid dysgenesis (Bregliano and Kidwell, 1983; Engels, 1989; Finnegan, 1989). As exemplified by *Drosophila*, genetic change in hybrid dysgenesis is limited to germline cells and involves both chromosome rearrangements and the accumulation of many new insertions into the genome (Engels, 1989). These rearrangements occur during the premeiotic development of the germline, with the consequence that several mitotic descendants of the original mutated cell can undergo meiosis to produce multiple gametes sharing the same constellation of distributed genetic changes. Thus, the progeny of a single dysgenic fly can form an interbreeding population characterized by shared multilocus alterations in the genome. In other words, hybrid dysgenesis illustrates the molecular and biological basis for so-called macromutations during evolution.

At a second level, molecular genetic analysis of all mobile genetic elements reveals many connections to the molecular events and signal transduction systems which regulate transcription and other cellular biochemical processes. For example, bacterial transposable elements respond to controls by the DnaA chromosome replication initiation protein and DNA methylation (Mahillon *et al.*, 1998), the stationary phase sigma factor RpoS (Gomez-Gomez *et al.*), regulatory proteases and the cAMP-catabolite activation system (Shapiro, 1993; Lamrani *et al.*, 1999); yeast retrotransposons respond to the mating pheromone protein kinase cascade and to mating-type transcriptional regulators (Errede *et al.*, 1981; Kinsey and Sandmeyer, 1995); Drosophila transposons are subject to cell-type specific splicing control (Rio,

1991); and both maize transposons and mammalian retroviruses are regulated by DNA methylation (Fedoroff, 1995; Simon *et al.*, 1983).

In addition to quantitative and temporal regulation, we are beginning to learn about a third level of regulation, *viz.* that molecular systems which control gene expression can also influence the targeting of mobile elements within the genome. One of the first indications for targeting was the specificity of yeast Ty retrotransposon insertions near the start sites of PolII (mRNA) or PolIII (stable RNA) transcripts (Eibel and Philippsen, 1984; Ji *et al.*, 1993). The beginnings of tRNA transcripts were particular hot spots for the Ty elements of *Saccharomyces cerevisiae*, and it was subsequently demonstrated in vitro that PolIII transcription factors could direct the insertion of Ty3 to a tRNA start site (Kirchner *et al.*, 1995). Transcription factor targeting in yeast is a general phenomenon, not a peculiarity of promoter regions, because the Ty5 element has quite a distinct specificity: rather than insert immediately upstream of actively transcribed sequences like Ty1–4, Ty5 has an almost total preference for transcriptionally silenced regions, indicating that its integration apparatus interacts with the negatively acting transcription factors involved in silencing (Zou *et al.*, 1996). *Drosophila* genetic engineering using the hybrid dysgenesis P factor transposon also provides evidence for transcriptional targeting. When transgenic flies are constructed using P factor vectors, a large number of genomic insertion sites are usually observed. However, incorporation of transcription factor binding sequences from the *white, engrailed,* or *polyhomeotic* loci results in a high frequency of insertions at sites near the cognate loci or other loci regulated by the same transcription factors (Hazelrigg *et al.*, 1984; Hama *et al.*, 1990; Kassis *et al.*, 1992; Fauvarque and Dura, 1993).

These observations of transcriptional guidance for mobile elements in yeast and *Drosophila* are extremely significant because they suggest a molecularly plausible answer to the question of how new functional multilocus systems may arise during evolution. The problem has always been to understand how coordinated changes can occur at several functionally related genomic locations. If we accept the proposition that the transcriptional control apparatus connects functionally related loci during differential gene expression (a widely accepted notion among biologists today), then the results quoted above indicate that this same apparatus can also connect distinct but functionally related loci during the insertion of mobile elements. Since mobile elements confer novel regulatory properties on the loci where they insert, transcriptional regulatory targeting can facilitate the creation of multilocus systems, determining new phenotypes.

7 Conclusion

The molecular genetic discoveries outlined above are significant for the concept of evolution as computation because they provide an observational basis for focusing our thinking on new possibilities in evolutionary theory:

- Evolution involves the reassembly of component pieces of complex distributed genomic systems.
- Key evolutionary changes result from the reorganization of regulatory sites and other repetitive DNA elements.
- Multiple changes can occur throughout the genome in a single cell generation.
- The timing and extent of genome rearrangements is subject to biological control in response to environmental and biological stimuli.
- Transcriptional and other cellular regulatory circuits can functionally coordinate multiple genomic changes.

In other words, it is now plausible at the molecular level to conceive of concerted, non-random changes in the genome guided by cellular computing networks during episodes of evolutionary change. Thus, just as the genome has come to be seen as a highly sophisticated information storage system, its evolution has become a matter of highly sophisticated information processing.

Acknowledgements. This research was supported by a grant from the National Science Foundation.

References

1. Avery, O.T., C.M. Macleod, and M. McCarty. 1944. Studies on the chemical nature of the substance inducing transformation of pneumococcal types. I. induction of transformation by a desoxyribonucleic acid fraction isolated from Pneumococcus type III. *J. Exptl. Med.* 79: 137.
2. Beerman, S. 1977. The diminution of heterochromatic chromosomal segments in Cyclops (Crustacea, Copepoda). *Chromosoma* 60: 297–344.
3. Berg, D.E. and M.M. Howe (eds). 1989. *Mobile DNA*. ASM Press, Washington, D.C.
4. Blackburn, E.H. 1991. Structure and function of telomeres. *Nature* 350: 569–573.
5. Blackwell, T.K. and F.W. Alt. 1989. Mechanism and developmental program of immunoglobulin gene rearrangement in mammals. *Ann. Rev. Genet.* 23: 605–636.
6. Bray, D. 1990. Intracellular signalling as a parallel distributed process. *J. theoret. Biol.* 143: 215–231.
7. Bregliano, J.-C. and M.G. Kidwell. 1983. Hybrid dysgenesis determinants. In Shapiro, 1983, pp. 363–410.
8. Britten, R.J. 1997. Mobile elements inserted in the distant past have taken on important functions. *Gene* 205: 177–182.

9. Britten, R.J. and D.E. Kohne. 1968. Repeated sequences in DNA. Hundreds of thousands of copies of DNA sequences have been incorporated into the genomes of higher organisms. *Science* 161: 529–540.

10. Brosius, J. 1991. Retroposons–Seeds of evolution. *Science* 251: 753.

11. Bukhari, A.I., J.A. Shapiro, and S.L. Adhya. 1977. *DNA Insertion Elements, Episomes and Plasmids.* Cold Spring Harbor Press, Cold Spring Harbor, N.Y.

12. Cato, A.C., J. Weinmann, S. Mink, H. Ponta, D. Henderson, and A. Sonnenberg. 1989. The regulation of expression of mouse mammary tumor virus DNA by steroid hormones and growth factors. *J. Steroid. Biochem.* 34: 139–143.

13. Cech T.R. 1983. RNA splicing: three themes with variations. *Cell* 34: 713–716.

14. Csink, A.K. and S. Henikoff. 1998. Something from nothing: the evolution and utility of satellite repeats. *Trends Genet.* 14: 200–204.

15. Darwin, C. 1859. *On the Origin of Species by Means of Natural Selection.* J. Murray: London.

16. Deininger, P.L. 1989. SINES: Short interspersed repeated DNA elements in higher eucaryotes. In Berg and Howe, 1989, pp. 619–636.

17. Demburg, A.F., J.W. Sedat, and R.S. Hawley. 1996. Direct evidence of a role for heterochromatin in meiotic chromosome segregation. *Cell* 86: 135–146.

18. Doolittle, R.F. 1995. The multiplicity of domains in proteins. *Ann. Rev. Biochem.* 64: 287–314.

19. Dover, G.A. 1982. Molecular drive: a cohesive mode of species evolution. *Nature* 299: 111–117.

20. Dowsett, A.P. 1983. Closely related species of *Drosophila* can contain different libraries of middle repetitive DNA sequences. *Chromosoma* 88: 104–108.

21. Duboule, D. and A.S. Wilkins. 1998. The evolution of "bricolage." *Trends Genet.* 14: 54–59.

22. Eibel, H. and P. Philippsen. 1984. Preferential integration of yeast transposable element Ty into a promoter region. *Nature* 307: 386–388.

23. Elder, J.F. and B.J. Turner. 1995. Concerted evolution of repetitive DNA sequences in eukaryotes. *Quart. Rev. Biol.* 70: 297–320.

24. Engels, W.R. 1989. P elements in Drosophila melanogaster. In Berg and Howe, 1989, pp. 437–484.

25. Epplen J.T., W. Maueler, and C. Epplen. 1994. Exploiting the informativity of "meaningless" repetitive DNA from indirect gene diagnosis to multilocus genome scanning. *Biol. Chem. Hoppe Seyler* 375: 795–801.

26. Errede, B., T.S. Cardillo, G. Wever, and F. Sherman. 1981. ROAM mutations causing increased expression of yeast genes: their activation by signals directed toward conjugation functions and their formation by insertions of Ty1 repetitive elements. *Cold Spr. Harb. Symp. Quant. Biol.* 45: 593–607.

27. Fauvarque, M.O. and J.M. Dura. 1993. Polyhomeotic regulatory sequences induce developmental regulator-dependent variegation and targeted P-element insertions in Drosophila. *Genes Dev.* 7: 1508–1520.

28. Fedoroff N. *et al.* 1995. Epigenetic regulation of the maize Spm transposon. *Bioessays* 17: 291–297.

29. Fields, S. and O. Song. 1989. A novel genetic system to detect protein–protein interactions. *Nature* 340: 245–246.

30. Finnegan, D.J. 1989. The I factor and I-R hybrid dysgenesis in Drosophila melanogaster. In Berg and Howe, 1989, pp. 503–518.

31. Foster, P.L. 1993. Adaptive mutation: the uses of adversity. *Ann. Rev. Microbiol.* 47: 467–504

32. Gómez-Gómez, J.M., J. Blázquez, F. Baquero, and J.L. Martinez. 1997. H-NS and RpoS regulate emergence of LacAra+ mutants of *Escherichia coli* MCS2. *J. Bacteriol.* 179: 4620–4622.

33. Grossman, A.D. 1995. Genetic networks controlling the initiation of sporulation and the development of genetic competence in *Bacillus subtilis. Ann. Rev. Genetics* 29: 477–508.

34. Gwinn, M.L., D. Yi, H.O. Smith, and J.F. Tomb. 1996. Role of the two-component signal transduction and the phosphoenolpyruvate: carbohydrate phosphotransferase systems in competence development of *Haemophilus influenzae* Rd. *J Bacteriol.* 178: 6366–6368.

35. Hama, C., Z. Ali, and T.B. Kornberg. 1990. Region-specific recombination and expression are directed by portions of the Drosophila engrailed promoter. *Genes Dev.* 4: 1079–1093.

36. Harrington, J.J., G. Van Bokkelen, R.W. Mays, K. Gustashaw, and H.F. Willard. 1997. Formation of de novo centromeres and construction of first-generation human artificial microchromosomes. *Nat. Genet.* 15: 345–355.

37. Hazelrigg, T., R. Levis, and G.M. Rubin. 1984. Transformation of white locus DNA in Drosophila: dosage compensation, zeste interaction, and position effects. *Cell* 36: 469–481.

38. Henikoff, S. 1992. Position effect and related phenomena. *Curr. Opin. Genet. Dev.* 2: 907–912.

39. Jacob, F. and J. Monod. 1961. Genetic regulatory mechanisms in the synthesis of proteins. *J. Mol. Biol.* 3: 318.

40. Ji H., D.P. Moore, M.A. Blomberg, L.T. Braiterman, D.F. Voytas, G. Natsoulis, and J.D. Boeke. 1993. Hot spots for unselected Ty1 transposition events on yeast chromosome III are near tRNA genes and LTR sequences. *Cell* 73: 1007.

41. Karpen, G.H. 1994. Position-effect variegation and the new biology of heterochromatin. *Curr. Opin. Genet. Dev.* 4: 281–291.

42. Karpen G.H., M.H. Le, and H. Le. 1996. Centric heterochromatin and the efficiency of achiasmate disjunction in Drosophila female meiosis. *Science* 273: 118–122.

43. Kassis, J.A., E. Noll, E.P. Vansickle, W.F. Odenwald, and N. Perrimon. 1992. Altering the insertional specificity of a Drosophila transposable element. *Proc. Nat. Acad. Sci. USA* 89: 1919–1923.

44. Kinsey, P.T. and S.B. Sandmeyer. 1995. Ty3 transposes in mating populations of yeast: a novel transposition assay for Ty3. *Genetics* 139: 81–94.

45. Kirchner, J., C.M. Connolly, and S.B. Sandmeyer. 1995. Requirement of RNA polymerase III transcription factors for in vitro position-specific integration of a retroviruslike element. *Science* 267: 1488–1491.

46. Lamrani, S., C. Ranquet, M.-J. Gama, H. Nakai, J.A. Shapiro, A. Toussaint and G. Maenhaut-Michel. 1999. Starvation-induced Mucts62-mediated Coding Sequence Fusion: Roles for ClpXP, Lon, RpoS and Crp. *Molec. Microbiol.* 32: 327–343.

47. Mahillon, J. and M. Chandler. 1998. Insertion sequences. *Microbiol. Mol. Biol. Rev.* 62: 725.

48. McClintock, B. 1951. Chromosome organization and genic expression. *Cold spr. Harb. Symp. Quant. Biol.* 16: 13.

49. McClintock, B. 1956. Intranuclear systems controlling gene action and mutation. *Brookhaven Symp. Biol.* 8: 58–74.

50. McClintock, B. 1965. The control of gene action in maize. *Brookhaven Symp. in Biol.* 18: 162.

51. McClintock, B. 1978. Mechanisms that rapidly reorganize the genome. *Stadler Genetics Symp.* 10: 25.

52. McClintock, B. 1984. Significance of responses of the genome to challenge. *Science* 226: 792–801.

53. McClintock, B. 1987. *The Discovery and Characterization of Transposable Elements: The Collected Papers of Barbara McClintock.* Garland, New York.

54. Mckee B.D., L. Habera, and J.A. Vrana. 1992. Evidence that intergenic spacer repeats of Drosophila melanogaster rRNA genes function as X-Y pairing sites in male meiosis, and a general model for achiasmatic pairing. *Genetics* 132: 529–544.

55. Miller, J.H. 1996. Structure of a paradigm. *Nat. Struct. Biol.* 3: 310–312.

56. Monod, J. and F. Jacob. 1961. Teleonomic mechanisms in cellular metabolism, growth and differentiation. *Cold Spr. Harb. Sympo. Quant. Biol.* 26: 389.

57. Prescott, D.M. 1992. The unusual organization and processing of genomic DNA in hypotrichous ciliates. *Trends Genet.* 8: 439–445.

58. Prescott, D.M. 1997. Origin, evolution, and excision of internal elimination segments in germline genes of ciliates. *Curr. Opin. Genet. Dev.* 7: 807–813.

59. Reznikoff, W.S. 1992. The lactose operon-controlling elements: a complex paradigm. *Mol. Microbiol.* 6: 2419–2422.

60. Rio, D.C. 1991. Regulation of Drosophila P element transposition. *Trends Genet.* 7: 282.

61. Sagot, M.F. and E.W. Myers. 1998. Identifying satellites and periodic repetitions in biological sequences. *J. Comput. Biol.* 5: 539.

62. Saier, M.H., Jr., T.M. Ramseier, and J. Reizer. 1996. Regulation of carbon utilization. In *Escherichia coli and Salmonella Cellular and Molecular Biology*, 2nd ed. F.C. Neidhardt *et al.* (eds), ASM Press, Washington, D.C. pp. 1325–1343.

63. Shapiro, J.A. (ed) 1983. *Mobile Genetic Elements* Academic Press, New York.

64. Shapiro, J.A. 1984. Observations on the formation of clones containing *araB-lacZ* cistron fusions. *Molec. Gen. Genet.* 194: 79–90.

65. Shapiro, J.A. 1991. Genomes as smart systems. *Genetica* 84: 3–4.

66. Shapiro, J.A. 1992. Natural genetic engineering in evolution. *Genetica* 86: 99–111.

67. Shapiro, J.A. 1993. A role for the Clp protease in activating Mu-mediated DNA rearrangements. *J. Bacteriol.* 175: 2625–2631.

68. Shapiro, J.A. 1995. The discovery and significance of mobile genetic elements. In Sherratt 1995. pp. 1–17.

69. Shapiro, J.A. 1997. Genome organization, natural genetic engineering, and adaptive mutation. *Trends Genet.* 13: 98–104.

70. Shapiro, J.A. 1999. Genome system architecture and natural genetic engineering in evolution. In *Molecular Strategies for Biological Evolution*. L. Caporale and W. Arber, (eds), *Annal. NY Acad. Sci.*, 870: 23–35.

71. Sherratt, D.J. (ed) 1995. *Mobile Genetic Elements — Frontiers in Molecular Biology*. IRL Press, Oxford.

72. Simon, D., H. Stuhlmann, D. Jahner, H. Wagner, E. Werner, and R. Jaenisch. 1983. Retrovirus genomes methylated by mammalian but not bacterial methylase are non-infectious. *Nature* 304: 275–277.

73. Trifonov, E.N. and V. Brendel. 1986. GNOMIC: A Dictionary of Genetics Codes. Balaban, Philadelphia.

74. van der Lelie, D., A. Sadouk, A. Ferhat, S. Taghavi, A. Toussaint, and M. Mergeay. 1992. Stress and survival in *Alacaligenes eutrophus* CH34: effects of temperature and genetic rearrangements. In *Gene Transfers and Environment*, M.J. Gauthier (ed), Springer-Verlag. pp. 27–32.

75. Walker, G. 1996. The SOS response of Escherichia coli. In *Escherichia coli and Salmonella Cellular and Molecular Biology*, 2nd ed. F.C. Neidhardt *et al.* (eds), ASM Press, Washington, D.C., pp. 1400–1416.

76. Watson, J.D. and F.H.C. Crick. 1953. Genetic implications of the structure of deoxyribonucleic acid. *Nature* 171: 964.

77. Weiler, K.S. and B.T. Wakimoto. 1995. Heterochromatin and gene expression in Drosophila. *Ann. Rev. Genet.* 29: 577–605.

78. Willard HF. 1990. Centromeres of mammalian chromosomes. *Trends Genet.* 6: 410–416.

79. Zou, S., N. Ke, J. M. Kim and D. F. Voytas. 1996. The Saccharomyces retrotransposon Ty5 integrates preferentially into regions of silent chromatin at the telomeres and mating loci. *Genes Dev.* 10: 634–645.

Evolutionary Computation as a Paradigm for DNA-Based Computing

Thomas Bäck, Joost N. Kok, and Grzegorz Rozenberg

Abstract. *Evolutionary Computation* focuses on probabilistic search and optimization methods gleaned from the model of organic evolution. *Genetic algorithms*, *evolution strategies*, and *evolutionary programming* are three independently developed representatives of this class of algorithms, with genetic programming and classifier systems as additional paradigms in the field.

This paper focuses on the link between evolutionary computation and DNA-based computing by discussing the relevant aspects of evolutionary computation both from a practical and a theoretical point of view. In particular, theoretical results concerning the calculation of convergence velocities and the derivation of optimal schedules for mutation rates, respectively steps sizes, are presented.

The potential for cross-fertilization between the fields of DNA-based computing and evolutionary computation is outlined both from a fundamental point of view and by means of an experimental investigation concerning the NP-hard maximum clique problem. A simple evolutionary approach to maximum clique is introduced and the hypothesis whether the increase in population size possible by realizing evolutionary computation with DNA yields the expected improvement in solution quality is tested. Results obtained for a limited range of population sizes up to 10^5 indicate that the hypothesis holds for about two-thirds of the investigated problem instances (which were taken from the DIMACS library).

1 Introduction

The goal of this paper is to bring together some of the main concepts from the areas of evolutionary computation and DNA-based (or, more general, molecular) computing. We argue that this is beneficial for both areas. For evolutionary computation, we outline how the massive parallelism of DNA-based computing can be exploited, and for DNA-based computing an evolutionary approach would offer the possibility of scaling the method to problems of increasing dimensionality by substituting the usual filtering approach with an evolutionary one. Therefore, we outline the main ideas, notions, and trends of evolutionary computation to the extent we feel is needed for creating a solid bridge to the area of DNA-based computing.

In particular, Sect. 2 presents an introduction to evolutionary computation that focuses on a general view of evolutionary algorithms, with genetic algorithms and evolution strategies as two rather different instances of the general algorithm. Especially, the theoretical results about evolution strategies and the transfer of this kind of theory to genetic algorithms provide the main topic of the theoretical overview given in this section. Section 3 gives a brief overview of the basic aspects of DNA-based computing, and Sect. 4 explains the relationship we see between the two areas.

In Sect. 5, we propose an application example for evolutionary DNA-based computing, namely the maximum clique problem, which has already been tackled in vitro with a filtering approach [55]. After reviewing this filtering approach, we outline an evolutionary DNA-based algorithm and its most closely related evolutionary algorithm. The latter is then applied with increasing population sizes up to 10^5 to maximum clique problem instances to check our hypothesis that the increase in population size improves the quality of solutions found by the algorithm. This should be seen as a first step on our way towards implementing the evolutionary DNA-based algorithm in vitro.

2   Evolutionary Computation

As was recently pointed out by Dennett [24], the identification of evolution as an *algorithmic* process is one of the fundamental insights offered by Darwin — an insight that evolutionary algorithms have tried to exploit for more than 35 years, with remarkable success for practical problem solving in a variety of application fields. With strong simplifications, evolutionary algorithms implement the process as an interplay between the creation of new information (by means of randomized mutation and recombination operators) and its evaluation and selection, where a single individual of the population is affected by other individuals as well as by the environment. The better an individual performs under such conditions, the greater its chance to survive for a longer time and to pass down its information to its offspring. The link between this abstract view on natural evolution and an optimization problem

$$f(\boldsymbol{x}) \to \max \qquad\qquad (1)$$

is then established by identifying individuals of a population with candidate solutions (i.e., vectors $\boldsymbol{x} \in M$ from the search space $M = M_1 \times \ldots \times M_n$) of the optimization problem, the fitness of an individual with the measured (according to the objective function $f : M \to I\!\!R$) quality of a solution, the evolutionary variation operators with probabilistic search operators, and the selection operator with a preference rule that better solutions be kept in the actual pool of solutions.

The general outline of an evolutionary algorithm forms the basis for all known instances of the idea, namely *genetic algorithms* [33,40], *evolutionary programming* [31,30], *evolution strategies* [59,60,65,67], and the variants *genetic programming* [46] and *learning classifier systems* [41]. Recent overviews of evolutionary algorithms can also be found in [7,30] and the introductory article [9]. The basic evolutionary algorithm can be summarized by the following pseudocode:

Algorithm 1

```
t := 0;
initialize(P(t)) randomly, |P(t)| = μ;
evaluate(P(t));
while not terminate do
    P'(t) := variation(P(t)), |P'(t)| = λ;
    evaluate(P'(t));
    P(t + 1) := select_μ(P'(t) ∪ Q);
    t := t + 1;
od
```

In Algorithm 1, $P(t)$ denotes a multiset at iteration t of candidate solutions to a given problem, and we write such a multiset of μ vectors (candidate solutions to the optimization problem) $\boldsymbol{x}_i \in M$ as $P(t) == \{\boldsymbol{x}_1, \ldots, \boldsymbol{x}_\mu\}$. In analogy with the underlying biological intuition, $P(t)$ is usually referred to as the *population* of size μ at *generation* t, and the members of $P(t)$ are called *individuals*. As a matter of fact, in some instances of evolutionary algorithms individuals in a population might be more complex than just a representation of a solution, i.e., $I = M \times S$ then denotes the set of individuals, M denotes the search space, and S summarizes the additional components of an individual, the so-called *strategy parameters*. In the following, we will use the notation $\boldsymbol{a} = (\boldsymbol{x}, \boldsymbol{s}) \in I$ for members of I as well as the notation $\boldsymbol{x} \in M$ to denote individuals whenever $S = \emptyset$. The *genetic load* Q denotes a special multiset of candidate solutions that might be considered for selection, typical choices being given by $Q = \emptyset$, $Q = P(t)$, or $Q = \{\boldsymbol{x}'\}$ where $\boldsymbol{x}'$ denotes the best candidate solution from $P(t)$.

An intermediary *offspring* population $P'(t)$ of size λ of candidate solutions is generated by sequentially applying variation operators such as *recombination* and/or *mutation* to members of the population $P(t)$. The elements of $P'(t)$ are then evaluated by calculating their objective function values $f(\boldsymbol{x}_k)$ for each of the solutions $\boldsymbol{x}_k$ represented by individuals in $P'(t)$, and selection based on the individual's *fitness* values drives the process toward better solutions. Typically, the fitness value F directly relates to the objective function value by means of some mapping g, i.e., $F(\boldsymbol{a}) = g(f(\boldsymbol{x}))$ for some individual $\boldsymbol{a} = (\boldsymbol{x}, \boldsymbol{s})$.

A variety of instances of the general evolutionary algorithm (Algorithm 1) are presently under active investigation in the field of evolutionary computation, but the majority of current implementations descend from the *genetic algorithm* [40,33], *evolution strategy* [59,60,65,67], and *evolutionary programming* [31,30] approaches. Roughly, the main distinguishing features of these different instances of the evolutionary algorithm are summarized in the following sections.

2.1   Genetic Algorithms

The classical genetic algorithm as introduced by Holland [40] uses a binary representation of individuals as vectors over the alphabet $\{0, 1\}$, such that the representation is well suited for combinatorial optimization problems of the general form $f(\boldsymbol{x}) \to \max$ with $f : \{0, 1\}^n \to \mathbb{R}$, i.e., $M = \{0, 1\}^n$. The initial population $P(0)$ is typically created by randomly assigning each bit of the members of $P(0)$ a value of either 0 or 1. The strategy parameter set S is normally empty in genetic algorithms such that $I = M$.

Concerning the variation operators, both recombination (crossover) and mutation are applied in genetic algorithms, with a preference on crossover as the main variation operator. In the simplest case of *one-point crossover*, illustrated in Fig. 1, two individuals are randomly chosen from the population and two new individuals are created by exchanging segments starting at a randomly determined crossover point between the two parents.

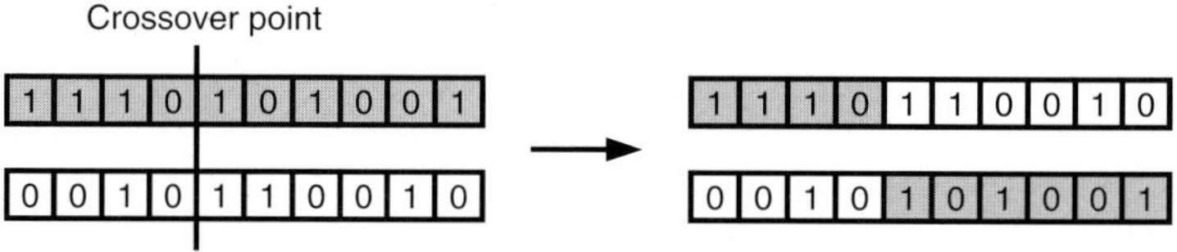

Fig. 1. Illustration of one-point crossover in genetic algorithms. The crossover point is located at position 4.

The crossover operator has been generalized in various ways, including an increased number of crossover points [27], a uniform crossover where each single bit is chosen randomly from one of the corresponding parent bits [71], and even multi-parent extensions involving more than two parent individuals have been proposed [25]. The crossover operator in a genetic algorithm is applied with a certain probability p_c, with typical values of $p_c = 0.6$ [42] or $p_c \in [0.75, 0.95]$ [64].

The mutation operator in genetic algorithms is implemented by inverting bits with small probability such as $p_m = 0.001$ [42] or $p_m \in [0.005, 0.01]$ [64]. Although mutation was originally introduced as a background operator of minor importance, recent studies have indicated that this operator deserves more attention. Theoretical investigations for simple objective functions resulted in the finding that $p_m = 1/n$ reflects the relationship between p_m and other parameters of the algorithm better than a constant value of p_m [4,5,51]. Moreover, much larger mutation rates which decrease over the course of evolution [29] as well as so-called *self-adaptive* mutation rates [12,69] (see also Section 2.2) are often helpful to increase the quality of solutions found by genetic algorithms.

The selection operator in genetic algorithms is usually a probabilistic operator, randomly sampling individuals from[1] $P'(t) \cup Q$ (typically, $Q = \emptyset$ is used) with a bias that individuals of better quality tend to be passed into $P(t+1)$. The original *proportional selection* method [33,40] determines the selection probability $p(\boldsymbol{x}_k)$ of an individual $\boldsymbol{x}_k$ by the relative fitness $f(\boldsymbol{x}_k)/\sum_{i=1}^{\lambda} f(\boldsymbol{x}_i)$. As this method expects positive fitness values and a maximization task, *scaling functions* have been introduced to perform appropriate fitness transformations (e.g., see [7,33]).

As an alternative to proportional selection, *tournament selection* has recently gained an increasing popularity because it avoids the need for a scaling function and allows for fine-tuning of the selective pressure by increasing or decreasing the *tournament size* q [54]. The method works by repeatedly taking a random uniform sample of size $q > 1$ from the population and selecting the best of these q individuals to be copied into the new population $P(t+1)$, until it is filled with μ individuals.

2.2   Evolution Strategies

In contrast to genetic algorithms, evolution strategies typically use a representation based on vectors of real values[2], such that they are most commonly applied to continuous parameter optimization problems of the general form $f(\boldsymbol{x}) \to \max$ with $f : \mathbb{R}^n \to \mathbb{R}$ and problem dimension n, i.e., $M = \mathbb{R}^n$.

Concerning the variation operators, mutation clearly plays a dominant role in evolution strategies, using normally distributed variations with expectation zero and a certain standard deviation to model the observation that in nature, small phenotypical changes are more likely to occur than large ones. More specifically, mutation is performed in an evolution strategy by adding a normally distributed random value to each of the object variables as follows[3]:

$$x_i' = x_i + \sigma \cdot N_i(0,1) \, . \tag{2}$$

As a specific property of evolution strategies, the standard deviation σ is no exogenous parameter of the algorithm (such as e.g., the mutation rate in genetic algorithms), but is instead endogenously controlled by means of a so-called *self-adaptation* principle. Essentially, this concept incorporates the mutational strategy parameters into the representation of individuals and applies evolutionary variation to object variables and strategy parameters at the same time. In other words, rather than employing an exogenous control, the mutational parameters undergo evolution in the same way as strategy parameters do. Putting this more formally, an individual $\boldsymbol{a} = (\boldsymbol{x}, \boldsymbol{\sigma})$ now

---

[1] For the meaning of Q, we refer to Algorithm 1.

[2] But they originally were developed and are still used for combinatorial optimization problems with an integer representation of solutions [62].

[3] The notation $N_i(\cdot, \cdot)$ indicates that the random variable is sampled anew for each value of the index i.

consists of the object variable vector $\boldsymbol{x}$ (i.e., the candidate solution to the problem f) and the strategy parameters $\boldsymbol{\sigma} \in \mathbb{R}^n_+$ (i.e., now $S = \mathbb{R}^n_+$). Technically, the mutation operator is then defined as

$$\sigma'_i = \sigma_i \cdot \exp(\tau' \cdot N(0,1) + \tau \cdot N_i(0,1)) \,, \tag{3}$$

$$x'_i = x_i + \sigma'_i \cdot N_i(0,1) \,, \tag{4}$$

where $\tau' \propto (\sqrt{2n})^{-1}$ and $\tau \propto (\sqrt{2\sqrt{n}})^{-1}$. In summary, this defines a mutation operator which works by adding a normally distributed random vector $\boldsymbol{z} \in \mathbb{R}^n$ with $z_i \sim N(0, \sigma'^2_i)$ to the individual's object variable vector $\boldsymbol{x}$ (i.e., the components of this vector are normally distributed with expectation zero and variance σ'^2_i, where σ'_i is the *mutated* standard deviation associated with variable x_i).

The factors τ and τ' define the speed of adaptation on the level of strategy parameters and can be regarded as *learning rates* for the self-adaptation process. Their settings indicated above are fairly robust upper bounds and can be used as defaults, although their optimal values depend on the topological characteristics of the objective function.

In addition to Eqns. (3) and (4), at least two modifications of the scheme have to be mentioned: A simplified version uses just one standard deviation σ, identical for all of the object variables, thus reducing the degree of freedom for the mutation operator to spherically symmetric mutations. On the other hand, the more elaborated *correlated mutation* operator also includes covariances into the self-adaptive framework, thus expanding the amount of information stored in the individuals even further and generalizing the mutation operator to arbitrarily rotated hyperellipses. For details on these two variants of self-adaptation, the reader is referred to [7,8,67].

The recombination operator in evolution strategies is incorporated into the main loop of the algorithm as the first operator in the variation step (see Algorithm 1) and generates a new intermediate population of λ individuals by λ-fold application to the parent population, creating one individual per application from ϱ $(1 \leq \varrho \leq \mu)$ individuals. Normally, $\varrho = 2$ or $\varrho = \mu$ (so-called global recombination) are chosen. The recombination types for object variables and strategy parameters in evolution strategies often differ from each other, and typical examples are *discrete recombination* (random choices of single variables from parents, comparable to uniform crossover in genetic algorithms) and *intermediary recombination* (arithmetic averaging). For further details about these operators, see [7].

Finally, selection in evolution strategies works by a deterministic choice of the μ best individuals from $P'(t)$ (i.e., $Q = \emptyset$ in Algorithm 1) or $P'(t) \cup P(t)$ (i.e., $Q = P(t)$). The former is referred to as (μ,λ)-selection, and the latter as $(\mu+\lambda)$-selection. The (μ,λ)-selection is typically recommended in evolution strategies for two main reasons: it allows the strategy to accept temporary deteriorations that might help it to leave the region of attraction of an optimum point to reach a better one, and it improves the self-adaptation capability of the algorithm as opposed to $(\mu+\lambda)$-selection. As in nature, where a reproduction surplus is the normal case, a setting of $\lambda \gg \mu$ is also required in evolution strategies (e.g., a setting of $\mu/\lambda \approx 1/7$ is often reasonable). The terminology for $(\mu+\lambda)$- and (μ,λ)-strategy given above is often used to denote the whole evolution strategy as the characterizing feature. This is due to the fact that the population structure of the different historical variants of the modern evolution strategy can be captured by the notation. In particular, we would like to mention briefly some of the special variants that have been described in the literature:

- $(1+1)$-strategy: One parent creates one offspring by mutation, and the better[4] of parent and offspring survives [59].
- $(1,\lambda)$- and $(1+\lambda)$-strategy: λ offspring individuals are created from just one parent by mutation [65,66].
- $(\mu+1)$-strategy: One offspring individual is created from μ parents by recombination and mutation and eventually substitutes the worst parent [59].

Thus, the selection scheme allows for a wide range of different algorithmic instances, especially simplified ones where the parent or the offspring population are reduced to just one individual. In the following sections, we will briefly review modern aspects of the theory of evolutionary algorithms and demonstrate some techniques for analyzing these simplified variants.

2.3   Some Theory

Traditionally, the strong preference for using binary representations in the genetic algorithm framework is explained using the concept of *schema theory* [33,40], which essentially analyzes genetic algorithms in terms of their expected schema sampling behavior under the assumption that mutation and crossover are detrimental. In this notation, a *schema* denotes a similarity template representing a subset of $\{0,1\}^n$, and the *schema theorem* of genetic algorithms says that the algorithm provides a near-optimal sampling strategy for schemata by increasing the number of well-performing, short (i.e., with small distance between their left-most and right-most specified bits) and low-order (i.e., with few specified bits) schemata (so-called *building blocks*) over subsequent generations (see [33]).

Although this theory seems to explain the claim for using a binary alphabet because this maximizes the number of schemata for a given number of search points in the search space, it does not allow for a derivation of *constructive* results concerning operator choices, parameter settings, or expected time behavior of the algorithm. These aspects of the theoretical analysis, however, have already been present in evolution strategy research for more than 30 years. A transfer of these concepts to genetic algorithms turned out to

---

[4] In case of identical fitness values, the offspring is chosen.

be beneficial and facilitated a number of successful steps in their analysis in the past few years. Some of these benefits are also reported in the following sections.

Convergence Velocity. Traditionally, evolution strategies have been analyzed in terms of their *convergence velocity*, i.e., the expected improvement in fitness (5) or distance from the optimum $\boldsymbol{x}^*$ (6) between two consecutive generations:

$$\varphi = \mathbf{E}(|f(\boldsymbol{x}^*) - f(\boldsymbol{x}_t)| - |f(\boldsymbol{x}^*) - f(\boldsymbol{x}_{t+1})|) \tag{5}$$

$$\varphi = \mathbf{E}(\|\boldsymbol{x}^* - \boldsymbol{x}_t\| - \|\boldsymbol{x}^* - \boldsymbol{x}_{t+1}\|) \tag{6}$$

The second alternative, distance to the optimum, has been used in evolution strategies with enormous success to analyze algorithmic instances such as the (1+1)- [59], (1+λ)- [65] and (1,λ)- [65], (μ,λ)-[5] [14] and (μ/μ,λ)-strategies[6] [15] on simple quadratic objective functions, namely the sphere model $f(\boldsymbol{x}) = \sum(x_i - x_i^*)^2 = R^2$ for the assumption of one fixed mutation step size σ and sufficiently large dimensionality n. Without going into the details here, it is worthwhile to mention that for comma-strategies the results can usually be expressed as simple quadratic equations such as

$$\varphi'_{(1,\lambda)} = c_{1,\lambda} \cdot \sigma' - \frac{\sigma'^2}{2} \tag{7}$$

for a (1,λ)-strategy, where $\varphi' = \varphi n/R$ abd $\sigma' = \sigma n/R$ denote normalized variables, R is the current distance of the parent $\boldsymbol{x}$ to the optimum location $\boldsymbol{x}^*$, and $c_{1,\lambda}$ denotes the *progress coefficient* of the strategy [60]. This constant depends only on λ and characterizes the selective pressure of the (1,λ)-selection method. It is defined as the expectation of the largest of λ independent random variables Z_i with identical, standardized, and normalized Gaussian distribution $Z_i \sim N(0,1)$, which are rearranged in increasing order and relabeled by $Z_{1:\lambda} \leq Z_{2:\lambda} \leq \ldots \leq Z_{\lambda:\lambda}$:

$$c_{1,\lambda} = \mathbf{E}(Z_{\lambda:\lambda}) = \int_{-\infty}^{\infty} z \cdot \frac{d}{dz}[\Phi(z)]^\lambda dz . \tag{8}$$

Here, Φ denotes the distribution function of the Gaussian distribution. The values of $c_{1,\lambda}$ are extensively studied and tabulated in the theory of order statistics (see, e.g., [22]).

Based on the theoretical approach discussed so far, the corresponding result for the (μ,λ)-strategy without recombination is obtained as

$$\varphi'_{(\mu,\lambda)} = c_{\mu,\lambda} \cdot \sigma' - \frac{\sigma'^2}{2} \tag{9}$$

---

[5] With $\mu > 1$, but without recombination for the sake of theoretical analysis.
[6] With global intermediary or global discrete recombination.

with a more complex definition of the generalized progress coefficient (which again characterizes the (μ,λ)-selection completely). Finally, depending on the recombination operator, the results for recombinative strategies are

$$\varphi'_{(\mu/\mu_I,\lambda)} = c_{\mu,\lambda} \cdot \sigma' - \frac{\sigma'^2}{2\mu} \tag{10}$$

for intermediary recombination and

$$\varphi'_{(\mu/\mu_D,\lambda)} = \sqrt{\mu} \cdot c_{\mu,\lambda} \cdot \sigma' - \frac{\sigma'^2}{2} \tag{11}$$

for global discrete recombination. It should be noted that, for all of these quadratic equations, it is straightforward to calculate the locally optimal standard deviation and the resulting maximum convergence velocity, e.g., resulting in $\sigma^* = c_{1,\lambda} \cdot R/n$ and $\varphi^* = c_{1,\lambda}^2 R/(2n)$ for the $(1,\lambda)$-strategy with $R = \|\boldsymbol{x}_t - \boldsymbol{x}^*\|$ denoting the current distance to the optimum.

To conclude, we would like to emphasize that the local convergence velocity analysis of evolution strategies is a constructive analysis facilitating statements about the optimal working regime of the algorithm. In particular, it allows evaluation of the performance of parameter control methods such as self-adaptation with respect to the optimal schedule for σ as obtained from the theory [16].

Starting in the early 90s, the constructive aspects of this theory were transferred to binary search spaces to facilitate the analysis of simple mutation-based variants of genetic algorithms, such as the $(1+1)$-GA[7]. The analysis presented in [4,51] focused on the bit-counting function $f : \{0,1\}^n \to \mathbb{R}$, $f(\boldsymbol{x}) = \sum_{i=1}^{n} x_i$ for simplicity, and a standard bit inversion mutation operator with a mutation rate p_m per bit. Assuming a maximization task, the convergence velocity is then defined by means of the expected improvement in fitness value as follows:

$$\varphi_{(1+1)} = \sum_{k=0}^{n-f(\boldsymbol{x})} k \cdot p(k) \, , \tag{12}$$

where $p(k)$ denotes the probability that mutation improves the fitness value of the vector $\boldsymbol{x}$ by a value of k, i.e., $p(k) = P\{f(\texttt{mutate}(\boldsymbol{x})) = f(\boldsymbol{x})+k\}$. Knowing these transition probabilities, the $(1+1)$-algorithm can then be modeled as a finite homogeneous Markov chain with a special triangular transition matrix and n different states corresponding with the different fitness values. The exact transition probabilities are given by the expression [4]

$$p(k) = \sum_{i=0}^{f(\boldsymbol{x})} \binom{f(\boldsymbol{x})}{i}\binom{n-f(\boldsymbol{x})}{i+k}p^{2i+k}(1-p)^{n-2i-k} \tag{13}$$

---

[7] One parent generates one offspring by mutation, and the better of the two individuals survives to become parent of the next generation.

and facilitate numerical investigations of the Markov chain with respect to properties such as the optimal mutation rate (maximizing convergence velocity) and the expected time to absorption (i.e., time to find the optimum; see [7] for details).

The results reported in [4,5,7] say that the optimal mutation rate for the bit counting problem is inversely proportional to the dimensionality of the problem and depends on the actual fitness value $f(\boldsymbol{x})$, i.e.,

$$p^* \approx \frac{1}{2(f(\boldsymbol{x})+1)-n} \ . \tag{14}$$

Moreover, an analytical treatment of the problem becomes possible by neglecting backward mutations, i.e., those mutations that change correct bits back to incorrect ones, resulting in an approximation of p^* as $p^* \approx 1/n$, which is a reasonable general-purpose heuristic for genetic algorithms [51].

While the generalization of this kind of convergence velocity analysis to $(1,\lambda)$- and (μ,λ)-genetic algorithms was formalized in [7] without obtaining a closed expression for the optimal mutation rate, Beyer recently presented an approximation for the $(1,\lambda)$-genetic algorithm applied to the bit counting function [17]:

$$p^* \approx \frac{c_{1,\lambda}^2}{4 \cdot (2 \cdot f(\boldsymbol{x})/f^* - 1)^2} \cdot \frac{1}{n} \ , \tag{15}$$

where $f^* = n$ denotes the optimal fitness value, $f(\boldsymbol{x}) > n/2$, and $n \to \infty$ are assumed.

Even though these results were derived for the simple bit counting problem, they can serve well as lower bounds for the mutation rate in case of more complex objective functions and are therefore of interest for any application of genetic algorithms to binary optimization problems. Consequently, the experimental investigation presented in Sect. 5 will be based on the results summarized in this section.

Large Populations. Still, relatively few results are available for large (but not infinite) populations in evolutionary algorithms. In calculating the optimal step size or mutation rate of evolutionary algorithms, the asymptotic approximation [11]

$$c_{1,\lambda} \approx \sqrt{2 \cdot \ln \lambda} \tag{16}$$

for $\lambda \to \infty$ is very helpful, because it allows for a further simplification of the expressions for optimal step size σ^* and mutation rate p^* for the $(1,\lambda)$-evolution strategy and $(1,\lambda)$-genetic algorithm, respectively. For the latter case, equation (15) yields

$$p^* \approx \frac{\ln \lambda}{2 \cdot (2 \cdot f(\boldsymbol{x})/f^* - 1)^2} \cdot \frac{1}{n} \ , \tag{17}$$

which can further be approximated in the final stage of the search, when $f(\boldsymbol{x})$ is relatively close to its optimum value f^*, as

$$p^* \approx \frac{1}{2n} \cdot \ln \lambda \ . \tag{18}$$

Obviously, these values can easily be used to parameterize the corresponding genetic algorithm instance.

Selection Operators. The selection operator in evolutionary algorithms is clearly independent of the solution representation, such that selection operators can easily be transferred between different evolutionary algorithms such as evolution strategies and genetic algorithms. Especially, the (μ,λ)-selection (sometimes also called *truncation selection* [52,53]) has been used successfully in the context of genetic algorithms [10,52,53]. In the past years, all selection methods have been intensively studied and analyzed within the general framework of their selective pressure or *selection intensity* under the assumption of normally distributed fitness values within the population selection is acting on. Assuming a standard deviation σ of the fitness values, the selection intensity I characterizes the difference in mean between the selected population, $\bar{f}_{\mathrm{sel}}$, and the parent population, $\bar{f}$: $\bar{f}_{\mathrm{sel}} - \bar{f} = I \cdot \sigma$ [52,53].

The selection intensity is analytically known for all of the most important selection methods (see, e.g., [6,18,50]), namely for (μ,λ)-selection

$$I = c_{\mu,\lambda} = \frac{1}{\mu} \cdot \sum_{i=\lambda-\mu+1}^{\lambda} \mathbf{E}(Z_{i:\lambda}) \ , \tag{19}$$

for proportional selection

$$I = \sigma/\bar{f} \ , \tag{20}$$

for tournament selection

$$I = \mathbf{E}(Z_{q:q}) \ , \tag{21}$$

and for linear ranking selection, where a linear function with slope $1 \leq \eta^+ \leq 2$ defines the selection probabilities

$$I = (\eta^+ - 1) \cdot \mathbf{E}(Z_{2:2}) \ . \tag{22}$$

These results facilitate a rigorous understanding of the effects of the various parameters of these selection methods and an analytical treatment of certain variants of the convergence velocity analysis including the calculation of absorption times (e.g., see [53]).

3   DNA-Based Computation

The idea of computing with biological macromolecules like DNA (deoxyribonucleic acid) and RNA (ribonucleic acid) is fascinating because of its potentially tremendous advantages concerning *speed* (realized by massively parallel operations on DNA strands), *energy consumption*, and *information storage density* (see [57] for an overview).

These potential advantages of DNA computing over the traditional approach and the seminal experimental work of Adleman, demonstrating the practical in vitro implementation of a DNA algorithm for solving an instance of the Hamiltonian path problem [2], caused a strong increase of interest in DNA computing over the past years. Although the set of "bio-operations" that can be executed on DNA strands in a laboratory (including operators such as *synthesizing, mixing, annealing, melting, amplifying, separating, extracting, cutting*, and *ligating* DNA strands; see, e.g., [3]) seems fundamentally different from traditional programming languages, theoretical work on the computational power of various models of DNA computing (see, e.g., [57]) demonstrates that certain subsets of these operators are computationally complete. In other words, everything that is Turing-computable can also be computed by these DNA models of computation. Furthermore, it has also been shown that *universal systems* exist, so that the *programmable DNA computer* is theoretically possible.

The algorithms for DNA computing that have been presented in the literature (e.g., for the satisfiability problem [48], the shortest common superstring problem [43], and the Hamiltonian path problem [2]) use an approach that will not work for NP-complete problems of realistic size, because these algorithms are all based on extracting an existing solution from a sufficiently large initial population of solutions [35,36]. Although a huge number ($\approx 10^{12}$) of DNA molecules (i.e., potential solutions to a given problem) can be manipulated in parallel, this so-called *filtering approach* (i.e., generate and test) quickly becomes infeasible as problem sizes grow (e.g., a 500-node instance of the traveling salesman problem has $> 10^{1000}$ potential solutions).

Using a particular subset of these fundamental operators, an example for a typical DNA computing algorithm as used for solving the Hamiltonian path problem can be formulated [2]. Given a directed graph $G = (V, E)$ with vertices $V = \{v_1, \ldots, v_n\}$ and edges $E \subseteq V \times V$, where v_{in} and v_{out} denote designated vertices, the graph is said to have a Hamiltonian path if and only if there exists a sequence of edges that begins at v_{in}, ends at v_{out}, and enters every other vertex exactly once [32]. The DNA computing algorithm as developed and executed by Adleman solves the Hamiltonian path problem as follows:

Algorithm 2

```
1:  Generate random paths through G;
2:  Keep only those paths that begin with v_in and end with v_out;
3:  Keep only those paths that enter exactly n vertices;
4:  Keep only those paths that enter all vertices at least once;
5:  if any paths remain then
        output YES;
    else
        output NO;
```

It is obvious that this algorithm follows the filtering approach: A random population of candidate solutions is generated, and a solution to the problem is obtained by filtering out all DNA molecules not representing a solution. By means of the biological operations, the steps of Algorithm 2 can be realized as follows: For step 1, each vertex of the graph is encoded into a 20-nucleotide strand of DNA. For each edge $e = (v_i, v_j)$ of the graph, a DNA sequence is then created, consisting of the second half of the sequence encoding v_i and the first half of the sequence encoding v_j. A solution containing multiple copies of the sequences encoding edges is then mixed with a solution containing a population of sequences complementary to the vertex-encoding strands (the solution contains also ligase). The complements of vertices act as splints, keeping together DNA sequences corresponding to compatible edges so that they can be ligated. The reaction results in the formation of DNA molecules encoding random paths through the graph.

Step 2 is implemented by amplifying the product of step 1 by PCR, however, by amplifying only the molecules encoding paths that begin with v_{in} and end with v_{out}.

Step 3 is facilitated by gel electrophoresis for separating the strands by length, which allows filtering out of those strands encoding paths of length n.

Step 4 is accomplished by an iterative process of extraction by means of affinity purification. In an iterative loop, strands containing each of the vertices are subsequently extracted as indicated in the following algorithm:

Algorithm 3

```
4:  S_0 := all DNA strands remaining after step 3;
    for i := 1 to n do
        S_i := extract(S_{i-1}, v_i);
    od
```

Here, the notation $\mathtt{extract}(S, v)$ indicates that all strands containing (a subsequence encoding) vertex v are extracted from the set S of strands.

Step 5 can then be applied to S_n by means of amplifying the result by PCR and determining the DNA sequence of the amplified molecules (i.e., detecting and reading of sequences). For a complete overview of this algorithm, see also [57].

4   Evolutionary DNA-Based Computation

In this paper we discuss an alternative to the filtering approach: a solution should *evolve* rather than be extracted. We propose to *combine DNA computing with the general approach of evolutionary algorithms*. The proposed approach aims at implementing so-called evolutionary algorithms by using DNA strands and biological operators to encode and evolve potential solutions. In contrast to a previous proposal that focuses on genetic algorithms [13,21], we clarify that concepts from both evolution strategies and genetic algorithms are beneficial for implementing evolutionary DNA-based computing.

Knowing that the filtering approach in DNA computing works only up to a certain dimensionality of the problem, and also knowing that evolutionary algorithms still yield good approximate solutions even for problems of high dimensionality, we propose to combine both approaches by *realizing an evolutionary algorithm with DNA: evolutionary DNA computing*. This approach combines the capability of evolutionary algorithms to construct a solution to a given problem by evolving it with the massive population sizes that become possible by using DNA. Moreover, the approach does not suffer from the erroneous behavior of the "bio-operations", but rather it exploits the errors as a source of diversity in the population in the sense of mutation in an evolutionary algorithm (the general principle is outlined in [23]).

The *selection* operator would require a detection of gradual quality differences between DNA strands and might impose larger difficulties than the two other operators mentioned, but some ideas towards solving this problem have already been indicated in the literature [23].

Moreover, the aspect of selection of gradually improving molecules is one of the topics especially emphasized in the field of *molecular evolution* [26], one of its goals being the evolutionary design of biopolymers such as drugs, vaccines, biosensors, etc. [45]. Kauffman clearly outlines how molecular evolution for these purposes could be realized [45], and it is clear that the ideas from molecular evolution can well be transferred to evolutionary DNA computing as proposed here. A large amount of theoretical work exists in the field of molecular evolution and can certainly be exploited to guide the design of evolutionary DNA computing algorithms for finding good solutions to NP-complete combinatorial optimization problems (see, e.g., [58]). New techniques from molecular evolution can be of particular help in the in vitro implementation of the selection operator required for evolutionary DNA computing. In the following, operators are discussed only to the extent that is required for outlining the idea of evolutionary DNA computing.

With these ideas in mind, the genetic algorithm concept with its discrete, binary alphabet seems closest to DNA computing, relying on the quaternary nucleotide base alphabet consisting of adenine (A), guanine (G), cytosine (C), and thymine (T). Due to the complementarity A–T and C–G of the double-stranded DNA, no loss of information occurs when interpreting the DNA as a string of *symbols* (encoding a potential solution) rather than a double-stranded structure.

As in all evolutionary algorithms, the *initial population* of an evolutionary DNA computing algorithm would be generated at random, which is also the standard method employed in conventional approaches to DNA computing (e.g. [2,48]). In contrast to the latter, however, evolutionary DNA computing would not typically generate the optimal solution as a member of the initial population, because the population size is still tiny compared to the size of the search space of combinatorial optimization problems.

While mutation in genetic algorithms is typically realized by inverting bits with a small probability p_m (see Section 2.1), the DNA computing version of an evolutionary algorithm would exploit the replication errors inherently occurring in biological operations on DNA such as amplification by means of PCR. A technique for realizing mutation by means of a controlled mutagenesis technique involving the enzymes *uvrABC* and *polymerase* has already been outlined [23].

Concerning *recombination*, the classical one-point crossover operator originally proposed for genetic algorithms [40], with its random choice of a crossover point and its exchange of information to the right of the crossover point (see Sect. 2.1), corresponds directly to the *splicing operation* on double-stranded DNA molecules as introduced by Head [38,39]. Experimental investigations have already been performed to confirm that a set of *restriction enzymes* (DraIII and BglII, as proposed in [39]) and a *ligase enzyme* (Taq) can be used successfully to perform the recombination operation in the test tube [47].

As demonstrated in Section 2.2, the (μ,λ)-selection operator in evolution strategies extracts the μ best solutions from a large offspring surplus of $\lambda \gg \mu$ solutions. In terms of DNA computing, this approach is somewhat similar to the filtering approach employed in [2], but rather than filtering out a single best solution, a set of best partial solutions (e.g., containing a large number of different nodes for the Hamiltonian path problem) is extracted. Of course, the "population sizes" μ and λ are no longer fixed in the evolutionary DNA computing algorithm, but are subject to variations caused by the noisy character of the biological DNA operations. These variations, however, do not matter at all as long as the ratio λ/μ remains sufficiently large (> 10).

The *evaluation* of candidate solutions with respect to a particular optimization problem raises problems for the DNA variant of evolutionary algorithms, because an automatic exact assessment of quality as in the case of numerical computer implementations is not possible. It is well known for evolutionary algorithms in general, however, that they do not suffer from noisy function evaluations [34,50] and require no more than a method to judge whether a candidate solution performs better or worse than another one to implement the selection operator. In the particular case of the Hamiltonian path problem, a combined selection / evaluation operator could be realized in the laboratory by using a sequence of affinity purification steps to extract those DNA molecules that contain the largest number of different vertices (rather than all n different vertices as in the original experiment). What seems more suitable for evolutionary DNA-based computing, however, are techniques based on separating strands by length using gel electrophresis, where the length of a strand directly reflects the fitness of the solution represented by the strand. An example of such an application is discussed in the next section.

5 A Potential Application Example

In this section, we focus on a candidate problem for evolutionary DNA-based computation that has recently been implemented in the laboratory using the filtering method, namely the *maximum clique problem* [55]. The maximum clique problem is introduced in Section 5.1, and the filtering approach used in [55] is outlined in Section 5.2. As an alternative, a naive evolutionary algorithm for maximum clique is outlined in Section 5.4, and experimental results are given in Section 5.5. It is important here to emphasize that the goal of these experimental investigations is *not* to compete with any of the specialized heuristics described in the literature for the maximum clique problem (see, e.g., [56] for classical approaches, and [19,20,49] for evolutionary approaches combined with local search heuristics). Rather, the goal is to investigate the working hypothesis that the evolutionary DNA-based computation approach can exploit the benefits of massive population sizes which become possible in the framework of molecular computation.

5.1 The Maximum Clique Problem

The maximum clique problem consists of finding a largest subset of vertices of a graph such that all of those vertices are connected by edges with each other. Thus, if $G = (V, E)$ denotes a graph where V is the set of nodes and E the set of edges, the problem is to determine a subset $V' \subseteq V$ such that $\forall i, j \in V' : \langle i, j \rangle \in E$ where $|V'|$ is maximal for all such subsets. This problem is encountered in many different real-life applications such as cluster analysis, information retrieval, mobile networks, and computer vision (see, e.g., the survey paper [56]).

In the following, we reformulate the problem so as to use Stinson's terminology for combinatorial optimization problems [70]:

Problem instance: A graph $G = (V, E)$, where $V = \{1, 2, \ldots, n\}$ is the set of vertices and $E \subseteq V \times V$ the set of edges. An edge between vertices i, j is denoted by the pair $\langle i, j \rangle \in E$, and we define the *adjacency matrix* (e_{ij}) according to

$$e_{ij} = \begin{cases} 1 & , \quad \text{if } \langle i, j \rangle \in E \\ 0 & , \quad \text{otherwise} \end{cases} .$$

Feasible solution: A set V' of nodes such that $\forall i, j \in V' : \langle i, j \rangle \in E$ (i.e., $e_{ij} = 1$).

Objective function: The size $|V'|$ of the clique V'.

Optimal solution: A clique V' that maximizes $|V'|$.

The maximum clique problem is highly intractable. It is one of the first problems that has been proven to be NP-complete [44]. Moreover, even its approximations within a constant factor are NP-hard [28]. In particular, a recent result [37] states that if $NP \neq P$, then no polynomial time algorithm can approximate the maximum clique to within a factor of $n^{1-\epsilon}$ for any $\epsilon > 0$. Due to these strong negative results on the computational complexity of the maximum clique problem, it is very desirable to design efficient heuristics that yield satisfactory sub-optimal solutions for this problem.

5.2   A DNA-Based Algorithm

The DNA-based algorithm for maximum clique as presented and executed in the laboratory by [55] is based on the filtering approach, i.e., the optimal solution is filtered out from the set of all candidate solutions. Assuming the existence of a suitable method to represent binary strings by DNA strands, the approach can be summarized as follows [55]:

Algorithm 4

```
1:   X := randomly generate DNA strands representing
         all candidates;
2:   Remove the set Y of all non-cliques from X: C = X − Y;
3:   Identify x* ∈ C with maximum value of Σⁿᵢ₌₁ xᵢ*;
         /* meaning smallest length; see below */
```

The double stranded DNA encoding approach presented in [55] uses value-position pairs $P_n(V_{n-1}, P_{n-1}) \ldots (V_0, P_0)$ (P_n is needed for PCR amplification) with $|P_i| = 20$bp (base pairs) and value sections of length $|V_i| = 0 \Leftrightarrow x_i = 1$ ($i \in V'$) and $|V_i| = 10$bp $\Leftrightarrow x_i = 0$ ($i \notin V'$). This encoding method implies that the fitness evaluation of candidate solutions can be done by determining the length of DNA strands, e.g., by means of gel electrophoresis. According to this encoding method, the length of strands can in principle vary between $20 \cdot (n + 1)$ and $20 \cdot (n + 1) + 10 \cdot n$ base pairs, and the optimal solution is represented by the shortest DNA strand found in C in step 3 of the algorithm.

Step 2 of Algorithm 4, the removal of strands not representing cliques, can be achieved by executing the following iterative procedure which removes sequentially all strands containing edges in E^C ($E^C = \{\langle i, j\rangle \mid \langle i, j\rangle \notin E\,\}$ denotes the set of all edges *not* in G) from the initial pool X [55]:

Algorithm 5

```
2:   for all ⟨i,j⟩ ∈ E^C do
          Divide test tube X into X_0 and X_1;
          Destroy all strings containing x_i = 1 in X_0;
          Destroy all strings containing x_j = 1 in X_1;
          X := X_0 ∪ X_1 does not contain ⟨i,j⟩ ∈ E^C;
     od
```

The destruction of DNA strands can be achieved by using restriction enzymes and excluding the cut strands from PCR amplification. The initial set of all possible strands necessary in step 1 of Algorithm 4 is generated by the parallel overlap assembly method briefly described in [55].

This method was successfully applied in the laboratory to a problem of size $n = 6$ with $|E^C| = 4$, i.e., four iterations of the loop given in Algorithm 5 were necessary (in general, the number of iterations of this loop depends quadratically on n). With the final length separation in step 3 of Algorithm 4, the length of the shortest DNA present was found to be 160bp, corresponding to the correct solution with $|V'| = 4$. However, as discussed in [55], the method is limited to problem sizes up to $n = 27$ nodes with picomole and $n = 36$ nodes with nanomole operations, and the authors also conclude that an evolutionary approach is needed to overcome this limitation. In the next section, we outline such an evolutionary approach.

5.3   An Evolutionary DNA-Based Algorithm

An evolutionary DNA computing algorithm for the maximum clique problem could be based on the fact that, by means of iteratively amplifying and mutating a population of strands, removing non-cliques from the population according to Algorithm 5, and selecting for shortest strands by means of gel electrophoresis, an optimal or near-optimal solution can be evolved rather than extracted from the initial population. To guarantee the absence of an optimal solution from the initial population, either the problem dimension has to be sufficiently large or the initial population has to be limited to contain multiple copies of only a few different sorts of DNA strands. The amplification of strands is easily achieved by the polymerase chain reaction, and the removal of non-cliques as well as the selection of shortest DNA strands are already available. Utilizing the current representation, the mutation of bit x_i from $x_i = 0$ to $x_i = 1$ in the binary representation would correspond to a deletion of the corresponding 10bp segment V_i from the DNA strand, while the inverse mutation would correspond to an insertion of V_i into the strand. Putting all these operators together, the evolutionary DNA computing algorithm for maximum clique is given by Algorithm 6:

Algorithm 6

```
1:  Generate an initial random population P, |P| ≪ 2ⁿ;
2:  while not terminate do
3:      P := amplify and mutate P;
4:      Remove the set Y of all non-cliques from P: P := P − Y;
5:      P := select shortest DNA strands from P;
6:  od
```

Although the algorithm presents only the mutation operator as an evolutionary variation method, it should be noticed that the algorithm can be extended to include recombination by means of the splicing operator.

It should also be noted that Algorithm 6 strongly resembles a (μ,λ)-genetic algorithm with a small μ ($\mu = 1$ would probably be possible) and a certain mutation rate, which depends on the practical implementation of mutation. The next section presents a computer algorithm as close as possible to a potential DNA-based realization so as to benefit from the insights obtained from studying experiments with increasing population sizes.

5.4 A Silicium-Based Evolutionary Algorithm for Maximum Clique

In order to encode the problem to use an evolutionary algorithm, we choose the following representation of a candidate solution as a binary string $(x_1, x_2, \ldots, x_n)$: $x_i = 1 \Leftrightarrow i \in V'$. This way, the ith bit indicates the presence ($x_i = 1$) or absence ($x_i = 0$) of vertex i in the candidate solution. Note that a particular bit string may (and will often happen to) represent an infeasible solution. Instead of trying to prevent this, we allow infeasible strings to join the population and use a penalty function approach to guide the search towards the feasible region [61]. The penalty term in the objective function has to be graded in the sense that the farther away from feasibility the string is, the larger its penalty term should be. The exact nature of the penalty function, however, is not of high importance if it fulfills the property of being graded (see [68]).

Taking this design rule for a penalty function into consideration, we developed the following fitness function to be maximized by the evolutionary algorithm:

$$f(\boldsymbol{x}) = \sum_{i=1}^{n} x_i \cdot \left(1 - n \cdot \sum_{j=i+1}^{n} x_j \cdot (1 - e_{ij}) \right) \rightarrow \max . \tag{23}$$

This fitness function penalizes infeasible strings $\boldsymbol{x}$ by a penalty of n for every missing edge between two nodes i and j in the candidate solution V'

represented by $\boldsymbol{x}$ (i.e., whenever $x_i = 1 \,\wedge\, x_j = 1 \,\wedge\, e_{ij} = 0$, a penalty is added). For feasible strings $\boldsymbol{x}$, $f(\boldsymbol{x}) \geq 0$ and the fitness value is given just by the number of nodes in the independent set represented by $\boldsymbol{x}$. In other words, if all strings were guaranteed to be feasible, the fitness of $\boldsymbol{x}$ would be given by the number of ones in the string, i.e., $f(\boldsymbol{x}) = \sum x_i$, and f would be a function of a unitation or bit counting function.

For the purpose of transferring the evolutionary computation approach to DNA-based computation, it seems reasonable to use the most simple population-based algorithm for trying to exploit the massive parallelism of DNA-based computation, namely the $(1,\lambda)$-algorithm, which is an instance of Algorithm 7 for $\mu = 1$:

Algorithm 7

```
t := 0;
initialize P(t) randomly, |P(t)| = μ;
while not terminate do
    P'(t) := ∅;
    for  i := 1 to λ do
        P'(t) := P'(t) ∪ {mutate-member_p(P(t))};
    od
    evaluate(P'(t));
    P(t + 1) :=  select_μ(P'(t));
    t := t + 1;
od
```

The operator `mutate-member` with mutation probability p randomly chooses an element from $P(t)$ and mutates this element by inverting each bit with probability p. A population of λ mutant individuals is generated from $P(t)$ by repeating this process λ times. These mutants are then evaluated, i.e., their fitness is determined, and the selection operator `select` deterministically selects the μ best individuals from $P'(t)$ to survive for the next generation (individuals worse than those are discarded). As introduced in Section 2, this operator is called (μ,λ)-selection.

For our purposes, we simplify Algorithm 7 even further by using $\mu = 1$, i.e., reducing the parent population size to one. This implies that the whole intermediate population $P'(t)$ consists of mutants generated from a single individual. This approach roughly reflects the idea that the selection step in DNA-based evolutionary computation would be implemented by means of gel electrophoresis selecting for the longest or shortest DNA strands (see Sections 5.2 and 5.3).

5.5   Experimental Results

The number of parameters of the $(1,\lambda)$-evolutionary algorithm defined by algorithm 7 for $\mu = 1$ is now reduced to two, namely the population size λ

and the mutation rate p, which can be calculated according to equation (18) from population size and problem dimension.

The main working hypothesis to be investigated here is related to the massive parallelism which becomes possible by DNA-based computation, providing population sizes of about 10^{13} candidate solutions, which is many orders of magnitude larger than the usual evolutionary algorithm population with its $10^2 - 10^5$ candidate solutions. Therefore, our working hypothesis is that even for the simple $(1,\lambda)$-genetic algorithm, realized by DNA computing, it should be possible to find solutions of increasing quality as the population size increases. This working hypothesis is investigated by performing computer simulations, with population sizes limited by the computational power available to values from the range $\lambda \in \{10, 100, 1000, 10000\}$.

Because all of the operations of one single generation would be executable in parallel in the case of evolutionary DNA-based computation, the experimental runs are all performed for an identical number of generations, *regardless of the value of* λ (i.e., $\lambda = 10000$ is treated as if it would require the same computational effort as $\lambda = 10$). The number of generations for each single run of the algorithm is, after a few initial tests demonstrating the stagnation of the search in locally optimal solutions after this number of generations, fixed to 200. The problem instances are taken from the second DIMACS International Implementation Challenge[8] on NP-hard problems (1993), but due to the fact that the evaluation of the objective function is computationally expensive, only problem instances of moderate dimensionality up to $n = 300$ are chosen. For each problem instance tested, 10 independent runs are executed and the average (best) objective function values found within these 10 runs are reported in Table 5.5, together with the known optimum values.

The results in Table 5.5 do not allow for drawing the unique conclusion that the working hypothesis can be confirmed, because the results essentially split into three different classes:

Class 1: Test problems where optimal solutions are obtained in all or almost all of the runs do not give any information about the impact of λ on solution quality. The problems `c-fat200-1`, `c-fat200-5`, `hamming6-2`, `hamming6-4`, `johnson8-4-4`, and `johnson16-2-4` belong to this class.

Class 2: Test problems where, regarding the average and/or the best objective function value, an improvement is obtained with increasing value of λ. This class contains problems such as `brock200_1`, `brock200_2`, `brock200_3`, `brock200_4`, `hamming8-4`, `keller4` (regarding the average) `p_hat300-1`, `p_hat300-2`, `p_hat300-3` (regarding the best).

Class 3: Test problems where no clear improvement is obtained, including `hamming8-2`, `c-fat200-2`, and all `san...` and `sanr...` problems, where the algorithm always stagnates in local optima of bad quality.

In summary, 15 of 23 test problems belong to either class 1 or 2, such that for roughly one-third of the problems our hypothesis could not be confirmed.

---

[8] See `ftp://dimacs.rutgers.edu/challenge/graph/benchmarks`

Problem	Nodes	λ				Opt.
		10	100	1000	10000	
brock200_1	200	-66.9 (14)	14.6 (17)	15.3 (17)	15.9 (19)	21
brock200_2	200	-173.6 (7)	7.6 (9)	7.3 (8)	7.8 (9)	12
brock200_3	200	-151.3 (10)	9.7 (11)	9.9 (12)	10.5 (12)	15
brock200_4	200	-5 (11)	10.7 (12)	11.5 (13)	11.7 (14)	17
c-fat200-1	200	-676 (6)	10.3 (12)	11 (12)	10.8 (12)	12
c-fat200-2	200	-545.5 (10)	22.1 (23)	22.5 (24)	22.2 (23)	24
c-fat200-5	200	9.1 (41)	56.9 (58)	57.8 (58)	57.6 (58)	58
hamming6-2	64	19.6 (24)	23.4 (32)	30.8 (32)	30.8 (32)	32
hamming6-4	64	4 (4)	4 (4)	4 (4)	4 (4)	4
hamming8-2	256	37.4 (70)	68.8 (82)	73.9 (104)	76.3 (91)	128
hamming8-4	256	-1246.1 (-758)	9.3 (12)	10 (12)	10.8 (16)	16
johnson8-4-4	70	9 (10)	14 (14)	14 (14)	14 (14)	14
johnson16-2-4	120	8 (8)	8 (8)	8 (8)	8 (8)	8
keller4	171	-26.6 (9)	7.6 (8)	8.4 (10)	9 (9)	11
p_hat300-1	300	-16276 (-8989)	5.6 (6)	5.5 (7)	5.7 (7)	8
p_hat300-2	300	-7725 (-2091)	17.7 (19)	17.2 (19)	18.8 (20)	25
p_hat300-3	300	-3671 (-280)	25.9 (29)	25.9 (31)	26.3 (30)	36
san200_0.7_1	200	-85.1 (16)	15.2 (16)	15 (15)	15 (15)	30
san200_0.7_2	200	-7.9 (12)	12 (12)	12 (12)	12.1 (13)	18
san200_0.9_1	200	43.6 (46)	45.3 (46)	45.3 (46)	45.2 (46)	70
san200_0.9_2	200	11.2 (37)	35.4 (42)	35.9 (38)	33.7 (38)	60
sanr200_0.7	200	-28.4 (13)	12.6 (14)	13.1 (14)	13.1 (14)	18
sanr200_0.9	200	28.4 (31)	30 (35)	32.4 (38)	32.2 (35)	42

Table 1. Average (best) clique sizes found within 10 runs after 200 generations

Arguably, the structure of these problems is more difficult for the simple, fixed rate mutation operator used in the experiment.

6   Conclusions and Further Work

The major goal of this paper is to outline the idea of evolutionary DNA-based computing by making use of the variety of useful techniques and the theoretical background available in evolutionary computation. Moreover, a proposal for a test case was also outlined, and some simulation-based experimental evidence supporting our claim concerning the benefits of massive population sizes was given. While these results are encouraging for our investigations, the proof of principle from running the algorithm in vitro was not yet given. In order to facilitate the DNA-based implementation, a number of questions have to be answered in cooperation with molecular biologists. These questions concern the problem of implementing mutation so as to reflect the concept of bit inversion on the DNA level, the problem of length-based fitness evaluation in the presence of DNA strands representing infeasible solutions, the question of scalability to higher problem dimension, and the efficient execution of the main loop iterations of the evolutionary algorithm. Tackling these questions forms an important part of a research project at the Leiden Center for Natural Computing.

Acknowledgements. The research of Thomas Bäck is supported by a grant of the German *Federal Ministry of Education, Science, Research and Technology* (BMBF). The authors are responsible for the contents of this publication.

References

1. *Proceedings of the Fourth IEEE Conference on Evolutionary Computation, Indianapolis, IN.* IEEE Press, Piscataway, NJ, 1997.
2. L. M. Adleman. Molecular computation of solutions to combinatorial problems. *Science*, 266:1021–1024, November 1994.
3. M. Amos. *DNA Computing*. PhD thesis, The University of Warwick, Warwick, UK, September 1997.
4. T. Bäck. The interaction of mutation rate, selection, and self-adaptation within a genetic algorithm. In R. Männer and B. Manderick, editors, *Parallel Problem Solving from Nature 2*, pages 85–94. Elsevier, Amsterdam, 1992.
5. T. Bäck. Optimal mutation rates in genetic search. In S. Forrest, editor, *Proceedings of the Fifth International Conference on Genetic Algorithms*, pages 2–8. Morgan Kaufmann, San Mateo, CA, 1993.
6. T. Bäck. Generalized convergence models for tournament- and (μ,λ)-selection. In L. Eshelman, editor, *Proceedings of the 6th International Conference on Genetic Algorithms*, pages 2–8. Morgan Kaufmann, San Francisco, CA, 1995.
7. T. Bäck. *Evolutionary Algorithms in Theory and Practice*. Oxford University Press, New York, 1996.
8. T. Bäck. Self-adaptation. In T. Bäck, D. B. Fogel, and Z. Michalewicz, editors, *Handbook of Evolutionary Computation*, chapter C7.1. Oxford University Press, New York, and Institute of Physics Publishing, Bristol, 1997.
9. T. Bäck, U. Hammel, and H.-P. Schwefel. Evolutionary computation: History and current state. *IEEE Transactions on Evolutionary Computation*, 1(1):3–17, 1997.
10. T. Bäck and F. Hoffmeister. Extended selection mechanisms in genetic algorithms. In R. K. Belew and L. B. Booker, editors, *Proceedings of the Fourth International Conference on Genetic Algorithms*, pages 92–99. Morgan Kaufmann, San Mateo, CA, 1991.
11. T. Bäck, G. Rudolph, and H.-P. Schwefel. Evolutionary programming and evolution strategies: Similarities and differences. In D. B. Fogel and W. Atmar, editors, *Proceedings of the Second Annual Conference on Evolutionary Programming*, pages 11–22. Evolutionary Programming Society, San Diego, CA, 1993.
12. T. Bäck and M. Schütz. Intelligent mutation rate control in canonical genetic algorithms. In Z. W. Ras and M. Michalewicz, editors, *Foundations of Intelligent Systems, 9th International Symposium, ISMIS '96*, volume 1079 of *Lecture Notes in Artificial Intelligence*, pages 158–167. Springer, Berlin, 1996.
13. J. M. Barreiro, J. Rodrigo, and A. Rodríguez-Patón. Evolutionary biomolecular computing. *Romanian Journal of Information Science and Technology*, 1(1):201–206, 1998.

14. H.-G. Beyer. Toward a theory of evolution strategies: The (μ,λ)-theory. *Evolutionary Computation*, 2(4):381–408, 1994.
15. H.-G. Beyer. Toward a theory of evolution strategies: On the benefits of sex — the $(\mu/\mu, \lambda)$-theory. *Evolutionary Computation*, 3(1):81–111, 1995.
16. H.-G. Beyer. Toward a theory of evolution strategies: Self-adaptation. *Evolutionary Computation*, 3(3):311–348, 1995.
17. H.-G. Beyer. An alternative explanation for the manner in which genetic algorithms operate. *BioSystems*, 41:1–15, 1997.
18. T. Blickle and L. Thiele. A comparison of selection schemes used in evolutionary algorithms. *Evolutionary Computation*, 4(4):361–394.
19. I. M. Bomze, M. Pelillo, and R. Giacomini. Evolutionary approach to the maximum clique problem: Empirical evidence on a larger scale. In I. M. Bomze, editor, *Developments in Global Optimization*, pages 95–108. Kluwer, Amsterdam, 1997.
20. Thang Nguyen Bui and P. H. Eppley. A hybrid genetic algorithm for the maximum clique problem. In L. Eshelman, editor, *Genetic Algorithms: Proceedings of the 6th International Conference*, pages 478–484. Morgan Kaufmann, San Francisco, CA, 1995.
21. L. Castellanos, S. Leiva, J. Rodrigo, and A. Rodríguez-Patón. Molecular computation for genetic algorithms. In L. Polkowski and A. Skowron, editors, *Rough Sets and Current Trends in Computing, RSCTC '98, Warsaw, June 98*, volume 1424 of *Lecture Notes in Artificial Intelligence*, pages 91–98. Springer, Berlin, 1998.
22. H. A. David. *Order Statistics*. Wiley, New York, 2nd edition, 1981.
23. R. Deaton, R. C. Murphy, J. A. Rose, M. Garzon, D. R. Franceschetti, and Jr. S. E. Stevens. A DNA based implementation of an evolutionary search for good encodings for DNA computation. In [1], pages 267–271.
24. D. C. Dennett. *Darwin's Dangerous Idea*. Simon & Schuster, New York, 1995.
25. A. E. Eiben, P.-E. Raué, and Zs. Ruttkay. Genetic algorithms with multiparent recombination. In Y. Davidor, H.-P. Schwefel, and R. Männer, editors, *Parallel Problem Solving from Nature — PPSN III International Conference on Evolutionary Computation*, volume 866 of *Lecture Notes in Computer Science*, pages 78–87. Springer, Berlin, 1994.
26. M. Eigen. Macromolecular evolution: Dynamical ordering in sequence space. *Ber. Bunsenges. Phy. Chem.*, 89:658–667, 1985.
27. L. J. Eshelman, R. A. Caruna, and J. D. Schaffer. Biases in the crossover landscape. In [63], pages 10–19.
28. U. Feige, S. Goldwasser, S. Safra, L. Lovász, and M. Szegedy. Approximating clique is almost NP-complete. In *Proceedings of the 32nd Annual IEEE Symposium on the Foundations of Computer Science (FOCS)*, pages 2–12, 1991.
29. T. C. Fogarty. Varying the probability of mutation in the genetic algorithm. In [63], pages 104–109.
30. D. B. Fogel. *Evolutionary Computation: Toward a New Philosophy of Machine Intelligence*. IEEE Press, Piscataway, NJ, 1995.
31. L. J. Fogel, A. J. Owens, and M. J. Walsh. *Artificial Intelligence through Simulated Evolution*. Wiley, New York, 1966.
32. M. R. Garey and D. S. Johnson. *Computers and Intractability — A Guide to the Theory of NP–Completeness*. Freemann & Co., San Francisco, CA, 1979.
33. D. E. Goldberg. *Genetic Algorithms in Search, Optimization and Machine Learning*. Addison-Wesley, Reading, MA, 1989.

34. U. Hammel and T. Bäck. Evolution strategies on noisy functions: How to improve convergence properties. In Y. Davidor, H.-P. Schwefel, and R. Männer, editors, *Parallel Problem Solving from Nature — PPSN III, International Conference on Evolutionary Computation*, volume 866 of *Lecture Notes in Computer Science*, pages 159–168. Springer, Berlin, 1994.

35. J. Hartmanis. On the computing paradigm and computational complexity. In Jiří Wiederman and Petr Hájek, editors, *Mathematical Foundations of Computer Science 1995. 20th International Symposium, MFCS '95, Prague, Czech Republic*, volume 969 of *Lecture Notes in Computer Science*, pages 82–92, Springer, Berlin, 1995.

36. J. Hartmanis. On the weight of computations. *Bulletin of the European Association for Theoretical Computer Science*, 55:136–138, February 1995.

37. J. Hastad. Clique is hard to approximate within $n^{1-\epsilon}$. In *Proceedings of the 37th Annual IEEE Symposium on the Foundations of Computer Science (FOCS)*, pages 627–636, 1996.

38. T. Head. Formal language theory and DNA: An analysis of the generative capacity of recombinant behaviors. *Bulletin of Mathematical Biology*, 49:737–759, 1987.

39. T. Head. Splicing systems and molecular processes. In [1], pages 203–205.

40. J. H. Holland. *Adaptation in Natural and Artificial Systems*. University of Michigan Press, Ann Arbor, MI, 1975.

41. J. H. Holland, K. J. Holyoak, R. E. Nisbett, and P. R. Thagard. *Induction: Processes of Inference, Learning, and Discovery*. MIT Press, Cambridge, MA, 1986.

42. K. A. De Jong. *An analysis of the behaviour of a class of genetic adaptive systems*. PhD thesis, University of Michigan, 1975. Diss. Abstr. Int. 36(10), 5140B, University Microfilms No. 76–9381.

43. L. Kari. From micro-soft to bio-soft: computing with DNA. *TUCS General Publication 6*, Turku Centre for Computer Science, Turku, Finland, August 1997.

44. R. M. Karp. Reducibility among combinatorial problems. In *Complexity of Computer Computations*, pages 85–103. Plenum Press, New York, 1972.

45. S. A. Kauffman. *The Origins of Order. Self-Organization and Selection in Evolution*. Oxford University Press, New York, NY, 1993.

46. J. R. Koza. The genetic programming paradigm: Genetically breeding populations of computer programs to solve problems. In Branko Souček and the IRIS Group, editors, *Dynamic, Genetic, and Chaotic Programming*, chapter 10, pages 203–321. Wiley, 1992.

47. E. Laun and K. J. Reddy. Wet splicing systems. In *Proceedings of the Third Annual Meeting on DNA Based Computers*. University of Pennsylvania, 1997.

48. R. Lipton. DNA solution of hard computational problems. *Science*, 268:542–545, April 1995.

49. E. Marchiori. A simple heuristic based genetic algorithm for the maximum clique problem. In *ACM Symposium on Applied Computing (SAC)*, pages 366–373, 1998.

50. B. L. Miller and D. E. Goldberg. Genetic algorithms, selection schemes, and the varying effects of noise. *Evolutionary Computation*, 4(2):113–132, 1996.

51. H. Mühlenbein. How genetic algorithms really work: I. mutation and hillclimbing. In R. Männer and B. Manderick, editors, *Parallel Problem Solving from Nature 2*, pages 15–25. Elsevier, Amsterdam, 1992.

52. H. Mühlenbein and D. Schlierkamp-Voosen. Predictive models for the breeder genetic algorithm. *Evolutionary Computation*, 1(1):25–49, 1993.

53. H. Mühlenbein and D. Schlierkamp-Voosen. The science of breeding and its application to the breeder genetic algorithm (BGA). *Evolutionary Computation*, 1(4):335–360, 1993.

54. C. K. Oei, D. E. Goldberg, and S.-J. Chang. Tournament selection, niching, and the preservation of diversity. IlliGAL Report 91011, University of Illinois at Urbana-Champaign, December 1991.

55. Qi Ouyang, P. D. Kaplan, S. Liu, and A. Libchaber. DNA solution of the maximal clique problem. *Science*, 278:446–449, 1997.

56. P. M. Pardalos and J. Xue. The maximum clique problem. *Journal of Global Optimization*, 4:301–328, 1994.

57. G. Păun, G. Rozenberg, and A. Salomaa. *DNA Computing.* Texts in Theoretical Computer Science – An EATCS Series. Springer, Berlin, 1998.

58. V. A. Ratner, A. A. Zharkikh, N. Kolchanov, S. N. Rodin, V. V. Solovyov, and A. S. Antonov. *Molecular Evolution.* Springer, Berlin, 1996.

59. I. Rechenberg. *Evolutionsstrategie: Optimierung technischer Systeme nach Prinzipien der biologischen Evolution.* Frommann–Holzboog, Stuttgart, 1973.

60. I. Rechenberg. *Evolutionsstrategie '94*, volume 1 of *Werkstatt Bionik und Evolutionstechnik.* Frommann–Holzboog, Stuttgart, 1994.

61. J. T. Richardson, M. R. Palmer, G. Liepins, and M. Hilliard. Some guidelines for genetic algorithms with penalty functions. In [63], pages 191–197.

62. G. Rudolph. Convergence analysis of canonical genetic algorithms. *IEEE Transactions on Neural Networks, Special Issue on Evolutionary Computation*, 5(1):96–101, 1994.

63. J. D. Schaffer, editor. *Proceedings of the Third International Conference on Genetic Algorithms.* Morgan Kaufmann, San Mateo, CA, 1989.

64. J. D. Schaffer, R. A. Caruana, L. J. Eshelman, and R. Das. A study of control parameters affecting online performance of genetic algorithms for function optimization. In [63], pages 51–60.

65. H.-P. Schwefel. *Numerische Optimierung von Computer-Modellen mittels der Evolutionsstrategie*, volume 26 of *Interdisciplinary Systems Research.* Birkhäuser, Basel, 1977.

66. H.-P. Schwefel. *Numerical Optimization of Computer Models.* Wiley, Chichester, 1981.

67. H.-P. Schwefel. *Evolution and Optimum Seeking.* Sixth-Generation Computer Technology Series. Wiley, New York, 1995.

68. A. E. Smith and D. M. Tate. Genetic optimization using a penalty function. In S. Forrest, editor, *Proceedings of the Fifth International Conference on Genetic Algorithms*, pages 499–505. Morgan Kaufmann, San Mateo, CA, 1993.

69. J. Smith and T. C. Fogarty. Self adaptation of mutation rates in a steady state genetic algorithm. In *Proceedings of the Third IEEE Conference on Evolutionary Computation*, pages 318–323. IEEE Press, Piscataway, NJ, 1996.

70. D. R. Stinson. *An Introduction to the Design and Analysis of Algorithms*, 2nd edition, The Charles Babbage Research Center, Winnipeg, Manitoba, Canada, 1987.

71. G. Syswerda. Uniform crossover in genetic algorithms. In [63], pages 2–9.

Evolution at the Edge of Chaos: A Paradigm for the Maturation of the Humoral Immune Response

Patricia K. Theodosopoulos and Theodore V. Theodosopoulos

Abstract. We study the maturation of the antibody population following primary antigen presentation as a global optimization problem. Emphasis is placed on the trade-off between the safety of mutations that lead to *local* improvements to the antibody's affinity and the necessity of eventual mutations that result in *global* reconfigurations in the antibody's shape. The model described herein gives evidence of the underlying optimization process from which the rapidity and consistency of the biologic response could be derived.

1 Introduction

The study of the mechanisms underlying the physiology of the immune system has been a very promising area for applications of mathematical models. The spectacular success of the healthy immune system to recognize the combinatorial plethora of antigenic agents while being endowed with a substantially smaller repertoire of immunoglobulin (Ig) molecules is a problem that shares much in common with mathematical problems of combinatorial complexity and optimization.

This article focuses on the maturation of the humoral immune response as a conveniently fast-paced example of selection-based evolution. Through this example, we attempt to investigate the interplay between lessons learned from mathematical models of global optimization and the need for theoretical models in the biological sciences.

The paper begins by providing a brief overview of the B cell immune response. The following sections provide the biological context for the presentation of our model and results, and a summary of the techniques involved in our analysis. The presentation of our model and results proceeds in five steps of increasing complexity which assume a familiarity with the corresponding sections in the biological discussion of the model.

2 The Naive B Cell Repertoire

The ability of the specific immune response to recognize and respond to myriad foreign antigen challenges rests in the generation of diversity at different stages in the development of the immune cells. With regards to the B cell, diversity is first explored at the time of antibody naive maturation with somatic recombination from a pool of genes that construct the variable and constant domains of the heavy and light chains of the antibody molecule, along with their inherent combinatorial freedom. At this stage, there is additional diversity introduced via imprecise rearrangements and junctional nucleotide insertions. It is now believed that conformational isomerism of Ig molecules may also add additional diversity to the primary repertoire [16]. Conservatively, these processes have been calculated to generate a repertoire on the order of 10^9 [14].

3   B Cell Response

The introduction of foreign antigen into the host results in a complex response that occurs rapidly and effectively. The initial phase following antigen introduction involves elimination via innate immunity. The mediators of this response are nonspecific, including the phagocytic cells, complement, and NK cells. Subsequently, the mediators of specificity in the immune response, represented by the T cells and B cells, are activated following interactions with macrophages, and other soluble factors. The evolution of specificity into the immune repertoire greatly enhanced the organism's ability to respond to foreign invaders, and even more importantly, to develop memory of this invader that is protective upon reintroduction of the pathogen.

Antigen localization following exposure occurs within two compartments in the lymph node. The first is the primary follicles, which are comprised of antigen–antibody complexes and follicular dendritic cells that present antigen to a circulating population of B cells. The second compartment is the paracortex. Within minutes of introduction, the antigen is taken up by the phagocytic cells present in the paracortex. These cells will process the antigen for presentation via MHC to specific CD4+ T cells. The interaction of the T cells, antigen-presenting cells (APCs), and B cells will give rise to a population of low affinity B cells that can generate large amounts of antibody for a fixed period of time, called the plasmacytes of the primary response, and another population that will give rise to the secondary follicles and germinal centers (GC) where hypermutation will take place [14,2]. Usually only 3–5 of these antigen stimulated B cells enter a given follicle and there they undergo exponential growth that fills the follicular network in 3–4 days [2]. These cells, called centroblasts, are believed to lose surface Ig expression and undergo hypermutation. It is believed that these hypermutating cells undergo several rounds of mutation followed by selection within the microenvironment of the germinal center [23,6]. The selection process is believed to encompass both a positive selection for higher affinity and a negative selection barrier to remove clones that have developed self-recognizing phenotype or other detrimental mutations [20]. Those cells that pass the selection barriers enter the circulation as high affinity plasma cells or memory cells.

4   The Process of Hypermutation

The primary repertoire appears to be sufficient for the organism to recognize with a certain threshold affinity, and in some cases even high affinity, the antigenic challenges presented by the environment. The additional mechanism of somatic hypermutation, that in humans appears to be primarily antigen driven, improves the affinity of the antibody for the antigen by two orders of magnitude or more, with some expense of energy and cells, and some risk (i.e., autoimmunity and malignancy) to the organism. It has been hypothesized that this process is simply an evolutionary relic [3] that was initially needed to generate primary repertoire locally. However, it seems unlikely that this process which produces immunologic memory and high affinity effector antibody molecules at some expense and risk to the organism is redundant. It provides at the very least the security that in a world of countless and evolving pathogens, the organism can protect itself, but one might also suspect that the true utility of this process may even extend beyond antigen response and into a larger control mechanism for the organism, whose role may be appreciated more during early development, [21] or during the shaping and maintenance of memory.

Although the actual mechanism of hypermutation is still not understood, it appears to have some link to the transcription process, and several models have been suggested along these lines [17,32,11,3]. The mutations introduced are primarily point mutations, although deletions and insertions do occur, and more frequently than previously suspected [24]. Mutations seem to occur preferentially in the region bounded by the transcriptional promoter at the $5'$ end, and the C gene at the $3'$ end. The pattern displayed is that of a rapid peak in mutation frequency, followed by a slow decline out to about 1.5–2kb downstream [14,26]. The regions of both light and heavy chain V genes that are selected have an average of 3–13 mutations, but can have upwards of 20 [19]. The primary targets for hypermutation are the CDRs, or complementary determining regions, of which there are three in both heavy and light chain, separated from each other by intervening framework sequences(FRW). The CDRs are only a few residues in length but their position in the protein molecule and configuration in three-dimensional space make them crucial in the evolution of diverse antigen combining sites [14].

The substrates within each CDR that are frequently seen mutated are defined as "hotspots". They are described by preferences for purines, rather than pyrimidines, as well as for particular codons, or codon motifs within the sequence. The fact that mutation in a *hotspot* can create or delete other hotspots indicates a higher order structure to the mutation process than that which is currently observable [5]. In vitro random mutagenesis studies show loss of around 50% of clones that accumulate more than one mutation [42]. This is due to the effects of both diminished antigen binding, as well as loss of expression of a functional Ig molecule. These cells are believed to undergo apoptosis, perhaps mediated via T cells [27,1]. Despite the evidence suggesting high loss and apoptosis in germinal centers [27], true numbers of in vivo loss are not well documented. If mutation results in the production of a functional Ig molecule, then it is believed to be tested for affinity against the available antigen trapped in the follicular dendritic network of the particular germinal center. The role of competition for limited antigen, although figuring prominently in prior models [23], is still being elucidated [31]. Following this process of selection, which appears to have both a positive and a negative barrier as described above, the high affinity antibody-producing B cell may leave the germinal center and enter the circulation as a plasma cell or a memory cell [20].

5 The Model

The model presented here attempts to delineate a selection-based model of the evolution of the affinity-matured antibody. The contribution of the microenvironment in the germinal center as well as the intrinsic properties of primary repertoire antigen– antibody interactions versus affinity-matured interactions are considered. In addition, the desire to understand such observations as repertoire shift of variable region genes in the memory compartment, as well as to suggest an underlying mechanism of somatic hypermutation which is an unique adaptive evolutionary process in mature organisms are considered. The following sections give a more detailed biological context within which the methods and results can be interpreted.

5.1 Local Steps versus Global Jumps

This component attempts to model the biological "trade-off" that occurs during the mutation process and allows the rapid generation of high affinity antibodies. One might understand this trade-off in terms of the mutations that produce only local changes in the conformation and are therefore more likely, although not exclusively, to produce incremental changes in the affinity, versus those mutations that produce more global changes in conformation and therefore might be expected to produce rather large *jumps* in affinity.

In terms of affinity, the prior treatment of affinity changes and mutational studies lead to the idea that through the mutation process, the selection is for those clones that undergo a stepwise increase in affinity [33,5] – an additive effect of changes that create new H bonds or new weak electrostatic or hydrophobic interactions between the residues and associated solvent molecules [7,1,4]. However, it is observed that all codon changes cannot be translated into stepwise energetic changes [7]. In the literature, this affinity increase is often correlated with a lower K_{off} more so than a higher K_{on} for affinity measurements, although which one is more important for overall affinity increase of Igs is still unclear [41,1]. With regards to the changes in conformation, the nature of the affinity change secondary to the stepwise energetic changes in the selected antibodies has also led to the idea that the conformation is handled likewise. This progression to a *lock and key* conformation occurs at the expense of entropy in exchange for a decrease in free energy and a commensurate increase in affinity [41]. Since we cannot reliably observe the process, we cannot presume that this stepwise search is what is always functioning in the germinal center. We predict that the rapid elaboration of high affinity antibodies through the germinal center reaction may necessitate that the system occasionally make a large *jump* in order to better sample the affinity landscape.

The positions of positively selected mutations show that replacement mutations occur preferentially in the CDRs versus the intervening framework regions (FRWs). The FRW was often described as being very sensitive to replacement mutations, but it appears now that they too can tolerate a certain number of replacement mutations, and that the CDRs may alternately, possess a sensitivity to mutation through the coding structural elements as well [42]. The greatest diversity a priori is seen in the CDR 3, which appears to have the most contact residues with the antibody, while the other CDRs usually comprise the sides of the binding pocket [21]. During hypermutation, it is often in CDR 1 and 2 that one observes most of the mutations, whereas in CDR 3, there are relatively fewer, and they do not usually affect the existing contact residues [21,13,40]. We might then hypothesize that the local steps will result preferentially from mutations in CDR 1 and 2 and that global conformation changes might occur from CDR 3 or even FRW mutations. This is of course not absolute, as experimentally, CDR 2 regions have also been seen to contribute to the binding pocket, and to have long range interactions at certain residues that make mutations in them change the antibody conformation significantly [42]. Furthermore, certain base pair positions that are frequently mutated in the CDRs may create conservative local changes, while mutations outside of these positions may be more frequently associated with global changes [40,26]. The less frequently mutated codons are more common within the CDR 3, and this CDR experiences fewer mutations than the other CDRs or FRW regions, thus supporting the above generalization.

Growing evidence suggests that the primary repertoire is composed of multivalent, and highly flexible Igs that *conform* to the antigen, but through hypermutation they generate a rigid *lock and key* fit. Studies comparing germline diversity with hypermutated V genes showed that the amino acid differences introduced by mutation were fewer than the underlying diversity of the primary repertoire, and further suggested that through mutation, the more conserved residues of the CDR 1 and 2 that often create the periphery of the binding site are favored for mutation, whereas the more diverse residues are generally not mutated [13]. Although exceptions were cited for both of these generalizations, this supports a notion that the mutation mechanism has evolved to focus mutations primarily in those residues which, in the three-dimensional geometry of the CDR loops, will create a tighter fit for the antigen, and decrease flexibility of CDR 1 and CDR 2 at the periphery of the binding pocket [40]. The diversity of the primary repertoire can rarely achieve this energetic feat.

An additional mechanism for global jumps may be appreciated from the relative frequency of deletions and insertions. Recent work has identified deletions and insertions in single cell analysis of GC derived B cells at a frequency much higher than previously suspected, in the range of 4–16 percent of in-frame rearrangements [24]. These types of alterations would be suspected to contribute greatly to the occurrence of global changes in conformation and affinity.

5.2   Consistency of Response

Another component of the model will be the robustness of this evolutionary optimization to the diverse population of antigens that the organism faces. How do the mutation and selection processes ensure consistency in their response? The resulting affinities as well as the timing of response exhibit remarkable consistency both within the response to a given antigen and across responses to diverse antigens. The naive repertoire usually produces antibodies with affinities on the order of $10^5 M^{-1}$, and the somatic hypermutation process produces antibodies with a range of affinities from 10^6 to $10^8 M^{-1}$ within a period of days from a finite number of clones.

Additional consistency arises within a response to a given antigen, as the high affinity clones share many favorable mutations [31]. Although other amino acid substitutions in that same position also confer high affinity, they are not selected for or observed in the mature response [31]. This is considered as evidence of a negative selection barrier in the process that might be protecting against harmful mutations [31,20], although it may be an intrinsic feature of the mutating sequences. There is also evidence of consistency in specific base pair positions selected for mutation across different responses in both productive and non-productive rearrangements [13]. The codons in the CDRs seem, by their nature, to be predisposed in favor of replacement mutations during hypermutation. How does the system work within the constraints of the available number of clones, the timing observed for response, and even the physical or energetic limitations of the mechanism to ensure a consistent response? [30]. The potential benefits of the germinal center may be discussed in this respect.

The compartmentalization of the germinal center provides ease of interaction of the necessary components, ability to segregate beneficial from detrimental mutations in a controlled fashion, perhaps decreased energy expenditure for the organism, and improved diversity of antibodies, since each germinal center appears to function autonomously [31]. If we treat our model according to this compartmentalized representation, the parameters are defined independently within each germinal center. The initial number of clones entering a given GC is observed to be 3–5 antigen stimulated B cells that have received helper T cell signals [14]. Previous models have suggested that these cells enter a division phase that fills the GC over a period of 3–4 days [28,23], stop expressing surface Ig, and begin hypermutation. This would be followed by selection against the antigen trapped within the individual GC follicular dendritic network. Current evidence favors several rounds of mutation and selection [23], and more than one mutation per cycle of hypermutation [30,18]. In between these rounds of selection and mutation, one might presume that the positively selected clones would have to be given an advantage, i.e., by generating more offspring, but how this is translated into a number of cell divisions is not clear, nor is how the final decision is made to allow the cell to exit the GC when and if it attains a high enough affinity. Maybe it is a time-related phenomenon, or perhaps an affinity threshold controlled process. Even with the benefit of the fastest observed generation time in the GC, which is 6 hours [30], the 3–4 days of generation time gives a population of 10^4-10^5. From this starting point, we want to understand the probability of generating a high affinity clone, and how many mutations it will take to get us there.

5.3   The Affinity Threshold

A third component will be to treat the dependence of response time on the affinity threshold level. Following the initial antigen presentation, at around day 5, there appear in circulation low affinity plasma cells. The germinal center begins to form a few days after antigen exposure, and the first mutated V genes are detected around day 5–7 [30,37]. Samples of V genes throughout this primary GC reaction show increasing numbers of mutations over time [30]. Subsequent immunization produces a fast response of high affinity antibodies, usually within 1–3 days [14]. The V genes of the high affinity antibodies of secondary and tertiary immune responses are also seen to have more accumulated mutations, but the incremental increase in affinity following the primary response is low [14,15]. This implies a point of diminishing returns for this process.

Interestingly, the high affinity antibody molecules of the secondary response frequently use different V genes than the primary response population. This is referred to as *repertoire shift* and there does not appear to be enough time for these cells to be created de novo from the newly forming germinal centers. Therefore, these cells must have evolved into memory either late in the primary germinal center response or during the interval between. This of course presumes adequate time between inoculations, as too short or long an interval between exposures will produce a diminished response. In humans, hypermutation seems to be largely confined to the germinal center. The reaction in the GC lasts about 2–3 weeks, although antigen has been shown to remain on follicular dendritic cells for years following the primary immune response [14] and Ig-expressing B cell blasts can be evident for months following initial exposure [28]. It has been suggested that this remaining antigen is the force behind the shaping of the memory immune response between exposures, and that it also accounts for the repertoire shift. In addition, the memory compartment for a given epitope is often oligoclonal, whereas the high affinity late primary stage clonotypes can be numerous [31]. It is possible that following the primary exposure, there are two populations of cells generated through the germinal center– a fast response effector population and a slow response memory population. It is interesting to note that there may be an energetic advantage to some of these repertoire shifted memory cells–that they do not necessarily have higher affinity, but perhaps they cross an energy barrier more easily [15]. This would imply kinetic selection as opposed to affinity selection during this phase of the immune response. Attempting to understand these other selection parameters in shaping the memory response would be an interesting prospect for future models.

5.4 Evolution of the Mutation Dynamics

The fourth component deals with the evolution of the above outlined trade-off and how it is optimized in the organism. The observed mutation distribution is not random, because if it were, the degeneracy rate predicted would be too high. Instead it has evolved to target certain sites preferentially [31,13]. Intrinsic hotspots are usually characterized as 3–4 base pair sequences such as AGC and TAC and their inverted repeats and RGYW motifs (R=purine, Y=pyrimidine and W=A,T) [43,12]. These motifs are shown to be hotspots independent of antigenic selection, and concentrated in the CDRs, often in an overlapping fashion [12,14]. Other characteristics of this mutator mechanism include a preference for point mutations over deletions/insertions, a bias against mutations in thymidine and in favor of mutations in purines instead of pyrimidines. Mutations are favored in non-degenerate sites, and the replacement/silent mutation (R/S) ratios are higher in the CDRs in both selected and non-selected V genes, also suggesting intrinsic targeting of these areas [12]. Most significantly, in experiments with a light chain transgene, silent mutation in one part of the gene resulted in loss of a hotspot motif and in the appearance and loss of hotspots in other areas [18]. This argues for a higher order template as well as an evolving dynamic with loss and acquisition of mutations. This higher order structure may be conferred by DNA folding, or perhaps DNA–protein interactions [18]. Nevertheless, this dynamic might be expected to exhibit some convergence in order to maintain the consistency observed between individual responses.

6 The Evolutionary Landscape

As described in the previous sections, we treat the affinity maturation of the primary humoral immune response as a problem of global optimization. This paradigm should be contrasted with the "population dynamics" approach. The latter class of models entails the tallying of individual immune cell types and the investigation of the transition dynamics between their allowable states. Such models stress the emergence of affinity optimization as a result of these cell population dynamics. In this vein, it is generally the evolution of the average affinity in the population that is the dominant variable.

The paradigm we employ here begins instead by considering the problem of affinity optimization. We study the sources of complexity in this problem and infer general principles that this optimization must adhere to, avoiding whenever possible making ad hoc assumptions about the particular mechanics at play. In this regard, our model concentrates on the hypermutation process inside the germinal centers.

Moreover, the dominant variable in our model is the maximum affinity level that has been achieved at any stage of the hypermutation process. Thus, the size of the B cell population is exogenous to our model. We assume that the hypermutation process is initiated via a mechanism outside the scope of our model. Our treatment of the hypermutation process terminates upon the development of a clone with sufficiently high affinity. Finally, as a result of our focus on the affinity improvement steps, we measure time in a discrete fashion by the inter-mutation periods.

Specifically, we begin with the space of all DNA sequences encoding the variable regions of the Ig molecules and a function on that space that models the likelihood that the resulting Ig molecule becomes attached to a particular antigen. This *affinity function* is conceptualized in a series of mappings which portray the biochemical mechanisms involved.

To begin with, the gene in question is transcribed into RNA and subsequently translated into the primary Ig sequence. This step describes the mapping from the genotype (a 4–letter alphabet per site) to the sequence of amino acids making up the Ig molecule (a 20–letter alphabet per site). The next step is the folding of the resulting protein into its *ground state* in the presence of the antigen under consideration. This step is modeled as a mapping from the space of amino acid sequences to the three-dimensional geometry of the resulting Ig molecule[1].

Finally, the protein shape gives rise to the free energy of the Ig molecule in the presence of the antigen. The free energy in turn is used to define the association/dissociation constants and the Gibbs measure which determines the likelihood of attachment. The resulting affinity is visualized as a high-dimensional landscape, where the peaks represent DNA sequences that encode Ig molecules with high affinity to the particular antigen.

The process outlined above for modeling the affinity landscape depends on detailed knowledge which is often unavailable. Furthermore, our interest in the universality of the immune response has led us to a model that does not assume a detailed knowledge of the particular invading antigen. We will return to this critical theme in the section on Performance Robustness. For

---

[1] This concept is analogous to that of *shape space* in Chap. 13 of [35].

the time being, we acknowledge the need to reduce the series of mappings described above to a set of *affinity classes*, as in [23].

The underlying theme of our modelling effort has been the delineation of the trade-off between the safety of mutations leading to *local steps* in shape space (and consequently incremental affinity improvements) and the eventual necessity of mutations that result in *global jumps* in shape space as discussed in section 5.1. The main ingredient of our affinity landscape model reflects this trade-off by associating to each point in the space of DNA sequences the unique sequence to which it converges following a discrete gradient ascent algorithm.

Let $\mathcal{X}$ denote the space of DNA sequences encoding the V_H and V_L regions of an Ig molecule. Let f be a positive, real-valued function on $\mathcal{X}$, which denotes the affinity function, as described above. Finally, consider the gradient operator $\mathcal{D}f(x) = \arg\min_{y \in \mathcal{N}(x)} f(y)$, where $\{\mathcal{N}(x) \subseteq \mathcal{X},\, x \in \mathcal{X}\}$ describes the neighborhood structure[2] in $\mathcal{X}$. With this notation, for each sequence $x \in \mathcal{X}$, successive applications of the gradient operator converge to the closest local optimum, i.e., there is a finite positive integer $d(x)$ and a sequence $\mathcal{F}^*(x) \in \mathcal{X}$ such that for all $n \geq d(x)$,

$$\mathcal{D}^n f(x) = \mathcal{D}^{d(x)} f(x) = \mathcal{F}^*(x).$$

Using this association, we partition $\mathcal{X}$ into subsets that map to the same integer under $d(\cdot)$. These level sets contain all sequences that are a fixed number of point mutations away from their closest local optimum.

A further ingredient of our model for the affinity landscape is the relative nature of the separation between strictly local and global optima. In practice, the global optimum is not necessarily the goal. Instead, some sufficient level of affinity is desired. This affinity threshold is generally unknown a priori. Our model allows us to view the landscape as a function of the desired affinity threshold. As we show in section 9.4, we are able to study the dependence of our model's performance for a variety of affinity thresholds and thus investigate the trade-off between the desired affinity and the required time.

The consideration of strictly local versus global optima as a variable characteristic necessitates a finer partition of the level sets. Specifically, for each level of affinity threshold, some of the local optima in $\mathcal{X}$ are below it and therefore are considered strictly local, while others are above it and are therefore considered global. This leads to a finer decomposition of each level set into the part containing sequences a certain number of steps below a strictly local optimum versus sequences whose closest local optimum is also global because its affinity is above the desired threshold.

---

[2] In this paper we concentrate on point mutations as the mechanism for local steps and thus the neighborhood we consider consists of all 1–mutant sequences. It has been suggested [30,18] that more than one point mutation may occur before the resulting Ig molecule is tested against an antigen presenting cell to determine its affinity. Our model can capture such an eventuality by appropriately modifying the neighborhood structure to include the 2– or generally k–mutant sequences.

Let $\mu \in \mathcal{M}_1(\mathcal{X})$ be a probability distribution[3] on $\mathcal{X}$. Let

$$q(j) \overset{\Delta}{=} \mu\left(\mathcal{W}\left(\mathcal{L}(M)\right) \cap d^{-1}(j)\right)$$

and

$$p(j) \overset{\Delta}{=} \mu\left(\left[\mathcal{X} \setminus \mathcal{W}\left(\mathcal{L}(M)\right] \cap d^{-1}(j)\right)\right),$$

where M is the affinity threshold, $\mathcal{L}(M) \overset{\Delta}{=} f^{-1}\left([M, \infty)\right)$ and

$$\mathcal{W}(A) \overset{\Delta}{=} \left\{ x \in \mathcal{X} : \lim_{k \to \infty} \left[\mathcal{D}^k f\right](x) \in A \right\}$$

is the set of sequences that converge to a member of A after sufficient iterations of the gradient operator. Using this notation, let

$$a \overset{\Delta}{=} \max\left\{j \geq 0 : p(j) > 0\right\}$$

and

$$b \overset{\Delta}{=} \max\left\{j \geq 0 : q(j) > 0\right\}$$

denote the height of the strictly local and the global optima, respectively. Notice that for notational convenience, we have suppressed the dependence of these measures on the affinity threshold M. Unless stated explicitly otherwise, these measures of the affinity landscape will always depend on the affinity threshold as discussed above.

Despite the lack of detailed knowledge of the biologically relevant affinity landscapes, there is broad agreement that the size of the level sets decreases rapidly. A popular model used previously in [23] asserts that the level sets are decreasing in size at an exponential rate controlled by a parameter β (in the notation of [23] this parameter corresponds to Λ^{-1}). In this paradigm, one has $q(j) = Z_q^{-1}\beta^{-j}$ and $p(j) = Z_p^{-1}\beta^{-j}$ with $\sum_{j=0}^{a} p(j) = c\sum_{j=0}^{b} q(j)$, Z_p, Z_q the appropriate normalization constants, and c, a parameter that determines the relative size of the set of points for which discrete gradient ascent traps them at strictly local optima versus those points for which this *greedy* algorithm is sufficient to take them to a global optimum.

While the biologically meaningful range of values for the parameters a, b, c, and β is uncertain, the values we use are in agreement with the values for similar parameters used by other authors. Specifically, we restrict our attention to $a, b \in [4, 20]$, $c \in [10^4, 10^5]$, and $\beta \in [0.05, 0.3]$.

---

[3] Often, as is the case in this paper, the uniform distribution is used. If there are prior preferences observed for particular alleles at certain base pairs, they can be modeled by altering μ accordingly.

7   Optimization Dynamics

We model the dynamics of evolutionary optimization on the affinity landscape as a Markov chain. Specifically, the chain may take one of two actions at each time step: it may search locally to find the gradient direction, and take one step in that direction or it may perform a global jump, which effectively randomizes the chain. The decision between the two available actions is taken based on a Bernoulli trial with probability p: when $p = 0$, the chain performs global jumps all the time while $p = 1$ prohibits any global jumps. Thus, the parameter p controls the degree of randomization in the Markov chain. The biological distinction between local search versus global jumps is realized by means of at least two mechanisms described earlier: deletions/insertions and point replacements that lead to large changes in the resulting geometry.

The mathematical description of the Markov chain model described above uses the following generator:

$$[\mathcal{G}\phi]\,(x) \stackrel{\Delta}{=} p\phi(\mathcal{D}f(x)) + (1 - p)\mathbf{E}^{\mu}[\phi] - \phi(x).$$

We are interested in estimating the extreme left tail of the distribution of the exit times for the resulting Markov chain. In particular, let

$$\tau(M) \stackrel{\Delta}{=} \inf\left\{k \geq 0 | X_k \in f^{-1}\left([M, \infty)\right)\right\},$$

where X_k denotes the Markov chain under consideration. We are interested in estimating the likelihood that at least one out of a population of n identical, non-interacting replicas of the Markov chain will reach an affinity level higher than M before time y. It should be noted that, by virtue of the discrete nature of the Markov chain, time in this context is measured by the number of mutation cycles experienced by the system. The probability we are looking for takes the form

$$\mathrm{P}^*\left(\inf_{i \leq n} \tau_i(M) \leq y\right) = 1 - (1 - \mathrm{P}^*\left(\tau_1(M) \leq y\right))^n,$$

where P^* denotes the path measure induced by the Markov chain and the index i tallies the replica under consideration. Since we are focusing our attention to the GC reaction, n is approximately 10^3-10^5.

8   Methodology

The study of the evolutionary optimization process outlined in the previous section uses results by the second author on the convergence rates of exit times of Markov chains [38]. The general approach for estimating the desired tails of the exit time distributions consists of the following steps:

(i) We formulate a Dirichlet problem for $\mathcal{G}$ on $f^{-1}\left([M,\infty)\right)$ whose solution provides a martingale representation of the Laplace transform of the exit time $\tau(M)$.

(ii) We solve the resulting Dirichlet problem and compute the desired Laplace transform as

$$\psi(\xi) \triangleq \mathbf{E}^*\left[e^{\xi\tau(\epsilon)}\right] = \frac{\left(1 - pe^\xi\right)\sum_{j=0}^b q(j)p^j e^{j\xi}}{1 - e^\xi + (1-p)e^\xi \sum_{j=0}^b q(j)p^j e^{j\xi}}$$

where $\mathbf{E}^*$ denotes the expectation starting from a μ–distributed initial sequence.

(iii) We compute the Legendre-Fenchel transform $I(y)$ of the cumulant of $\tau(M)$ as

$$I(y) = \begin{cases} \int_1^{\frac{y}{\mathbf{E}^*[\tau]}} \Xi(t)dt, & \text{if } y \geq \mathbf{E}^*[\tau] \\ \int_{\frac{y}{\mathbf{E}^*[\tau]}}^1 \Xi(t)dt, & \text{otherwise} \end{cases},$$

where $\Xi(t)$ is the (positive or negative depending on whether $y \geq \mathbf{E}^*[\tau]$ or not) solution to

$$\frac{d\psi}{d\xi}\left(\frac{\Xi(t)}{\mathbf{E}^*[\tau]}\right) = t\mathbf{E}^*[\tau]\psi\left(\frac{\Xi(t)}{\mathbf{E}^*[\tau]}\right).$$

It turns out [38] that, for $y \leq \mathbf{E}^*[\tau]$,

$$\mathbf{P}^*\left(\tau_I(M) \leq y\right) \approx \exp\left\{I(y)\right\}.$$

(iv) We estimate $\Xi(t)$ by performing a Taylor expansion of the cumulant of $\tau(M)$ at $-\infty$, yielding

$$\frac{d\psi}{d\xi}\left(\log\lambda\right) = \sum_{i=1}^\infty \frac{c_i\lambda^i}{i!},$$

as $\lambda \searrow 0^+$. Inverting the polynomial on the right-hand side we obtain the general form

$$\Xi(t) \approx c_1\log\left(c_2 t\right).$$

Specifically, if we stop the Taylor series after the linear term we have $c_1 = 1$ and $c_2 = \frac{p+\beta(1-p)[1-q(0)]}{\beta q(0)}$.

9   Results

Our goal in the modelling exercise described in the previous few sections was not to produce exact quantitative predictions for the behavior of the immune system. Instead, our main goal has been to elucidate the drivers behind the apparently complex behavior of the immune system and develop a qualitative understanding of what makes the system work so efficiently. With this in mind, we acknowledge that our model is purposely kept simple enough to allow a thorough analysis and simulation.

Our results are presented in this section in order of increasing complexity. First we discuss the surprisingly quick response of the system which is governed by a variant of the recently discovered *cutoff phenomenon* in many Markov chains. Then, we proceed to exhibit the dependence of the system's response time on the value of the trade-off parameter p. A relatively narrow band of values for p are shown to significantly outperform all others.

Until this point, we consider one antigen and resulting affinity landscape at a time. In order to appreciate the general applicability of the trade-off discussed above, we proceed next to describe the response dynamics as a function of p for a sequence of landscapes, randomly generated from the biologically motivated range of parameters presented earlier. This investigation yields a sharp transition between two regimes: a frozen, ordered regime with too little randomization and a liquid, chaotic regime with too much randomization. The corresponding *phase transition* occurs within the same range of p that strikes the right balance in the trade-off studied before.

The next step is to investigate the trade-off between the diminishing incremental benefit in affinity against the increasing time burden as the system attempts to reach higher affinity peaks. This trade-off may cast some light on the mechanism that ends the hypermutation process.

Once we have established the existence of the narrow band of desired values of the parameter p, we finally turn our attention to its biological implementation. How does the system know to set the parameter p at the right level? It turns out that it doesn't need to know. In fact, with some mild assumptions, we show that the system cannot help but adapt to exhibit a value of p within the narrow desired range despite exogenous shocks.

9.1   The Cutoff Phenomenon

This phenomenon has been studied by Diaconis and his collaborators in [8–10]. In the context of the Markov chain modeling the affinity maturation dynamics, Fig. 1 shows a typical realization of the observed cutoff. Methodologically, our approach differs from that of Diaconis in the measure of convergence used. The more traditional approach employs the total variation distance between the sample distribution of the Markov chain after a finite number of steps and the stationary distribution. Instead, we concentrate our attention on the left tail of a family of exit times. At a deeper level, the two approaches are not as dissimilar as they may appear. The technique for estimating the total variation distance often relies on coupling arguments which reduce to the distribution of the coupling time, a stopping time not unlike the ones underpinning our approach.

In the immunological literature, this behavior has been captured by previous models of the humoral response maturation process [6,23,34]. Qualita-

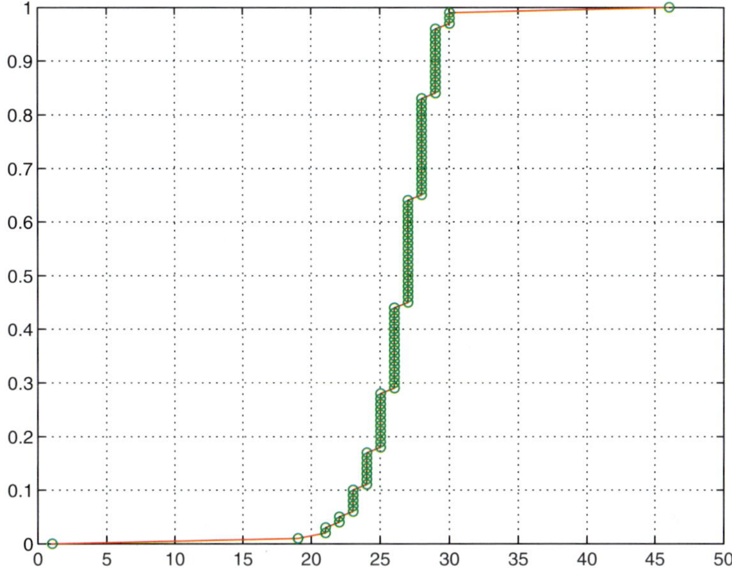

Fig. 1. Convergence of $P^*(\tau(M) \leq y)$ as a function of y

tively, this result corroborates the observed speed of the affinity maturation in the immune system. Within a few mutation cycles spanning 3–4 days, the specific immune response gains 2–3 orders of magnitude in affinity. This time, measured in the number of mutational cycles, is at least 2 orders of magnitude less than the expected time to equilibrium of the Markov chain. Thus, the observed affinity maturation is a decidedly disequilibrium effect. This is certainly one of the sources of complexity in the system that defies interpretation using traditional equilibrium-oriented techniques.

9.2   The Optimal Value of p

The result described in this section is an outgrowth of previous work by the second author. In [39] similar techniques as the ones described above were applied in the study of the asymptotic convergence rate of a class of Markov chains encompassing the one employed here. In that context, it was shown that under very mild conditions on the landscape, there is a nonzero level of *randomization by design* (i.e., beyond the minimum randomization required to avoid remaining trapped in strictly local optima asymptotically) that substantially increases the convergence rate.

Here we study the more complex problem of transient behavior of the Markov chain. Nevertheless, we obtain a qualitatively equivalent result. A fine-tuning of the parameter p in a narrow range of values confers a remarkable performance improvement (Fig. 2). This qualitative behavior appears

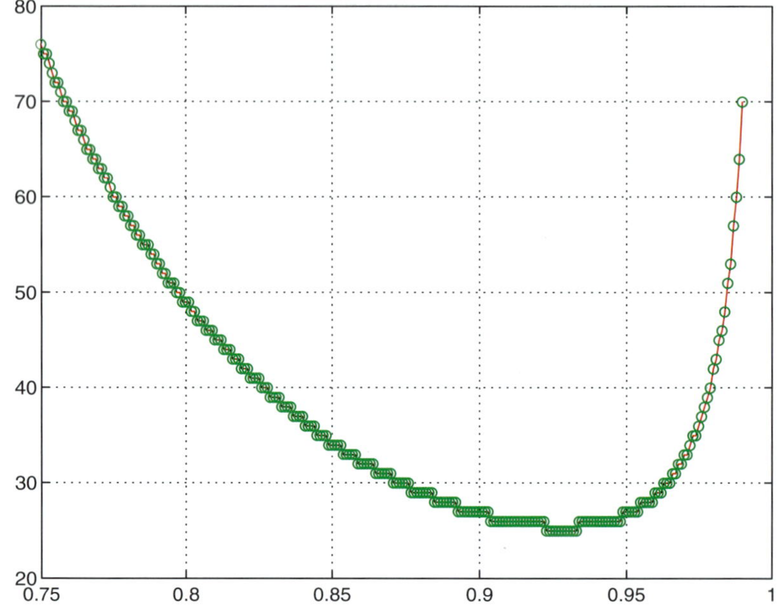

Fig. 2. $\tau(M)$ as a function of p

to be surprisingly insensitive to the structure of the landscape. In [39] this was investigated for exponential landscapes similar to the ones studied here, as well as broader classes of polynomial, logarithmic, and uniformly random landscapes. In all cases, the optimal level of p was within a narrow overlapping range. It is this hint of universality that has led us to the robustness investigation that follows next and which uncovers a further characteristic of this class of complex systems.

9.3 Performance Robustness

As alluded to above, performance robustness refers to the surprising consistency in the efficiency of the immune system response to a combinatorially large set of invading antigens. We studied this problem by performing the trade-off analysis for the optimal value of p for a population of randomly generated landscapes from within the biologically justifiable range of parameters mentioned earlier. One thousand different antigens were tried, and the resulting mean and standard deviation of the response time were plotted for a series of p values. In Fig. 3 we have suppressed the third dimension (p) in order to illustrate more clearly the two observed regimes. For clarity, note that $p = 0.75$ corresponds to the point close to (80,80) in the graph, while $p = 0.99$ corresponds to the point close to (190,150), with equally-spaced, increasing p values in between. We observe that for very high values of p, the

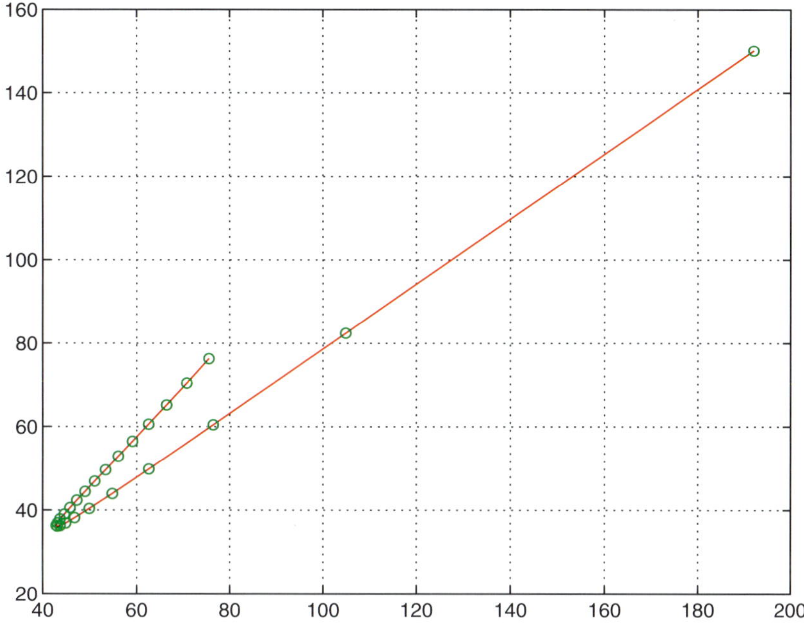

Fig. 3. Expected value of $\tau(M)$ versus standard deviation of $\tau(M)$

system's expected performance suffers a rapid decrease accompanied by an increase in the variability of that response across different antigens. As we lower the p value, there is a narrow range between about 0.85 and 0.91 for which the system attains the best response time and the lowest variability of that performance as it faces varying antigens. Past that point, there is a sharp change in the system's behavior. The expected response time deteriorates and the variability of that response grows even faster. Using the same approach as in [39], we identify the first of these two regimes with a *solid* phase which is too ordered to escape the strictly local optima that abound in a randomly generated landscape. Similarly, the second regime is seen as a *liquid* phase with too little structure to effect the desired progression towards higher affinity peaks. Instead, systems in this regime appear to diffuse aimlessly in sequence space. The observed sharp transition between these two regimes is analogous to the concept of *the edge of chaos* introduced by Kauffman [22] as well as the notion of a critical level of parallelism investigated in [29].

9.4   The Affinity Threshold

So far, our results related to the time it takes the system to achieve a fixed desired level of affinity. In this subsection we ask the question how the system

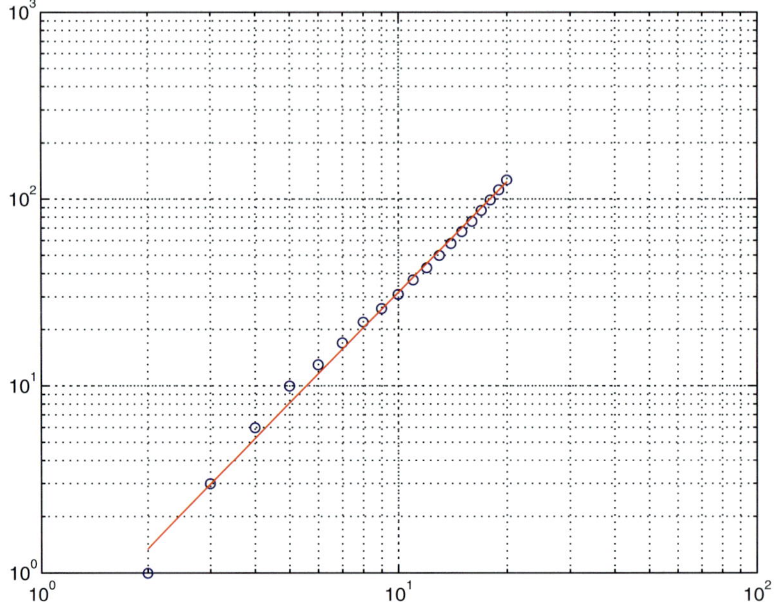

Fig. 4. $\tau(M)$ as a function of the desired affinity threshold M

knows when to terminate the hypermutation process. Clearly, a new trade-off becomes relevant between the energetic costs and immunologic risks (e.g. autoimmunity or malignancy) of continuing the hypermutation process for longer and the expected incremental affinity gains as discussed in detail in Sect. 9.4.

As one might expect, diminishing incremental affinity gains are a rule. Figure 4 exhibits the power-law relationship that holds between the response time and the resulting affinity for a fixed value of p. We are interested to investigate how this power law varies as p moves across the two regimes outlined above. In general, we obtain an approximate power law of the form $M(\tau) \sim \tau^{\alpha}$. Figure 5 shows the behavior of $\alpha(p)$. We observe that the rate at which incremental gains diminish increases as p is lowered deeper into the liquid phase. It is worthwhile to note that the value of $\alpha = 0.5$, which corresponds to the scaling behavior of Brownian motion, is attained close to the value of p that leads to the phase transition described in the previous paragraph.

One biological interpretation of this observed variation of the scaling law as a function of p is that lower values of p are intrinsically riskier, in that they entail more global jumps. Thus, it is not surprising that lower values of p exacerbate the trade-off between time and diminishing affinity gains in favor of stopping the hypermutation process sooner. We conjecture that this type of behavior leads to a two-tiered response:

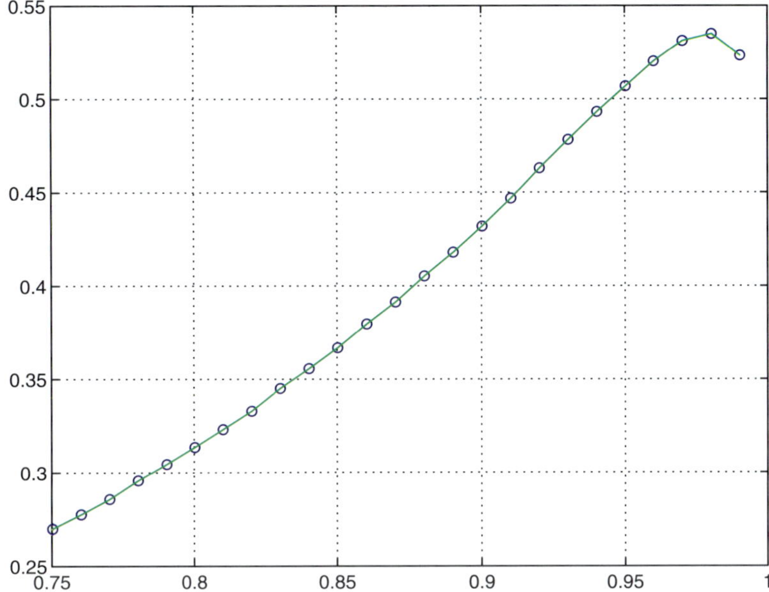

Fig. 5. α as a function of p

(i) After a relatively modest amount of time, a sufficient affinity improvement has been gained to allow the mature plasma cells to exit the GC and mount a highly specific and effective attack on the invading antigen.

(ii) The hypermutation process continues in the background and now strives to generate an even more specialized population of memory cells to maintain long-term immunity to the antigen in subsequent reinfections.

The above results lead to the consideration of the repertoire shift phenomenon. The emergence of this phenomenon has been studied in earlier models. Specifically, within the population dynamics paradigm, Shannon and Mehr in [36] use the "destructive effect of hypermutation on the primary B cell repertoire" to show that the memory cell population converges to a moderately high affinity level: lower than the constituents of the fast primary response, which die out due to the overwhelming likelihood of affinity-reducing (and thus lethal) subsequent mutations, but higher than the background average due to the hypermutation and antigen-constrained selection processes in the germinal centers.

The global optimization paradigm presented in this paper offers a further, direct explanation for the occurrence of repertoire shift. While the result is not new, we present it mainly to illustrate the usefulness of the proposed paradigm. During the ongoing hypermutation process suggested in step (ii) above, new peaks in the affinity landscape eventually emerge, arising from V genes not present in the primary response. Specifically, while some sufficiently high affinity peaks are reached relatively fast during the primary response, they sometimes lead to lineages whose affinity plateaus shortly thereafter. At the same time, other clones, which found themselves in more modestly rising affinity "hills" during the primary response, eventually experience more rapid affinity improvement, which leads them to surpass the performance of the lineages that dominated the primary response.

9.5   Evolving to the Edge of Chaos

Having exhibited the critical dependence of the response time to the value of p, we address the question of how the immune system knows to fine-tune the value of p within the narrow desired band. We hypothesize that the nature of the evolutionary dynamics is such that the system cannot avoid being *drawn* to operate within the desired band of p values.

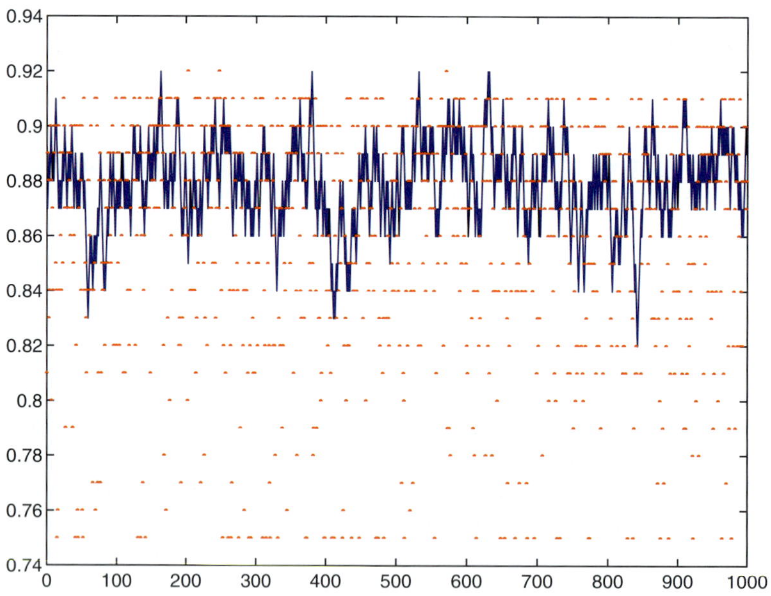

Fig. 6. Stability of evolving p inside robust band

In order to examine this hypothesis, we conjectured that the mutator control mechanism which biologically instantiates the parameter p is itself coded in the genome along with the rest of the Ig molecule. Even though that part of the genome is not involved directly in the affinity-based selection as it doesn't code for the Ig molecule, it may be subject to a longer scale mutation and selection process.

Such a stipulated evolution may theoretically occur at the level of individual cells (as is the case with the evolution of Ig affinity during the hypermutation process), or at the level of organisms across more traditional evolutionary time scales. The mechanics of mutation control, which account for the p value, are not understood well enough to determine this point with certainty. In any case, the simulations we present below of an evolving p value apply equally well to either interpretation. The idea is that, as a population of individuals is facing ever changing antigens, p values in the optimal range dominate the population. The main difference between the two interpretations would be in the meaning of individual (cell vs. organism).

For the rest of this subsection we assume that such a process exists. We further assume that we can model the process as a reversible nearest neighbor random walk with biased transitions towards the neighboring p value that reduces the immune response time. We justify this bias by the strong selectional advantage of clones with faster immune response times.

Using this hierarchical evolution model, we simulated one thousand successive infections with different antigens. During each infection the system has a fixed value of p and behaves as described in the sections above. Between infections, the system performs one step of the biased random walk, motivated by the selectional advantage of the clones with the faster response time to the previous infection.

Fig. 6 depicts the evolving value of p in the population (solid line) and the optimal value of p corresponding to the landscape for each new infection (dashes). This optimal p value for each new landscape acts as an exogenous shock to the system. Figure 6 shows that despite these periodic exogenous shocks (which would have the system decrease its p value into the liquid phase), the system stays within the desired narrow band, without explicit instruction.

In Fig. 7 we show the histograms associated with the p values in Figure 6. Once again, we see the system converging to a much tighter distribution of p values than the exogenous shocks. Finally, Fig. 8 shows the transient behavior of this hierarchical evolution model. We purposely started the system at $p = 0.75$, distinctly outside the desired range, and let it evolve autonomously. Once again, without any instruction, the system converges rapidly to the desired band of p values and is stabilized within that band.

10   Conclusions and Directions of Further Study

Our model gives evidence that the mutator mechanism functioning during somatic hypermutation has evolved to a trade-off value of p that gives a fast, efficient, and consistent response. While this mechanism is still not understood, certain transcriptional elements appear to be necessary for mutation

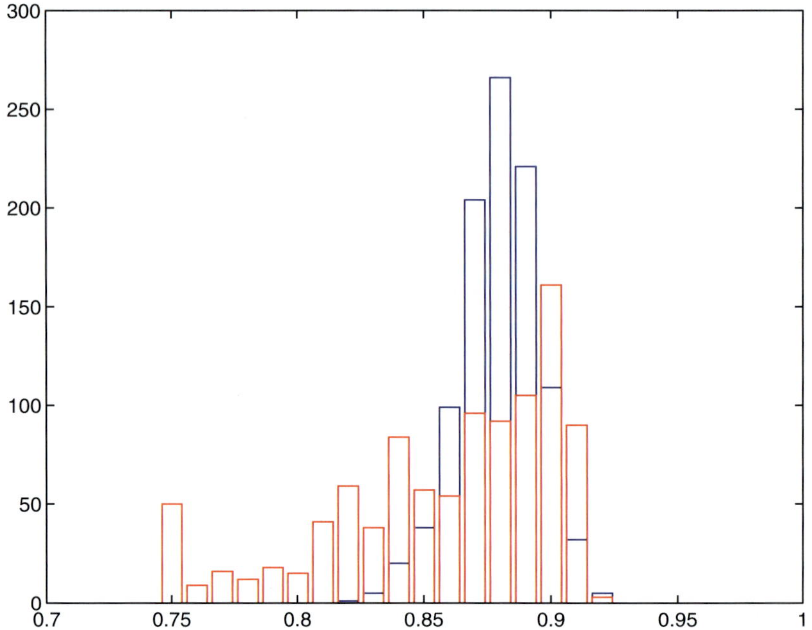

Fig. 7. Comparison of the distribution of evolving p versus the distribution of optimal p for the sequence of landscapes that drive the system

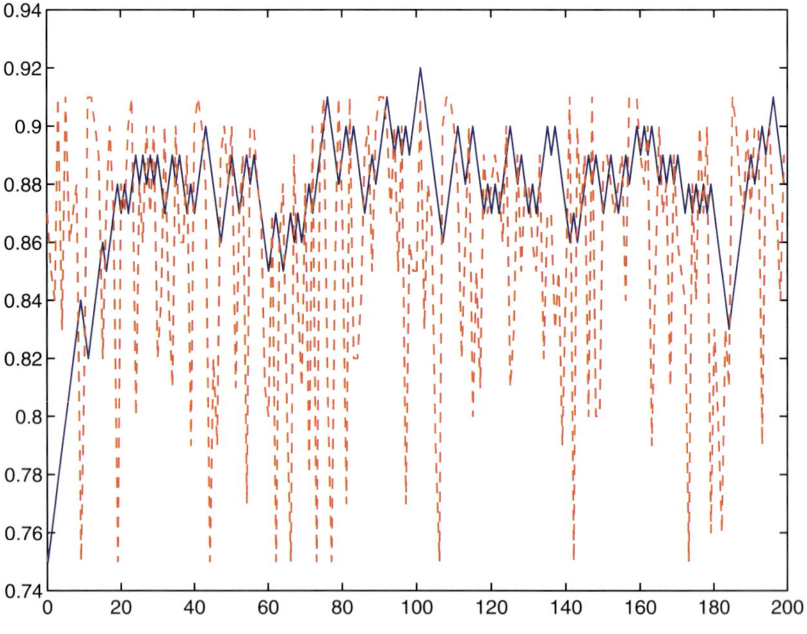

Fig. 8. Convergence of evolving p to robust band

in both heavy and light chain genes, although additional, novel molecular mechanisms need to be considered. Our model shows that *global jumps* need to occur during the process, and from a biologic standpoint, these may be understood in two ways. They may be encompassed by the less frequent, but necessary mutations in the more diverse regions of the CDRs or in the FRW regions, or in the form of deletions and insertions. This latter characteristic in particular implies the occurrence of double strand breaks in the DNA and may support evidence of recombination [25] or reverse transcription models [3].

With regards to the affinity threshold, our model demonstrates the diminishing incremental gains in affinity as a function of an increasing number of mutations. Future investigation might include analysis of the selection process during the ongoing shaping of the memory compartment that likely encompasses alternative selection parameters, kinetics, for example, being one of them [15].

Our model also shows that the evolution of mutations seems to have an internal driver that points us toward the higher order template underlying the observable hotspots. This finding emphasizes the evolutionary efficiency with which nature approaches this problem.

As was mentioned during the description of the model, the process of hypermutation has been suggested to be redundant or unnecessary [3]. We proposed larger control mechanisms underlying the changes between primary and secondary repertoire that might explain the persistence of this process. A next step might include an attempt to understand these mechanisms, perhaps in regards to tolerance and early development, or in regards to the dynamics of the antibody repertoire acting beyond affinity selection. Hypermutation may be a necessary step to set the stage for these other processes to occur.

References

1. Kerstin Andersson, Jens Wrammert, and Tomas Leanderson. Affinity selection and repertoire shift: paradoxes as a consequence of somatic mutation. *Immunological Reviews*, 162:172–182, 1998.
2. C. Berek, A. Berger, and M. Apel. Maturation of the immune response in germinal centers. *Cell*, 67:1121–1129, December 1991.
3. Robert V. Blanden, Harald S. Rothenfluh, Paula Zylstra, Georg F. Weiller, and Edward J. Steele. The signature of hypermutation appears to be written into the germline IgV segment. *Immunological Reviews*, 162:117–132, 1998.
4. Bradford C. Braden, Ana Cauerhff, William Dall'Acqua, Barry A. Fields, Fernando A. Goldbaum, Emilio L Malchiodi, Roy A. Mariuzza, Roberto J Poljak, Frederick P. Schwarz, Xavier Ysern, and T.N. Bhat. Structure and thermodynamics of antigen recognition by antibodies. *Annals of NY Academy of Science*, 764:315–327, 1995.
5. McKay Brown, Mary Stenzel-Poore, Susan Stevens, Sophia K. Kondoleon, James Ng, Hans Peter Bachinger, and Marvin B. Rittenberg. Immunologic

memory to phosphocholine keyhole limpet hemocyanin. *Journal of Immunology*, 148(2):339–346, January 1992.

6. Franco Celada and Philip Seiden. Affinity maturation and hypermutation in a simulation of the humoral immune response. *European Journal of Immunology*, 26:1350–1358, 1996.

7. D.G. Covell and A. Wallqvist. Analysis of protein-protein interactions and the effects of amino acid mutations on their energetics. *Journal of Molecular Biology*, pages 281–297, 1997.

8. P. Diaconis and L. Saloff-Coste. Walks on generating sets of abelian groups. *Probability Theory and Related Fields*, 105:393–421, 1996.

9. Persi Diaconis. The cutoff phenomenon in finite markov chains. *Proceedings of the National Academy of Science*, 93:1659–1664, February 1996.

10. Persi Diaconis. From shuffling cards to walking around the building: an introduction to modern markov chain thery. *Documenta Mathematica*, Extra Volume ICM 1998(I):47–64, August 1998.

11. Marilyn Diaz and Martin F. Flajnik. Evolution of somatic hypermutation and gene conversion in adaptive immunity. *Immunological Reviews*, 162:13–24, 1998.

12. Thomas Dorner, Hans-Peter Brezinschek, Ruth I. Brezinschek, Sandra J. Foster, Rana Domiati-Saad, and Peter E. Lipsky. Analysis of the frequency and pattern of somatic mutations within nonproductively rearranged human variable heavy chain genes. *Journal of Immunology*, 158:2779–2789, 1997.

13. Thomas Dorner, Sandra J. Foster, Hans-Peter Brezinschek, and Peter E. Lipsky. Analysis of the targeting of the hypermutational machinery and the impact of subsequent selection on the distribution of nucleotide changes in human VDJH rearrangements. *Immunological Reviews*, 162:161–171, 1998.

14. K. Elgert. *Immunology: Understanding the Immune System.* Wiley-Liss, 1996.

15. Jefferson Foote and Cesar Milstein. Kinetic maturation of an immune response. *Nature*, 352:530–532, August 1991.

16. Jefferson Foote and Cesar Milstein. Conformational isomerism and the diversity of antibodies. *Proc. Natl. Acad. Sci.*, 91:10370–10374, October 1994.

17. Y. Fukita, H. Jacobs, and K. Rajewsky. Somatic hypermutation in the heavy chain locus correlates with transcription. *Immunity*, 9:105–114, July 1998.

18. Beatriz Goyenechea and Cesar Milstein. Modifying the sequence of an immunoglobulin V gene alters the resulting pattern of hypermutation. *Immunology*, 93:13979–13984, 1996.

19. Nancy S. Green, Mark M. Lin, and Matthew D. Schraff. Somatic hypermutation of antibody genes: a hot spot warms up. *BioEssays*, 20:227–234, 1998.

20. Shailaja Hande, Evangelia Notidis, and Tim Manser. Bcl-2 obstructs negative selection of autoreactive, hypermutated antibody V regions during memory B cell development. *Immunity*, 8:189–198, February 1998.

21. Harry W. Schroeder Jr, Frank Mortari, Satoshi Shiokawa, Perry Kirkham, Rotem A. Elgavish, and F.E. Bertrand III. Developmental regulation of the human antibody repertoire. *Annals of NY Acad. of Science*, pages 242–260, September 1995.

22. Stuart Kauffman. *At Home in the Universe: The Search for Laws of Self-Organization and Complexity.* Oxford University Press, 1995.

23. Thomas B. Kepler and Alan S. Perelson. Somatic hypermutation in B cells: An optimal control treatment. *Journal of Theoretical Biology*, 164:37–64, 1993.

24. Ulf Klein, Tina Goossens, Matthias Fischer, Holger Kanzler, Andreas Brae-uninger, Klaus Rajewsky, and Ralf Kuppers. Somatic hypermutation in normal and transformed B cells. *Immunological Reviews*, 162:261–280, 1998.

25. Qingzhong Kong, Reuben S. Harris, and Nancy Maizels. Recombination-based mechanisms for somatic hypermutation. *Immunological Reviews*, 162:67–76, 1998.

26. Serge Lebecque and Patricia J. Gearhart. Boundaries of somatic mutation in re-arranged immunoglobulin genes. *Journal of Experimental Medicine*, 172:1717–1727, December 1990.

27. Y-J Liu, D.E. Joshua, G.T. Williams, C.A. Smith, J. Gordon, and I.C.M. MacLennan. Mechanism of antigen-driven selection in germinal centres. *Nature*, 342:929–931, December 1989.

28. Ian MacLennan. The centre of hypermutation. *Nature*, 354:352–353, 1991.

29. William G. Macready, Athanassios G. Siapas, and Stuart A. Kauffman. Criti-cality and parallelism in combinatorial optimization. *Science*, 271:56–59, Jan-uary 1996.

30. Tim Manser. The efficiency of antibody affinity maturation: can the rate of B-cell division be limiting. *Immunology Today*, 11(9):305–308, 1990.

31. Tim Manser, Kathleen M. Tumas-Brundage, Lawrence P. Casson, Angela M. Giusti, Shailaja Hande, Evangelia Notidis, and Kalpit A. Vora. The roles of antibody variable region hypermutation and selection in the development of the memory B cell compartment. *Immunological Reviews*, 162:182–196, 1998.

32. M. Neuberger, Norman Klix, Christopher J. Jolly, Jose Yelamos, Cristina Rada, and Cesar Milstein. The intrinsic features of somatic hypermutation. *Immuno-logical Reviews*, 162:107–116, 1998.

33. G.J.V. Nossal. The molecular and cellular basis of affinity maturation in the antibody response. *Cell*, 68:1–2, 1992.

34. Mihaela Oprea and Alan S. Perelson. Somatic mutation leads to efficient affinity maturation when centrocytes recycle back to centroblasts. *Journal of Immunol-ogy*, 158:5155–5162, 1997.

35. G.W. Rowe. *Theoretical Models in Biology*. Oxford University Press, 1994.

36. Michele Shannon and Ramit Mehr. Reconciling repertoire shift with affin-ity maturation: the role of deleterious mutations. *Journal of Immunology*, 162(7):3950–3956, April 1999.

37. Haifeng Song, Xiaobo Nie, Subhendu Basu, and Jan Cerny. Antibody feedback and somatic mutation in B cells: regulation of mutation by immune complexes with IgG antibody. *Immunological Reviews*, 162:211–218, 1998.

38. T.V. Theodosopoulos. *Stochastic Models for Global Optimization*. PhD thesis, MIT, May 1995.

39. T.V. Theodosopoulos. Some remarks on the optimal level of randomization in global optimization. In P.M. Pardalos, S. Rajasekaran, and J. Rolim, editors, *Randomization Methods in Algorithm Design*. American Mathematical Society, 1999.

40. Ian Tomlinson, Gerald Walter, Peter T. Jones, Paul H. Dear, Erik L.L. Sonnhammer, and Greg Winter. The imprint of somatic hypermutation on the repertoire of human germline V genes. *Journal of Molecular Biology*, 256:813–817, 1996.

41. Gary J. Wedemayer, Phillip A. Patten, Leo H. Wang, Peter G. Schultz, and Raymond C. Stevens. Structural insights into the evolution of an antibody combining site. *Science*, 276(5319):1421–1423, June 1997.

42. Gregory D. Wiens, Victoria A. Roberts, Elizabeth A. Whitcomb, Thomas O'Hare, Mary P. Stenzel-Poore, and Marvin B. Rittenberg. Harmful somatic mutations: lessons from the dark side. *Immunological Reviews*, 162:197–209, 1998.

43. J. Yelamos, N. Klix, B. Goyenechea, F. Lozano, Y.L. Chui, A. Gonzalez Fernandez, R. Pannell, M.S. Neuberger, and C. Milstein. Targeting of non-Ig sequences in place of the V segment by somatic hypermutation. *Nature*, 376:225–229, July 20, 1995.

The Evolutionary Unfolding of Complexity

James P. Crutchfield and Erik van Nimwegen

Abstract. We analyze the population dynamics of a broad class of fitness functions that exhibit epochal evolution—a dynamical behavior, commonly observed in both natural and artificial evolutionary processes, in which long periods of stasis in an evolving population are punctuated by sudden bursts of change. Our approach—*statistical dynamics*—combines methods from both statistical mechanics and dynamical systems theory in a way that offers an alternative to current "landscape" models of evolutionary optimization. We describe the population dynamics on the macroscopic level of fitness classes or phenotype subbasins, while averaging out the genotypic variation that is consistent with a macroscopic state. Metastability in epochal evolution occurs solely at the macroscopic level of the fitness distribution. While a balance between selection and mutation maintains a quasistationary distribution of fitness, individuals diffuse randomly through selectively neutral subbasins in genotype space. Sudden innovations occur when, through this diffusion, a genotypic portal is discovered that connects to a new subbasin of higher fitness genotypes. In this way, we identify evolutionary innovations with the unfolding and stabilization of a new dimension in the macroscopic state space. The architectural view of subbasins and portals in genotype space clarifies how frozen accidents and the resulting phenotypic constraints guide the evolution to higher complexity.

1 Evolutionary Computation Theory

The recent mixing of evolutionary biology and theoretical computer science has resulted in the phrase "evolutionary computation" taking on a variety of related but clearly distinct meanings.

In one view of evolutionary computation we ask whether Neo-Darwinian evolution can be productively analyzed in terms of how biological information is stored, transmitted, and manipulated. That is, is it helpful to see the evolutionary process as a computation?

Instead of regarding evolution itself as a computation, one might ask if evolution has produced organisms whose internal architecture and dynamics are capable *in principle* of supporting arbitrarily complex computations. Landweber and Kari argue that, yes, the information processing embedded in the reassembly of fragmented gene components by unicellular organisms is quite sophisticated; perhaps these organisms are even capable of universal computation [31]. It would appear, then, that evolved systems themselves must be analyzed from a computational point of view.

Alternatively, from an engineering view we can ask, does Neo-Darwinian evolution suggest new approaches to solving computationally difficult problems? This question drives much recent work in evolutionary search—a class of stochastic optimization algorithms, loosely based on processes believed to operate in biological evolution, that have been applied successfully to a variety of different problems; see, for example, [4,6,8,11,16,20,22,30,33] and references therein.

Naturally enough, there is a middle ground between the scientific desire to understand how evolution works and the engineering desire to use nature for human gain. If evolutionary processes do embed various kinds of computation, then one can ask, is this biological information processing of use to us? That is, can we use biological nature herself to perform computations that are of interest to us? A partial but affirmative answer was provided by Adleman, who mapped the combinatorial problem of Directed Hamiltonian Paths onto a macromolecular system that could be manipulated to solve this well known hard problem [2].

Whether we are interested in this middle ground or adopt a scientific or an engineering view, one still needs a mathematical framework with which to analyze how a population of individuals (or of candidate solutions) compete through replication and so, possibly, improve through natural (or artificial) selection. This type of evolutionary process is easy to describe. In the Neo-Darwinian view each individual is specified by a *genotype* and replicates (i) according to its *fitness* and (ii) subject to genetic variation. During the passage from the population at one *generation* to the next, an individual is translated from its genotypic specification into a form, the *phenotype*, that can be directly evaluated for fitness and thus selected for inclusion in the next generation. Despite the ease of describing the process qualitatively, the mechanisms constraining and driving the population dynamics of evolutionary adaptation are not well understood.

In mathematical terms, evolution is described as a nonlinear population-based stochastic dynamical system. The complicated dynamics exhibited by such systems has been appreciated for decades in the field of mathematical population genetics [24]. For example, the effects on evolutionary behavior of the rate of genetic variation, the population size, and the genotype-to-fitness mapping typically cannot be analyzed separately; there are strong, nonlinear interactions between them. These complications make an empirical approach to the question of whether and how to use evolutionary optimization in engineering problematic. They also make it difficult to identify the mechanisms that drive behavior observed in evolutionary experiments. In any case, one would like to start with the basic equations of motion describing the evolutionary process, as outlined in the previous paragraph, and then predict observable features—such as, the time to find an optimal individual—or, at a minimum, identify mechanisms that constrain and guide an evolving population.

Here we review our recent results that address these and similar questions about evolutionary dynamics. Our approach derives from an attempt to unify and extend theoretical work that has been done in the areas of evolutionary search theory, molecular evolution theory, and mathematical population genetics. The eventual goal is to obtain a more general and quantitative understanding of the emergent mechanisms that control the population dynamics of evolutionary adaptation and that govern other population-based dynamical systems.

2   Epochal Evolution

To date we have focused on a class of population-dynamical behavior that we refer to as *epochal evolution*. In epochal evolution, long periods of stasis (*epochs*) in the average fitness of the population are punctuated by rapid *innovations* to higher fitness. These innovations typically reflect an increase of complexity—that is, the appearance of new structures or novel functions at the level of the phenotype. One central question then is, how does epochal evolutionary population dynamics facilitate or impede the emergence of such complexity?

Engineering issues aside, there is a compelling biological motivation for a focus on epochal dynamics. There is the common occurrence in natural evolutionary systems of "punctuated equilibria"—a process first introduced to describe sudden morphological changes in the paleontological record [23]. Similar behavior has been recently observed experimentally in bacterial colonies [15] and in simulations of the evolution of t-RNA secondary structures [18]. This class of behavior appears sufficiently general that it occurs in artificial evolutionary systems, such as evolving cellular automata [10,34] and populations of competing self-replicating computer programs [1]. In addition to the increasing attention paid to this type of epochal evolution in the theoretical biology community [18,21,26,35,41,49], recently there has also been an increased interest by evolutionary search theorists [5,25]. More directly, Chen et al. recently proposed to test our original theoretical predictions in an experimental realization of a genetic algorithm that exhibits epochal evolution [9].

2.1   Local Optima versus Neutral Subbasins

How are we to think of the mechanisms that cause epochal evolutionary behavior? The evolutionary biologist Wright introduced the notion of "adaptive landscapes" to describe the (local) stochastic adaptation of populations to themselves and to environmental fluctuations and constraints [50]. This geographical metaphor has had a powerful influence on theorizing about natural and artificial evolutionary processes. The basic picture is that of a gradient-following dynamics moving over a "landscape" determined by a fitness "potential". In this view an evolving population stochastically crawls along a surface determined, perhaps dynamically, by the fitness of individuals, moving to peaks and very occasionally hopping across fitness "valleys" to nearby, and hopefully higher fitness, peaks.

More recently, it has been proposed that the typical fitness functions of combinatorial optimization and biological evolution can be modeled as "rugged landscapes" [28,32]. These are fitness functions with wildly fluctuating fitnesses even at the smallest scales of single-point mutations. Consequently, it is generally assumed that these "landscapes" possess a large number of local optima. With this picture in mind, the common interpretation of punctuated equilibria in evolving populations is that of a population being "stuck" at a local peak in the landscape, until a rare mutant crosses a valley of relatively low fitness to a higher peak, a picture more or less consistent with Wright's.

At the same time, an increasing appreciation has developed, in contrast to this rugged landscape view, that there are substantial degeneracies in the genotype-to-phenotype and the phenotype-to-fitness mappings. The history of this idea goes back to Kimura [29], who argued that on the genotypic level, most genetic variation occurring in evolution is adaptively *neutral* with respect to the phenotype. Even today, the crucial role played by neutrality continues to find important applications in molecular evolution, for example; see [19]. During neutral evolution, when degeneracies in the genotype-phenotype map are operating, different genotypes in a population fall into a relatively small number of distinct fitness classes of genotypes with approximately equal fitness. Due to the high dimensionality of genotype spaces, sets of genotypes with approximately equal fitness tend to form components in genotype space that are connected by paths made of single-mutation steps.

Additionally, due to intrinsic or even exogenous effects (e.g., environmental), there simply may not exist a deterministic "fitness" value for each genotype. In this case, fluctuations can induce variation in fitness such that genotypes with similar average fitness values are not distinct at the level of selection. Thus, genotype-to-fitness degeneracies can, to a certain extent, be induced by noise in the fitness evaluation of individuals.

When these biological facts are taken into account, we end up with an alternative view to both Wright's "adaptive landscapes" and the more recent "rugged landscapes". That is, the genotype space decomposes into a set of neutral networks, or *subbasins* of approximately isofitness genotypes, that are entangled with each other in a complicated fashion; see Fig. 1. As illustrated in Fig. 1, the space of genotypes is broken into strongly and weakly connected sets with respect to the genetic operators. Equal-fitness genotypes form strongly connected neutral subbasins. Moreover, since subbasins of high fitness are generally much smaller than subbasins of low fitness, a subbasin tends to be only weakly connected to subbasins of higher fitness.

Since the different genotypes within a neutral subbasin are not distinguished by selection, neutral evolution—consisting of the random sampling and genetic variation of individuals—dominates. This leads to a rather different interpretation of the processes underlying punctuated equilibria. Instead of the population being pinned at a local optimum in genotype space as sug-

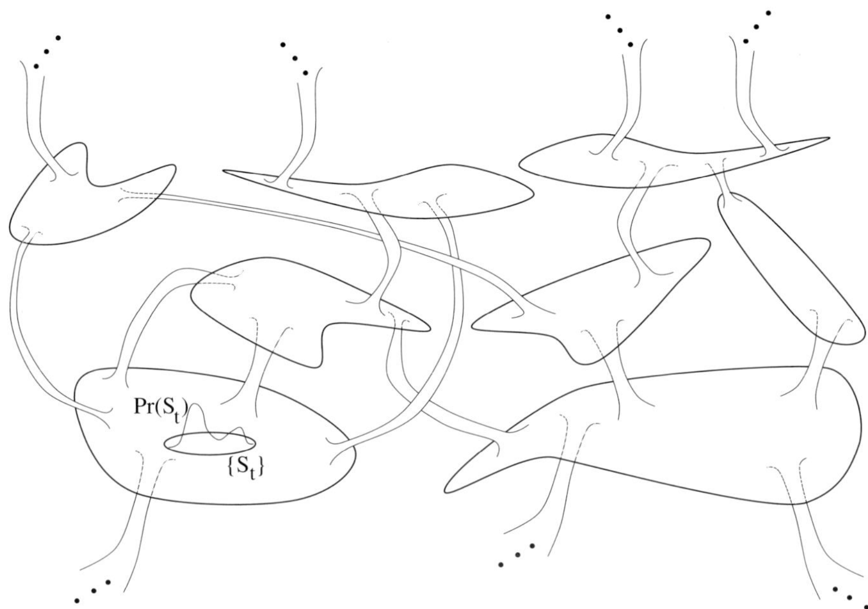

Fig. 1. Subbasin and portal architecture underlying epochal evolutionary dynamics. A population—a collection of individuals $\{S_t\}$ with distribution $\Pr(S_t)$—diffuses in the subbasins (large sets) until a portal (tube) to a higher-fitness subbasin is found.

gested by the "landscape" models, the population drifts randomly through neutral subbasins of isofitness genotypes. A balance between selection and deleterious mutations leads to a (meta-)stable distribution of fitness (or of phenotypes), while the population is searching through these spaces of neutral genotypic variants. Thus, there is no genotypic stasis during epochs. As was first pointed out in the context of molecular evolution in [27], through neutral mutations, the best individuals in the population diffuse over the neutral network of isofitness genotypes until one of them discovers a connection to a neutral network of even higher fitness. The fraction of individuals on this network then grows rapidly, reaching a new equilibrium between selection and deleterious mutations, after which the new subset of most-fit individuals diffuses again over the newly discovered neutral network.

Note that in epochal dynamics there is a natural separation of time scales. During an epoch, selection acts to establish an equilibrium in the proportions of individuals in the different neutral subspaces, but it does not induce *adaptations* in the population. Adaptation occurs only in a short burst during an innovation, after which equilibrium on the level of fitness is re-established in the population. On a time scale much faster than that between innovations, members of the population diffuse through subbasins of isofitness genotypes

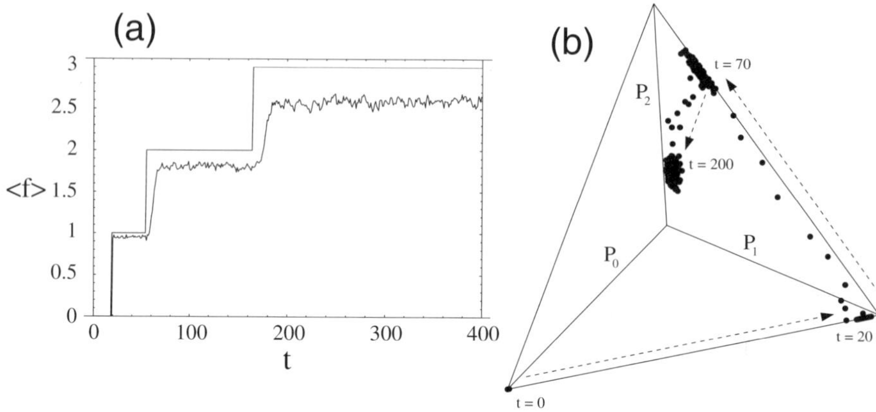

Fig. 2. Dynamics of (a) the average fitness (lower curve) and best fitness (upper curve), and (b) the fitness distribution for a population evolving under a Royal Road fitness function. The fitness function has $N = 3$ constellations of $K = 10$ bits each. The population size is $M = 250$ and the mutation rate $\mu = 0.005$. In (b) the location of the fitness distribution at each generation is shown by a dot. The dashed lines there indicate the direction in which the fitness distribution moves from metastable to metastable state through the population's fitness-distribution state space (a simplex). The times at which the different metastable states are first reached are indicated as well.

until a (typically rare) higher-fitness genotype is discovered. Long periods of stasis occur because the population has to search most of the neutral subspace before a portal to a higher fitness subspace is discovered.

In this way, we shift our view away from the geographic metaphor of evolutionary adaptation "crawling" along a "landscape" to the view of a diffusion process constrained by the subbasin-portal architecture induced by degeneracies in the genotype-to-phenotype and phenotype-to-fitness mappings. Moreover, our approach is not simply a shift towards an architectural view, but it also focuses on the *dynamics* of populations as they move through the subbasins to find portals to higher fitness.

2.2   Epochal Evolution—An Example

In our analysis [45,46], we view the subbasin-portal mechanism sketched above as the main source of epochal behavior in evolutionary dynamics. We will now discuss a simple example of epochal evolution that illustrates more specifically the mechanisms involved and allows us to introduce several concepts used in our analysis.

Figure 2 shows the fitness dynamics of an evolving population on a sample fitness function that exhibits large degeneracies in the genotype-fitness mapping. This fitness function is an example of the class of Royal Road fitness functions explained in Sect. 3 below. The genotype space consists of all bit-strings of length 30 and contains neutral subbasins of fitnesses 0, 1, 2, and 3. There is only one genotype with fitness 3, 3069 genotypes have fitness 2, 3.14×10^6 have fitness 1, and all others have fitness 0. The evolving population consists of 250 individuals that at each generation are selected in proportion to their fitness and then mutated with probability 0.005 per bit. Figure 2(a) shows the average fitness $\langle f \rangle$ in the population (lower curve) and the best fitness in the population (upper curve) as a function of generation t.

At time $t = 0$ the population starts out with 250 random genotypes. As can be seen from Fig. 2(a), during the first few generations all individuals are located in the largest subbasin with fitness 0, since both the average and the best fitness are 0. The population randomly diffuses through this subbasin until, around generation 20, a "portal" is discovered into the subbasin with fitness 1. The population is quickly taken over by genotypes of fitness 1, until a balance is established between selection and mutation: selection expanding and deleterious mutations (from fitness 1 to 0) decreasing the number of individuals with fitness 1. The individuals with fitness 1 continue to diffuse through the subbasin with fitness 1, until a portal is discovered connecting to the subbasin with fitness 2. This happens around generation $t = 60$ and by $t = 70$ a new selection-mutation equilibrium is established. Individuals with fitness 2 continue diffusing through their subbasin until the globally optimal genotype with fitness 3 is discovered sometime around generation $t = 170$. Descendants of this genotype then spread through the population until around $t = 200$, when a final equilibrium is reached.

The same dynamics is plotted in Fig. 2(b), but from the point of view of the population's fitness distribution $\vec{P} = (P_0, P_1, P_2, P_3)$. In the figure, the P_0 axis indicates the proportion of fitness 0-genotypes in the population, P_1 the proportion of fitness-1 genotypes, and P_2 the proportion of fitness-2 genotypes. Of course, since $\vec{P}$ is a distribution, $P_3 = 1 - P_0 - P_1 - P_2$. As a result, the space of possible fitness distributions forms a three-dimensional *simplex*. We see that initially $P_0 = 1$ and the population is located in the lower left corner of the simplex. Later, between $t = 20$ and $t = 60$, the population is located at a metastable fixed point on the line $P_0 + P_1 = 1$ and is dominated by fitness-1 genotypes ($P_1 \gg P_0$). Sometime around generation $t = 60$ a genotype with fitness 2 is discovered and the population moves into the plane $P_0 + P_1 + P_2 = 1$—the front plane of the simplex. From generation $t = 70$ until generation $t = 170$ the population fluctuates around a metastable fixed point in this plane. Finally, a genotype of fitness 3 is discovered and the population moves to the asymptotically stable fixed point in the interior of the simplex. It reaches this fixed point around $t = 200$ and remains there fluctuating around it for the rest of the evolution.

This example illustrates the general qualitative dynamics of epochal evolution. It is important to note that the architecture of neutral subbasins and portals is such that a higher-fitness subbasin is always reachable from the current best-fitness subbasin. Metastability is a result of the fact that the connections (portals) to higher-fitness subbasins are very rare. These portals are generally only discovered after the population has diffused through most of the subbasin. Additionally, at each innovation, the fitness distribution expands into a new dimension of the simplex. Initially, when all members have fitness 0, the population is restricted to a point. After the first innovation it moves on a one-dimensional line, after the second it moves within a two-dimensional plane, and finally it moves into the interior of the full three-dimensional simplex. One sees that, when summarizing the population with fitness distributions, the number of components needed to describe the population grows dynamically each time a higher-fitness subbasin is discovered. We will return to this observation when we describe the connection of our analytical approach to the theory of statistical mechanics.

3   The Terraced Labyrinth Fitness Functions

As just outlined, the intuitive view of phenotypically constrained, genotype-space architectures—as a relatively small number of weakly interconnected neutral subbasins—is the one we have adopted in our analyses. We will now define a broad class of fitness functions that captures these characteristics. The principal motivation for this is to illustrate the generality of our existing results via a wider range of fitness functions than previously analyzed.

We represent genotypes in the population as bit-strings of a fixed length L. For any genotype of a certain fitness, there is a subset of its bits that are *fitness constrained*. Mutations in any of the constrained bits lower an individual's fitness. All the other bits are considered *free* bits, in the sense that they may be changed without affecting fitness. Of all possible configurations of free bits, there is a small subset of *portal configurations* that lead to an increased fitness. A portal consists of a subset of free bits, called a *constellation*, that is set to a particular "correct" configuration. A constellation may have more than one correct configuration. When a constellation is set to a portal configuration, the fitness is increased, and the constellation's bits become constrained bits. That is, via a portal free bits of an incorrectly set constellation become the constrained bits of a correctly set constellation.

The general structure of the fitness functions we have in mind is that fitness is conferred on individuals by having a number of constellations set to their portal configurations. Mutations in the constrained bits of the correct constellations lower fitness, while setting an additional constellation to its portal configuration increases fitness. A fitness function is specified by choosing sets of constellations, portal configurations, and assigning the fitness that each constellation confers on a genotype when set to one of its portal configurations.

3.1   A Simple Example

Let's illustrate our class of fitness functions by a simple example that uses bit-strings of length $L = 15$. The example is illustrated in Fig. 3. Initially, when no constellation is set correctly the strings have fitness f. The first constellation, denoted c, consists of the bits 1 through 5. This constellation can be set to two different portal configurations: either $\pi_1 = 11111$ or $\pi_2 = 00000$. When $c = \pi_1$ or $c = \pi_2$ the genotypes obtain fitnesses f_1 and f_2, respectively. Once constellation $c = \pi_1$, say, there is a constellation c_1, consisting of bits 9 through 15, that can be set correctly to portal configuration $\pi_{1,1} = 1100010$, in which case the genotype obtains fitness $f_{1,1}$. The constellation c_1 might also be set to configuration $\pi_{1,2} = 0101101$, leading to a fitness of $f_{1,2}$. Finally, once constellation $c_1 = \pi_{1,1}$, there is a final configuration $c_{1,1}$, consisting of bits 6 through 8, that can be set correctly. With $c = \pi_1$ and $c_1 = \pi_{1,1}$ configuration $c_{1,1}$ needs to be set to configuration $\pi_{1,1,1} = 001$ in order to reach fitness $f_{1,1,1}$. If instead $c_1 = \pi_{1,2}$, the final constellation $c_{1,2}$ needs to be set to portal $\pi_{1,2,1} = 100$, giving fitness $f_{1,2,1}$.

Alternatively, if constellation $c = \pi_2$, the next constellation c_2 consists of bits 8 through 10, which have portal configuration $\pi_{2,1} = 111$. Setting c_2 to $\pi_{2,1}$ leads to fitness $f_{2,1}$. Once c_2 is set correctly, there is a constellation $c_{2,1}$ consisting of bits 13 through 15, which has portal configuration $\pi_{2,1,1} = 110$ and fitness $f_{2,1,1}$. Finally, there is the constellation $c_{2,1,1}$ consisting of bits 6, 7, 11, and 12. The portal configuration for this constellation is $\pi_{2,1,1,1} = 1000$, leading to fitness $f_{2,1,1,1}$.

Generally, the hierarchical ordering of constellations and their connections via portals can be most easily represented as a tree, as in Fig. 3. Each tree node represents a subbasin of equal-fitness genotypes. The tree branches represent the portals that connect a lower-fitness subbasin to a higher-fitness subbasin. The fitness and structure of genotypes within a subbasin are also shown at each node. Stars (*) indicate the free bits within a subbasin. The constellations at each node indicate which subset of bits needs to be set to a portal configuration in order to proceed further up the tree. Thus, setting a constellation to a portal configuration leads one level up the tree, while mutating one or more of the constrained bits leads down the tree. In fact, a single point-mutation might lead all the way back to the root node.

We assume that setting a new constellation correctly leads to an increase in fitness. That is, f_1 and f_2 are larger than f, $f_{1,1}$ is larger than f_1, and so on. For simplicity in this example, we chose the constellation bits contiguously, except for $c_{2,1,1}$. Since our genetic algorithm, introduced shortly, does not employ crossover, the population dynamics remains the same under arbitrary permutations of the bits in the genome. Note further that we chose the portal configurations rather arbitrarily. In cases where a constellation has only a single portal, this configuration can be chosen arbitrarily without affecting the dynamics. When a constellation has more than one portal, the evolutionary dynamics can be affected by the Hamming distances between

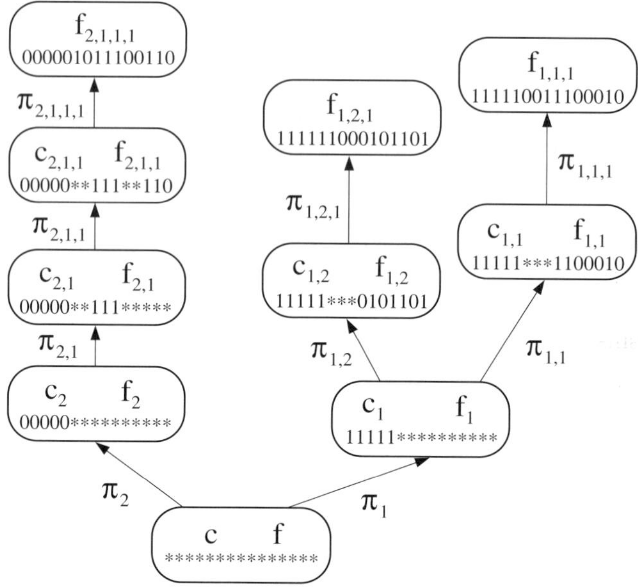

Fig. 3. Tree representation of a Terraced Labyrinth fitness function. The nodes of the tree represent subbasins of genotypes with equal fitness. They are represented by strings that have *'s for the free bits. The fitness f of the genotypes in the subbasins is indicated as well. The constellation c inside each node indicates the subset of bits that needs to be set correctly in order to move *up* a level in the tree to a higher-fitness subbasin. The portal configurations π that connect subbasins to higher-fitness subbasins are shown as branches.

the different portal configurations. A key assumption is that portal configurations such as π_1 and π_2 are mutually exclusive. Once evolution follows a certain branch up the tree, it is very unlikely to revert later on. We discuss in Sect. 8 how different evolutionary paths through the tree formalize such notions as *historical accident* and *structural phenotypic constraints*.

Finally, in this setting the genotype-to-phenotype map is nonexistent, since fitness is evaluated directly on the genotypes, without an intervening developmental process.

3.2   Definitions

We will now generalize this example by way of defining the class of *Terraced Labyrinth* fitness functions. As we saw in the example, constellations and portals form a hierarchy that can be most easily represented as a tree. Thus, we define Terraced Labyrinth fitness functions using trees, similar to the one illustrated in Fig. 3, as follows:

1. The *genotypes* are bit strings $\mathbf{s} = s_0 s_1 s_2 \cdots s_{L-1}$ of length L with bits $s_i \in \mathcal{A} \equiv \{0,1\}$.

2. The hierarchy of subbasins, constellations, and portals form a *tree*, consisting of nodes $\{\vec{\imath}\}$ and branches $\{\pi_{\vec{\imath}}\}$.
   (a) Tree *nodes* $\vec{\imath}$ are specified by a set of indices: $\vec{\imath} = \{i_1, i_2, i_3, \ldots, i_n\}$. The number n of indices denotes $\vec{\imath}$'s tree level. A particular setting of the indices labels the path from the root to $\vec{\imath}$. That is, one reaches $\vec{\imath}$ by taking branch i_1 at the root, branch i_2 at node i_1, and so on. The tree nodes represent both subbasins of genotypes with equal fitness and constellations of bits that, when set correctly, lead out of one subbasin to the next higher-fitness subbasin.
   (b) Tree *branches* represent portal configurations that connect the subbasins of equal-fitness genotypes to each other. Branch $\pi_{\vec{\imath}}$ points to node $\vec{\imath}$.
3. A *constellation* is a subset of **s**'s bits. Constellation $c_{\vec{\imath}}$ is located at node $\vec{\imath}$ and corresponds to the subset of bits that must be set to a portal configuration in order to move from subbasin $B_{\vec{\imath}}$ to a higher fitness subbasin. The number of bits in a constellation $c_{\vec{\imath}}$ is denoted $K_{\vec{\imath}}$.
4. A *portal* $\pi_{\vec{\imath},j}$ is one particular configuration of the $K_{\vec{\imath}}$ bits in constellation $c_{\vec{\imath}}$ out of the $2^{K_{\vec{\imath}}}$ possible configurations. The indices $\vec{\imath}, j$ of a portal $\pi_{\vec{\imath},j}$ indicate the node to which it points.
5. The *subbasin* $B_{i_1,i_2,\ldots,i_n}$ is the set of genotypes that have constellations c through $c_{i_1,\ldots,i_{n-1}}$ set to portals π_{i_1} through $\pi_{i_1,\ldots,i_n}$, respectively, but do *not* have constellation $c_{i_1,\ldots,i_n}$ set to any of its portal configurations.
6. All genotypes in the subbasin $B_{\vec{\imath}}$ have a fitness $f_{\vec{\imath}}$.
7. A leaf-node $\vec{\imath}$ in the tree represents a set of equal-fitness genotypes that form a local optimum of the fitness function. The fitness of these genotypes is $f_{\vec{\imath}}$.

The trees that define the hierarchy of constellations, subbasins, and portals are not entirely arbitrary. They have the following constraints.

1. The number of branches leaving node $\vec{\imath}$ is at most $2^{K_{\vec{\imath}}}$.
2. A constellation is *disjoint* from the root constellation c and all other constellations that connect it to the root. That is, the set $c_{i_1,i_2,\ldots,i_n}$ is disjoint from the sets c, c_{i_1}, c_{i_1,i_2}, and so on.

This class of Terraced Labyrinth fitness functions incorporates and extends the previously studied *Royal Road* fitness functions of [45] and [46] and the *Royal Staircase* fitness functions of [43]. In those fitness functions, all constellations had the same number of defining bits K, and there was only a single portal configuration $\pi = 1^K$ for each constellation. A Royal Staircase fitness function corresponds to a Terraced Labyrinth fitness function whose tree is a simple linear chain. Additionally, in the Royal Road fitness functions, constellations were allowed to be set in any arbitrary order.

The architectural approach we have taken here should be contrasted with the use of randomized fitness functions that have been modified to have neutral networks. These include the NKp landscapes of [5] and the discretized

NK fitness functions of [35]. The popularity of random fitness functions seems motivated by the idea that something as complicated as a biological genotype-phenotype mapping can only be statistically described using a randomized structure. Although this seems sensible in general, the results tend to be strongly dependent on the specific randomization procedure that is chosen; the results might be biologically misleading. For instance, NK models create random epistatic interactions between bits, mimicking spin-glass models in physics. In the context of spin glasses, this procedure is conceptually justified by the idea that the interactions between the spins were randomly frozen in when the magnetic material formed. However, in the context of genotype-phenotype mappings, the interactions between different genes are themselves the result of evolution. This can lead to very different kinds of "random" interactions, as shown in [3].

At a minimum, though, the most striking difference between our choice of fitness function class and randomized fitness functions is that the population dynamics of the randomized classes is very difficult, if not impossible, to analyze at present. In contrast, the population dynamics of the class of fitness functions just introduced can be analyzed in some detail. Moreover, for biological systems it could very well be that structured fitness functions, like the Terraced Labyrinth class, may contain all of the generality required to cover the phenomena claimed to be addressed by the randomized classes. Several limitations and generalizations of the Terraced Labyrinth fitness functions are discussed in Sect. 9.2.

4   A Simple Genetic Algorithm

For our analysis of epochal evolutionary dynamics, we chose a simplified form of a genetic algorithm (GA) that does not include crossover and that uses fitness-proportionate selection. A population of M individuals, each specified by a genotype of length L bits, reproduces in discrete non-overlapping generations. Each generation, M individuals are selected (with replacement) from the population in proportion to their genotype's fitness. Each selected individual is placed into the population at the next generation after mutating each genotype bit with probability μ.

This GA effectively has two parameters: the mutation rate μ and the population size M. A given evolutionary optimization problem is specified, of course, by the fitness function parameters as given by the constellations, portals, and their fitness values. Stated most prosaically, then, our central goal is to analyze the population dynamics, as a function of μ and M, for any given fitness function in the Terraced Labyrinth class. Here we review the essential aspects of the population dynamics analysis.

5   Statistical Dynamics of Evolutionary Search

In [45] and [46] we developed an approach, which we called *statistical dynamics*, to analyze the behavioral regimes of a GA searching fitness functions that lead to epochal dynamics. Here we can only briefly review the mathematical details of this approach to evolutionary dynamics, emphasizing the motivations and the main ideas and tools from statistical mechanics and dynamical systems theory. The reader is referred to [46] for an extensive and mathematically detailed exposition. There, the reader will also find a review of the connections and similarities of our work to the alternative methodologies for GA theory developed by Vose and collaborators [36,47,48], by Prügel-Bennett, Rattray, and Shapiro [37–39], in the theory of molecular evolution [13,14], and in mathematical population genetics [24].

5.1   Statistical Mechanics

Our approach builds on ideas from statistical mechanics [7,40,51] and adapts its equilibrium formulation to apply to the piecewise steady-state dynamics of epochal evolution. The microscopic state of systems that are typically studied in statistical mechanics—such as, a box of gas molecules—is described in terms of the positions and momenta of all particles. What is of physical interest, however, are observable (and reproducible) quantities, such as the gas's pressure P, temperature T, and volume V. The goal is to predict the relationships among these macroscopic variables, starting from knowledge of the equations of motion governing the particles and the space of the entire system's possible microscopic states. A given setting of macroscopic variables—e.g., a fixed P, V, and T—is often referred to as a *macrostate*; whereas a snapshot of the positions and momenta of all particles is called a *microstate*.

There are two kinds of assumptions that allow one to connect the microscopic description (a collection of microstates and equations of motion) to observed macroscopic behavior. The first is the assumption of *maximum entropy*, which states that all microscopic variables, unconstrained by a given macrostate, are as random as possible.

The second is the assumption of *self-averaging*. In the *thermodynamic limit* of an infinite number of particles, self-averaging means that the macroscopic variables are expressible only in terms of themselves. In other words, the macroscopic description does not require knowledge of detailed statistics of the microscopic variables. For example, at equilibrium the macroscopic variables of an *ideal gas* of noninteracting particles are related by the *equation of state*, $PV = kNT$, where k is a physical constant, and N is the total number of particles in the box. Knowing, for instance, the frequency with which molecules come within 100 nanometers of each other does not improve this macroscopic description.

Varying an experimental control parameter of a thermodynamic system can lead to a sudden change in its structure and in its macroscopic properties. This occurs, for example, as one lowers the temperature of liquid water below the freezing point. The liquid macrostate undergoes a *phase transition* and the water turns to solid ice. The macrostates (*phases*) on either side of the transition are distinguished by different sets of macroscopic variables. That is, the set of macrovariables that is needed to describe ice is not the same as the set of macrovariables that is needed to describe water. The difference between liquid water and solid ice is captured by a sudden reduction in the freedom of water molecules to move. While the water molecules move equally in all directions, the frozen molecules in the ice-crystal possess a relatively definite spatial location. Passing through a phase transition can be thought of as creating, or destroying, macroscopic variables and making or *breaking* the symmetries associated with them. In the liquid to solid transition, the rotational symmetry of the liquid phase is broken by the onset of the rigid lattice symmetry of the solid phase. As another example, in the Curie transition of a ferromagnet, the *magnetization* is the new macroscopic variable that is created with the onset of magnetic-spin alignment as the temperature is lowered.

5.2   Evolutionary Statistical Mechanics

The statistical mechanical description can also be applied to evolutionary processes. From a microscopic point of view, the exact state of an evolving population is only fully described when a list S of all genotypes with their frequencies of occurrence in the population is given. On the microscopic level, the evolutionary dynamics is implemented as a *Markov chain* with the conditional transition probabilities $\Pr(S'|S)$ that the population at the next generation will be the "microscopic" collection S'; see [17] and [36] for the microscopic formulation in the context of mathematical population genetics and genetic algorithms, respectively. For any reasonable genetic representation, however, there is an enormous number of these microscopic states S and so too of their transition probabilities. The large number of parameters, $\mathcal{O}(2^L!)$, makes it almost impossible to quantitatively study the dynamics at this microscopic level.

More practically, a full description of the dynamics on the level of microscopic states S is neither useful nor typically of interest. One is much more likely to be concerned with relatively coarse statistics of the dynamics, such as the evolution of the best and average fitness in the population or the waiting times for evolution to produce a genotype of a certain quality. The result is that quantitative mathematical analysis faces the task of finding a macroscopic description of the microscopic evolutionary dynamics that is simple enough to be tractable numerically or analytically and that, moreover, facilitates predicting the quantities of interest to an experimentalist.

With these issues in mind, we specify the macrostate of the population at each time t by some relatively small set of macroscopic variables $\{\mathcal{X}(t)\}$. Since this set of variables intentionally ignores vast amounts of detail in the microscopic variables $\{x(t)\}$, it is generally impossible to exactly describe the evolutionary dynamics in terms of these macroscopic variables. To achieve the benefits of a coarser description, we assume that the population has equal probabilities to be in any of the microscopic states consistent with a given macroscopic state. That is, we assume *maximum entropy* over all microstates $\{x(t)\}$ that are consistent with the specific macrostate $\{\mathcal{X}(t)\}$.

Additionally, in the limit of infinite population size, we assume that the resulting equations of motion for the macroscopic variables become closed. That is, for infinite populations, we assume that we can predict the state of the macroscopic variables at the next generation, given the present state of *only* the macroscopic variables. This infinite-population limit is analogous to the thermodynamic limit in statistical mechanics. The corresponding assumption is analogous to *self-averaging* of the macroscopic evolutionary dynamics in this limit.

We use the knowledge of the microscopic dynamics together with the maximum entropy assumption to predict the next macrostate $\{\mathcal{X}(t+1)\}$ from the current one $\{\mathcal{X}(t)\}$. Then we re-assume maximum entropy over the microstates $\{x(t+1)\}$ given the new macrostate $\{\mathcal{X}(t+1)\}$. Since this method allows one to relax the usual equilibrium constraints and so account for the dynamical change in macroscopic variables, we refer to this extension of statistical mechanics as *statistical dynamics*. A similar approach has been developed in some generality for non equilibrium statistical mechanics by Streater and, not surprisingly, it goes under the same name [42].

5.3 Evolutionary Macrostates

The key, and as yet unspecified, step in developing such a statistical dynamics framework of evolutionary processes is to find an appropriate set of macroscopic variables that satisfy the above assumptions of maximum entropy and self-averaging. In practice, this is difficult. Ultimately, the suitability of a set of macroscopic variables has to be verified by comparing theoretical predictions with experimental measurements. In choosing such a set of macroscopic variables, one is guided by knowledge of the fitness function and the genetic operators. Although not reduced to a procedure, this choice is not made in the dark.

First, there may be *symmetries* in the microscopic dynamics. Imagine, for instance, that genotypes can have only two possible values for fitness, f_A and f_B. Assume also that under mutation all genotypes of type A are equally likely to turn into type-B genotypes and that all genotypes of type B have equal probability to turn into genotypes of type A. In this situation, it is easy to see that we can take the macroscopic variables to be the relative proportions of A genotypes and B genotypes in the population. The reason one can do this is that *all* microstates with a certain proportion of A and B types give rise to exactly the same dynamics on the level of proportions of A and B types. That is, the dynamics is symmetric under any transformation of the microstates that leaves the proportions of A and B types unaltered. Neither selection nor mutation distinguish different genotypes within the sets A and B on the level of the proportions of A's and B's that they produce in the next generation. Obviously, one wants to take advantage of such symmetries in a macroscopic description. However, for realistic cases, such symmetries are not often abundant. Simply taking them into account, while important, does not typically reduce the complexity of the description sufficiently.

One tends to make more elaborate assumptions in developing a macroscopic description. Assume that the A and B genotypes are not all equally likely to turn from type A to B and vice versa, but do so only on *average*. For example, it might be the case that not all A types behave exactly the same under mutation, but that the dominant subset of A's that occurs in a population typically behaves like the *average* over the set of all A types. This is a much weaker symmetry than the exact one mentioned above. Importantly, it still leads to an accurate description of the dynamics on the level of A and B types under the maximum entropy assumption.

The Neo-Darwinian formalism of biological evolution suggests a natural decomposition of the microscopic population dynamics into a part that is guided by selection and a part that is driven by genetic diversification. Simply stated, selection is an ordering force induced by the environment that operates on the level of the phenotypic fitness in a population. In contrast, genetic diversification is a disordering and randomizing force that drives a population to an increased diversity of genotypes. Thus, it seems natural to choose as macrostates the proportion of genotypes in the different fitness classes (subbasins) and to assume that, due to random genetic diversification within each subbasin, genetic variation can be approximated by the maximum entropy distribution within each subbasin. This intuition is exactly the one we use in our statistical dynamics analysis of the Terraced Labyrinth fitness functions. Specifically, we describe the population in terms of the proportions $P_{\vec{i}}$ that are located in each of the subbasins $B_{\vec{i}}$. The maximum entropy assumption entails that within subbasin $B_{\vec{i}}$, individuals are equally likely to be any of the genotypes in $B_{\vec{i}}$. In other words, we assume that all free bits in a constellation are equally likely to be in any of their nonportal configurations.

The essence of our statistical dynamics approach is to describe the population state at any time during a GA run by a relatively small number of macroscopic variables—variables that (i) in the limit of infinite populations self-consistently describe the dynamics at their own level and (ii) can change over time. After obtaining the dynamics in the limit of infinite populations explicitly, one then uses this knowledge to solve for the GA's dynamical behaviors with finite populations.

6   Evolutionary Dynamical Systems

Up to this point we have described our approach in terms of its similarities with statistical mechanics. We appealed intuitively to macroscopic "dynamics", which can be derived in terms of the microscopic equations of motion (of selection and mutation on genotypes) and the maximum entropy assumption. Now we fill in the other half of the story, the half that clarifies what "dynamics" is and that draws out the similarities of our approach with dynamical systems theory.

As we just explained, we approximate the complete finite-population dynamics in two steps. First, we use the maximum entropy assumption together with the microscopic equations of motion to construct an infinite-population "flow" that describes the deterministic (macroscopic) dynamics of the subbasin distribution of an infinite population. Then, we construct the finite-population dynamics by accounting for the finite-population sampling at each generation. The net result is a stochastic nonlinear dynamical system. We now explain these two steps in more detail.

6.1   Infinite Populations

Consider an infinite population with subbasin distribution $\vec{P}$, where component $P_{\vec{i}} \in [0, 1]$ is the proportional of individuals in the subbasin $B_{\vec{i}}$. Note that the number of components in $\vec{P}$ is equal to the number of nodes in the constellation tree that describes the Terraced Labyrinth fitness function. Given this, the question is how selection and mutation, acting on the distribution $\vec{P}(t)$, create the distribution $\vec{P}(t+1)$ at the next generation.

The effects of selection are simple, since all genotypes in subbasin $B_{\vec{i}}$ have the same fitness. If $\langle f \rangle$ is the average fitness in the population, we simply have that after selection the components are $P_{\vec{i}}^{select} = f_{\vec{i}} P_{\vec{i}}(t)/\langle f \rangle$. To calculate the effects of mutation we have to use our maximum entropy assumption. The probability that a genotype in subbasin $B_{\vec{j}}$ turns into a genotype in subbasin $B_{\vec{i}}$ is simply given by the *average* probability of a mutation from a genotype in $B_{\vec{j}}$ to *any* genotype in $B_{\vec{i}}$. The average is taken with equal weights over all genotypes in $B_{\vec{j}}$. Putting the effects of selection and mutation together, we obtain a generation operator $\mathbf{G}$ that specifies the macroscopic evolutionary dynamical system:

$$\vec{P}(t+1) = \mathbf{G}[\vec{P}(t)] \ . \tag{1}$$

The infinite-population dynamics on the level of subbasin distributions is simply given by iterating the operator $\mathbf{G}$. Following the terminology introduced in molecular evolution theory we call $\vec{P}(t)$ the *phenotypic quasispecies*.

The expected[1] change $\langle d\vec{P} \rangle$ in the fitness distribution over one generation is given by:

$$\langle d\vec{P} \rangle = \mathbf{G}[\vec{P}] - \vec{P}. \tag{2}$$

---

[1] It will become clear shortly why we call this change an *expected* change.

We can visualize the *flow* induced by the macroscopic equations of motion by plotting $\langle d\vec{P} \rangle$ at a number of states in the simplex of populations. This is shown in Fig. 4; taken after [46]. The fitness function and evolution parameters of Fig. 4 are those of Fig. 2. The temporal behavior of the system, starting in an initial condition $\vec{P}(t = 0)$, is indicated by the flow arrows.

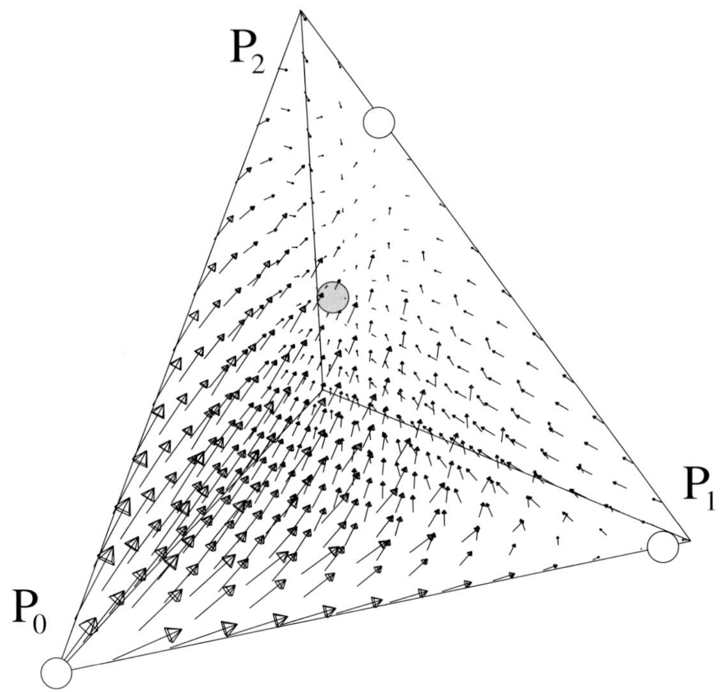

Fig. 4. Fitness distribution flow $\langle d\vec{P} \rangle$ in the simplex for the Royal Road fitness function with $N = 3$ constellations with $K = 10$ bits each and for the simple GA with mutation rate $\mu = 0.005$; cf. Fig. 2. Fixed points of the flow are shown as large balls. The grey ball is the stable, asymptotic fixed point inside the simplex. The white balls indicate the locations of the unstable fixed points that are outside the simplex. The latter do not represent valid populations, but nonetheless they can affect the dynamics of allowed populations within the simplex by slowing down (short arrows) the flow near them.

For large ($M > 2^L$) populations, the dynamics of the subbasin distribution is simple: $\langle f \rangle$ increases smoothly and monotonically to an asymptote over a small number of generations. (See Fig. 3 of [45].) That is, there are no epochs. The reason for this is simple: for an infinite population, *all* genotypes, and therefore all subbasins, are represented in the initial population. Instead of the evolutionary dynamics *discovering* fitter genotypes over time, it essentially only expands the proportion of globally optimal genotypes already present in the initial population at $t = 0$.

6.2   Finite Populations

In spite of the qualitatively different dynamics for infinite and finite populations, we showed in [46] that the (infinite population) operator $\mathbf{G}$ is the essential ingredient for describing the finite-population dynamics with its epochal dynamics as well. Beyond the differences in observed behavior, there are two important mathematical differences between the infinite-population dynamics and that with finite populations. The first is that with finite populations the components $P_{\vec{\imath}}$ cannot take on continuous values between 0 and 1. Since the number of individuals in subbasin $B_{\vec{\imath}}$ is necessarily an integer, the values of $P_{\vec{\imath}}$ are quantized in multiples of $1/M$. Thus, the continuous simplex of allowed infinite-population fitness distributions turns into a regular, discrete lattice with spacing of $1/M$. Second, due to finite-population sampling fluctuations, the dynamics of the subbasin distribution is no longer deterministic, as described by Eq. (1). In general, we can only determine the conditional probabilities $\Pr[\vec{Q}|\vec{P}]$ that a given fitness distribution $\vec{P}$ leads to another $\vec{Q}$ in the next generation.

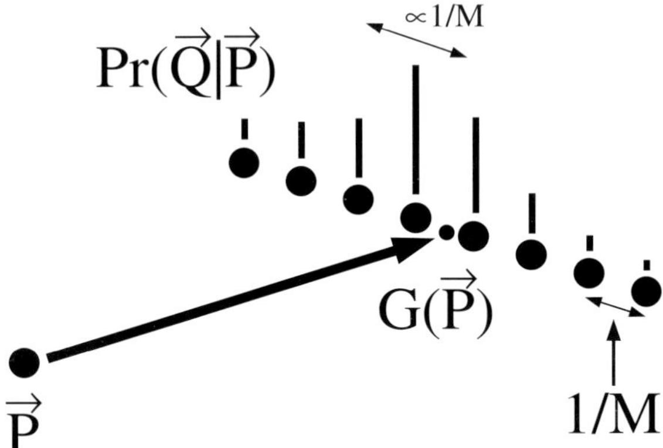

Fig. 5. Illustration of the stochastic dynamics that maps from one generation to the next. Starting with finite population $\vec{P}$, the arrow indicates the *expected* next population $\mathbf{G}[\vec{P}]$. Due to sampling, the probability that the actual next population is $\vec{Q}$ is given by a multinomial distribution $\Pr[\vec{Q}|\vec{P}]$, Eq. (3). Note that the underlying state space is a discrete lattice with spacing $1/M$.

The net result is that the probabilities $\Pr[\vec{Q}|\vec{P}]$ are determined by a multinomial distribution with mean $\mathbf{G}[\vec{P}]$:

$$\Pr[\vec{Q}|\vec{P}] = M! \prod_{\vec{\imath}} \frac{\left(\mathbf{G}_{\vec{\imath}}[\vec{P}]\right)^{m_{\vec{\imath}}}}{m_{\vec{\imath}}!} \ . \tag{3}$$

where $Q_{\vec{\imath}} = m_{\vec{\imath}}/M$, with $0 \leq m_{\vec{\imath}} \leq M$ integers, and the product runs over all subbasins $\vec{\imath}$. (The stochastic effects of finite-population sampling are illustrated in Fig. 5.) For any finite-population subbasin distribution $\vec{P}$, the operator $\mathbf{G}$ gives the evolution's *average* dynamics over one time step, since by Eq. (3) the *expected* subbasin distribution at the next time step is $\mathbf{G}[\vec{P}]$. Note that the components $\mathbf{G}_{\vec{\imath}}[\vec{P}]$ need not be multiples of $1/M$. Therefore, the actual subbasin distribution $\vec{Q}$ at the next time step is not $\mathbf{G}[\vec{P}]$, but is instead one of the allowed lattice points in the finite-population state space consistent with the distribution $\Pr[\vec{Q}|\vec{P}]$. Since the variance around the expected distribution $\mathbf{G}[\vec{P}]$ is proportional to $1/M$, $\vec{Q}$ tends to be one of the lattice points close to $\mathbf{G}[\vec{P}]$.

Putting both the infinite-population dynamical system and the stochastic sampling effects induced by finite populations together, we arrive at our basic model of evolutionary population dynamics. We can now begin to draw out some consequences.

7   Metastability and the Unfolding of Macrostates

Assume that there are no individuals in a certain subbasin $B_{\vec{\imath}}$ and that the component $\langle dP_{\vec{\imath}} \rangle$ is much smaller than $1/M$. In that case, the actual change in component $P_{\vec{\imath}}$ is likely to be $dP_{\vec{\imath}} = 0$ for a long succession of generations. That is, if there are no individuals in subbasin $B_{\vec{\imath}}$ and the rate of creation of such individuals is much smaller than $1/M$, then subbasin $B_{\vec{\imath}}$ is likely to stay empty for a considerable number of generations. Consequently, there is no movement to increase fitness to level $f_{\vec{\imath}}$ during this time. More generally, if the size of the flow $\langle dP_{\vec{\imath}} \rangle$ (and its variance) in some direction $\vec{\imath}$ is much smaller than the lattice spacing $(1/M)$ of allowed finite populations, we expect the subbasin distribution to not change in direction $\vec{\imath}$. In [45] and [46] we showed this is the mechanism that causes epochal dynamics for finite populations.

More formally, an epoch corresponds to the population being restricted to a region of an n-dimensional subsimplex of the macroscopic state space. Stasis occurs because the flow out of this subspace is much smaller than the finite-population–induced lattice spacing. In particular, for the Terraced Labyrinth fitness functions, an epoch corresponds to the time during which the highest-fitness individuals are located in subbasin $B_{i_1,i_2,\dots,i_n}$. During this time, an equilibrium subbasin distribution is established in the population. Its components are nonzero only for subbasins B, B_{i_1}, B_{i_1,i_2}, through $B_{i_1,\dots,i_n}$.

That is, they are nonzero for all of the lower-fitness subbasins that connect $B_{\vec{i}}$ to the root. Since the discovery of a portal configuration of constellation $c_{i_1,\ldots,i_n}$ is rare, the population remains in this n-dimensional subsimplex for a considerable number of generations. The number of generations it remains in this epoch is, of course, directly dependent on the number of portals out of the subbasin $B_{\vec{i}}$ and the number of bits $K_{\vec{i}}$ in constellation $c_{\vec{i}}$.

Recall the example of epochal behavior of Sect. 2.2 and Fig. 2. Initially, the population was located in the zero-dimensional macrostate corresponding to all genotypes located in the root subbasin. Then the first portal configuration was discovered and the population moved onto the line of population states that have some individuals in the root subbasin and some in the basin B_1. After this epoch, a genotype in subbasin $B_{1,1}$ was discovered and the population moved to a steady-state in the plane of proportions P, P_1, and $P_{1,1}$. (These were labeled according to their fitnesses—P_0, P_1, and P_2—in Fig. 2.) Finally, the global optimum string in subbasin $B_{1,1,1}$ was discovered, and the population moved to its final fixed point in the three-dimensional simplex.

The global evolutionary dynamics can be viewed as an incremental discovery (an unfolding) of successively more (macroscopic) dimensions of the subbasin distribution space. In most realistic settings, it is typically the case that population sizes M are much smaller than the number of possible genotypes 2^L. Initially, then, the population consists only of genotypes in subbasins of low fitness. Assume, for instance, that genotypes in subbasin $B_{1,2}$ are the highest fitness ones in the initial population. Mutation and selection establish an equilibrium phenotypic quasispecies $\vec{P}^{1,2}$, consisting of nonzero proportions of genotypes in the subbasin B, B_1, and $B_{1,2}$, and zero proportions of genotypes in all other subbasins. Individuals and their descendants drift through subbasin $B_{1,2}$. The subbasin distribution fluctuates around $\vec{P}^{1,2}$ until a portal configuration $\pi_{1,2,i}$ of the constellation $c_{1,2}$ is discovered and genotypes of (higher) fitness $f_{1,2,i}$ spread through the population. The population then settles into subbasin distribution $\vec{P}^{1,2,i}$ with average fitness $\langle f \rangle_{1,2,i}$ until a portal $\pi_{1,2,i,j}$ of constellation $c_{1,2,i}$ is discovered, and so on, until a local optimum corresponding to a leaf of the fitness function tree is found. In this way, the macroscopic dynamics can be seen as stochastically hopping between the different epoch distributions $\vec{P}^{\vec{i}}$ of subbasins $B_{\vec{i}}$ that are connected to each other in the fitness function tree.

Note that at each stage $\vec{P}^{i_1,\ldots,i_n}$ has only $n+1$ (nonzero) components, each corresponding to a subbasin connecting $B_{\vec{i}}$ to the tree root. All other subbasin components are zero. The selection-mutation balance maintains a constant proportion of genotypes with correct configurations in all constellations that define the epoch. By the maximum entropy assumption, the action of the generation operator $\mathbf{G}$ is symmetric with respect to all remaining nonportal constellation configurations. That is, $\mathbf{G}$'s action is indifferent to the various proportions of particular incorrect configurations in the constellation. The symmetry among constellation $c_{\vec{i}}$'s incorrect configurations is *broken dynamically* when a (typically rare) portal configuration is discovered. This symmetry breaking adds a new macroscopic variable—a new "active" dimension of the phenotype. This symmetry breaking and stabilization of a new phenotypic dimension is the dynamical analogue of a phase transition.

As alluded to earlier, much of the attractiveness of the Terraced Labyrinth class of fitness functions lies in the fact that, to a good approximation, analytical predictions can be obtained for observable quantities, such as average epoch fitness $\langle f \rangle_{\vec{i}}$ and the epoch subbasin distribution $\vec{P}^{\vec{i}}$ in terms of the evolutionary and fitness function parameters. For instance, assume that the highest fitness genotypes are in subbasin $B_{i_1, i_2, \ldots, i_n}$ and that the population resides in the steady-state distribution $\vec{P}^{i_1, i_2, \ldots, i_n}$. The number of constrained bits in each of the subbasins that have nonzero proportions during this epoch is denoted by

$$L_{i_1, i_2, \ldots, i_m} = K + K_{i_1} + K_{i_1, i_2} + \ldots + K_{i_1, i_2, \ldots, i_{m-1}}. \tag{4}$$

(Note that $L = 0$ for the root subbasin.) Then, up to some approximation,[2] the average epoch fitness is given simply by

$$\langle f \rangle_{\vec{i}} = f_{\vec{i}} (1 - \mu)^{L_{\vec{i}}}. \tag{5}$$

One can also derive the subbasin distribution $\vec{P}^{\vec{i}}$. In order to express the results most transparently, we introduce the epoch fitness-level ratio using Eq. (5):

$$\alpha_{\vec{i}\vec{j}} = \frac{f_{\vec{j}}}{f_{\vec{i}}} (1 - \mu)^{L_{\vec{j}} - L_{\vec{i}}}. \tag{6}$$

Then we have for the highest-fitness component of the subbasin distribution $P_{\vec{i}}$ that

$$P_{\vec{i}}^{\vec{i}} = \prod_{\vec{m} < \vec{i}} \frac{1 - \alpha_{\vec{i}\vec{m}}}{1 - \alpha_{\vec{i}\vec{m}}(1 - \mu)^{K_{\vec{m}}}}, \tag{7}$$

where $\vec{m} < \vec{i}$ indicates the set of all nodes lying along the path between $\vec{i}$ and the tree's root, including the root. For the other components of $P^{\vec{i}}$ we have that

$$P_{\vec{j}}^{\vec{i}} = \frac{(1 - \mu)^{L_{\vec{j}}} \left(1 - (1 - \mu)^{K_{\vec{j}}}\right)}{1 - \alpha_{\vec{i}\vec{j}}(1 - \mu)^{K_{\vec{j}}}} \prod_{\vec{m} < \vec{j}} \frac{1 - \alpha_{\vec{i}\vec{m}}}{1 - \alpha_{\vec{i}\vec{m}}(1 - \mu)^{K_{\vec{m}}}}. \tag{8}$$

Describing the dynamics in and between epoch distributions $\vec{P}^{\vec{i}}$ using diffusion approximations and then invoking (dynamical systems) concepts— such as stable and unstable manifolds, Jacobian eigenvalues, and their eigenvectors—a number of additional properties of epochal evolution can be derived analytically and predicted quantitatively. The reader is referred to [46]

---

[2] The approximation here is that, during an epoch, the *back mutations* from lower fitness subbasins to higher subbasins can be neglected. This assumption is generally valid for constellation lengths $K_{\vec{i}}$ that are not too small.

and [43] for a detailed analysis of the distribution of epoch fluctuations, the stability of epochs, and the average waiting times for portal discovery.

8   Frozen Accidents, Phenotypic Structural Constraints, and the Subbasin-Portal Architecture

The subbasin-portal architecture, whose population dynamics we are analyzing, suggests a natural explanation for the occurrence and longevity of *frozen accidents* in evolution. Generally speaking, frozen accidents refer to persistent phenotypic characters that are selected out of a range of possible, structurally distinct alternatives by specific random events in the evolutionary past. One imagines an arbitrary event, such as a sampling fluctuation, promoting one or another phenotype, which then comes to dominate the population and thereby excludes alternatives that could be equally or even more fit in the long term.

Within the class of Terraced Labyrinth fitness functions, *frozen accidents* occur via a simple mechanism. In particular, a given evolutionary path through the fitness-function tree can be regarded as a sequence of frozen accidents. Since different portals of the same constellation are mutually exclusive, their subbasins are separated by a *fitness barrier*. Across a wide range of parameter settings, the crossing of such fitness barriers takes much longer than the discovery of new portals, via neutral evolution, in the current subbasin. Once evolution has taken a certain branch up the tree, it is therefore unlikely that it will ever return. That is, once a subbasin $B_{\vec{\imath}}$ is discovered, the further course of evolution is restricted to the subtree with its root at $\vec{\imath}$. In this way, the genotypic constellations up to $\vec{\imath}$ become installed in the population.

The alternative evolutionary paths are not merely a case of genetic bookkeeping. Different portals of a constellation $c_{\vec{\imath}}$ may be associated with very different *phenotypic* innovations. Once a particular phenotypic innovation has occurred, the phenotype determines which range of future phenotypic innovations can occur. This contingency—how evolutionary futures depend on current phenotypic constraints—goes under the name of *structural phenotypic constraints*. In the Terraced Labyrinth, this phenomenon is reflected in the possibility that fitness-function trees have very dissimilar subtrees. For instance, the subtrees rooted at nodes 1 and 2 in Fig. 3 are very dissimilar. This dissimilarity reflects the fact that evolutionary futures starting from the phenotype corresponding to node 1 are very different from those starting from the phenotype associated with node 2.

Naturally, the Terraced Labyrinth class of fitness functions does not indicate which kind of tree structures, reflecting structural constraints, are appropriate or biologically realistic. This will ultimately be decided by experiment. The generality of this class of fitness functions, however, illustrates that qualitative concepts—such as frozen accidents and structural phenotypic constraints—are very easily represented and analyzed within the statistical dynamics framework.

9   Concluding Remarks

9.1   Summary

We introduced a generalized subbasin-portal architecture by way of defining a new class of fitness functions—the Terraced Labyrinth. The detailed mathematical analysis of the population dynamics that we introduced previously can be adapted straightforwardly to this generalized setting. In this way, statistical dynamics was shown to have a wider applicability, and its results on epochal evolution were seen to have wider ranging consequences than the first analyses in [45] and [46] might have suggested.

We described this more general view of epochal evolution, attempting to clarify the connections to both statistical mechanics and dynamical systems theory. The result is a dynamical picture of a succession of "phase transitions" in which microscopic symmetries are broken and new macroscopic dimensions are discovered and then stabilized. These new macroscopic dimensions then become the substrate and historical context for further evolution.

9.2   Extensions and Generalizations

There are a number of extensions to more complex evolutionary processes that should now be possible. Here we mention a few limitations of the class of fitness functions analyzed and make several generalizations.

First, constellations do not overlap constellations higher in the tree. Second, all the subbasins have a similar regular architecture: there is a set of constrained bits (in the constellations) that define the subbasin and all other bits are free. This is undoubtedly not the case generally. Different subbasins can have distinct irregular architectures and different kinds of portals. Moreover, the diffusion dynamics through distinct subbasins may be different. For instance, subbasins might also be defined with respect to more complicated genetic operations, such as gene duplication, unequal crossovers, and gene conversion.

Third, all of a subbasin's portals correspond to configurations of a single constellation. This ensures that the topology of the subbasin hierarchy forms a tree, as opposed to the more general topologies suggested by Fig. 1. Extending the analysis to more complicated subbasin architectures is formally straightforward, but becomes considerably more complicated to carry out. For very complicated architectures, the approximations in our analysis may have to be reworked.

Fourth, one would like to extend statistical dynamics to *open-ended* models in which (say) the genotype length can grow over time, allowing the tree to *dynamically* grow new branches as well, perhaps along the lines investigated in [3]. One would hope to see how the evolutionary dynamics adapts as the mutation-genome length error threshold is approached [13]. As long as such open-ended models adhere to the tree topology of the subbasin-portal hierarchy, it would appear that our analyses could easily be extended to them.

Finally, the maximum entropy assumption only holds to some degree of approximation. For instance, whenever a new macrodimension unfolds, the population is initially concentrated around the portal genotype in the subbasin; this is a type of *founder effect*. The population then spreads out randomly from there, but the genotypes never completely decorrelate due to finite-population sampling fluctuations [12]. Moreover, as we have shown in [44], the population members in lower-fitness subbasins are closely genetically related to members in the subbasin of currently highest fitness. These facts flatly contradict the maximum entropy assumption that individuals are randomly and independently spread through the subbasins. Since these complications do not generally alter the rate of deleterious mutations from subbasins to lower-fitness subbasins, theoretical predictions—such as the epoch distributions $\vec{P^i}$—are not much affected. However, as shown in [46], statistics—such as the average waiting time for the discovery of a portal—may be significantly affected. This leaves open the question of how to extend the set of macroscopic variables to account for these complications.

Acknowledgments. This work was partially supported under Keck Foundation Evolutionary Dynamics Program at SFI and under SFI's Computation, Dynamics, and Learning Program by NSF IRI-9705830, by AFOSR via NSF grant PHY-9970158, and by DARPA under contract F30602-00-2-0583.

References

1. C. Adami. Self-organized criticality in living systems. *Phys. Lett. A*, 203:29–32, 1995.
2. L. M. Adleman. Molecular computation of solutions to combinatorial problems. *Science*, 266:1021–1024, 1994.
3. L. Altenberg. Genome growth and the evolution of the genotype-phenotype map. In W. Banzhaf and F. H. Eeckman, editors, *Evolution and Biocomputation. Computational Models of Evolution, Monterey, California, July 1992*, pages 205–259, Springer-Verlag, Berlin, 1995.
4. T. Bäck. *Evolutionary algorithms in theory and practice: Evolution strategies, evolutionary programming, genetic algorithms*. Oxford University Press, New York, 1996.
5. L. Barnett. Tangled webs: Evolutionary dynamics on fitness landscapes with neutrality. Master's thesis, School of Cognitive Sciences, University of East Sussex, Brighton, 1997. http://www.cogs.susx.ac.uk/lab/adapt/nnbib.html.
6. R. K. Belew and L. B. Booker, editors. *Proceedings of the Fourth International Conference on Genetic Algorithms*. Morgan Kaufmann, San Mateo, CA, 1991.

7. J. J. Binney, N. J. Dowrick, A. J. Fisher, and M. E. J. Newman. *The Theory of Critical Phenomena: An Introduction to the Renormalization Group*. Oxford Science Publications, 1992.

8. L. Chambers, editor. *Practical Handbook of Genetic Algorithms*. CRC Press, Boca Raton, 1995.

9. J. Chen, E. Antipov, B. Lemieux, W. Cedeno, and D. H. Wood. *In vitro* selection for a OneMax DNA evolutionary computation. In E. Winfree and D.K. Gifford, editors, *DNA Based Computers V*. American Mathematical Society, Providence, RI, 2000.

10. J. P. Crutchfield and M. Mitchell. The evolution of emergent computation. *Proc. Natl. Acad. Sci. U.S.A.*, 92:10742–10746, 1995.

11. L. D. Davis, editor. *The Handbook of Genetic Algorithms*. Van Nostrand Reinhold, 1991.

12. B. Derrida and L. Peliti. Evolution in a flat fitness landscape. *Bull. Math. Bio.*, 53(3):355–382, 1991.

13. M. Eigen. Self-organization of matter and the evolution of biological macro-molecules. *Naturwissenschaften*, 58:465–523, 1971.

14. M. Eigen, J. McCaskill, and P. Schuster. The molecular quasispecies. *Adv. Chem. Phys.*, 75:149–263, 1989.

15. S. F. Elena, V. S. Cooper, and R. E. Lenski. Punctuated evolution caused by selection of rare beneficial mutations. *Science*, 272:1802–1804, 1996.

16. L. Eshelman, editor. *Proceedings of the Sixth International Conference on Genetic Algorithms*. Morgan Kaufmann, San Mateo, CA, 1995.

17. W. J. Ewens. *Mathematical Population Genetics*, volume 9 of *Biomathematics*. Springer-Verlag, Berlin, 1979.

18. W. Fontana and P. Schuster. Continuity in evolution: On the nature of transitions. *Science*, 280:1451–5, 1998.

19. W. Fontana, P. F. Stadler, E. G. Bornberg-Bauer, T. Griesmacher, I. L. Hofacker, M. Tacker, P. Tarazona, E. D. Weinberger, and P. Schuster. RNA folding and combinatory landscapes. *Phys. Rev. E*, 47:2083–2099, 1992.

20. S. Forrest, editor. *Proceedings of the Fifth International Conference on Genetic Algorithms*. Morgan Kaufmann, San Mateo, CA, 1993.

21. C. V. Forst, C. Reidys, and J. Weber. Evolutionary dynamics and optimizations: Neutral networks as model landscapes for RNA secondary-structure folding landscape. In F. Moran, A. Moreno, J. Merelo, and P. Chacon, editors, *Advances in Artificial Life*, volume 929 of *Lecture Notes in Artificial Intelligence*. Springer-Verlag, Berlin, 1995.

22. D. E. Goldberg. *Genetic Algorithms in Search, Optimization, and Machine Learning*. Addison-Wesley, Reading, MA, 1989.

23. S. J. Gould and N. Eldredge. Punctuated equilibria: The tempo and mode of evolution reconsidered. *Paleobiology*, 3:115–251, 1977.

24. D. L. Hartl and A. G. Clark. *Principles of population genetics*, 2nd edition. Sinauer Associates, 1989.

25. R. Haygood. The structure of Royal Road fitness epochs. *Evolutionary Computation*, submitted, 1997. ftp://ftp.itd.ucdavis.edu/pub/people/rch/StrucRoyRdFitEp.ps.gz.

26. M. Huynen. Exploring phenotype space through neutral evolution. *J. of Mol. Evol.*, 43:165–169, 1996.

27. M. Huynen, P. F. Stadler, and W. Fontana. Smoothness within ruggedness: the role of neutrality in adaptation. *Proc. Natl. Acad. Sci. USA*, 93:397–401, 1996.

28. S. A. Kauffman and S. Levin. Towards a general theory of adaptive walks in rugged fitness landscapes. *J. Theor. Bio.*, 128:11–45, 1987.

29. M. Kimura. *The Neutral Theory of Molecular Evolution.* Cambridge University Press, Cambridge, 1983.

30. J. R. Koza. *Genetic Programming: On the Programming of Computers by Means of Natural Selection.* MIT Press, Cambridge, MA, 1992.

31. L. F. Landweber and L. Kari. Universal molecular computation in ciliates. In L. F. Landweber, E. Winfree, editors, *Evolution as Computation*, this volume.

32. C. A. Macken and A. S. Perelson. Protein evolution in rugged fitness landscapes. *Proc. Nat. Acad. Sci. USA*, 86:6191–6195, 1989.

33. M. Mitchell. *An Introduction to Genetic Algorithms.* MIT Press, Cambridge, MA, 1996.

34. M. Mitchell, J. P. Crutchfield, and P. T. Hraber. Evolving cellular automata to perform computations: Mechanisms and impediments. *Physica D*, 75:361–391, 1994.

35. M. Newman and R. Engelhardt. Effect of neutral selection on the evolution of molecular species. *Proc. R. Soc. London B.*, 256:1333–1338, 1998.

36. A. E. Nix and M. D. Vose. Modeling genetic algorithms with Markov chains. *Ann. Math. Art. Intel.*, 5, 1991.

37. A. Prügel-Bennett. Modelling evolving populations. *J. Theor. Bio.*, 185:81–95, 1997.

38. A. Prügel-Bennett and J. L. Shapiro. Analysis of genetic algorithms using statistical mechanics. *Phys. Rev. Lett.*, 72(9):1305–1309, 1994.

39. M. Rattray and J. L. Shapiro. The dynamics of a genetic algorithm for a simple learning problem. *J. Phys. A*, 29(23):7451–7473, 1996.

40. L. E. Reichl. *A Modern Course in Statistical Physics.* University of Texas, Austin, 1980.

41. C. M. Reidys, C. V. Forst, and P. K. Schuster. Replication and mutation on neutral networks. *Bull. Math. Biol.*, 63(1):57–94, 2001.

42. R. F Streater. *Statistical Dynamics: A Stochastic Approach to Nonequilibrium Thermodynamics.* Imperial College Press, London, 1995.

43. E. van Nimwegen and J. P. Crutchfield. Optimizing epochal evolutionary search: Population-size dependent theory. *Machine Learning*, 45(1):77–114, 2001.

44. E. van Nimwegen and J. P. Crutchfield. Optimizing epochal evolutionary search: Population-size independent theory. *Computer Methods in Applied Mechanics and Engineering*, 186:171–194, 2000. Special issue on Evolutionary and Genetic Algorithms in Computational Mechanics and Engineering, D. Goldberg and K. Deb, editors.

45. E. van Nimwegen, J. P. Crutchfield, and M. Mitchell. Finite populations induce metastability in evolutionary search. *Phys. Lett. A*, 229:144–150, 1997.

46. E. van Nimwegen, J. P. Crutchfield, and M. Mitchell. Statistical dynamics of the Royal Road genetic algorithm. *Theoretical Computer Science*, 229:41–102, 1999. Special issue on Evolutionary Computation, A. Eiben and G. Rudolph, editors.

47. M. D. Vose. Modeling simple genetic algorithms. In L. D. Whitley, editor, *Foundations of Genetic Algorithms 2*, Morgan Kauffman, San Mateo, CA, 1993.

48. M. D. Vose and G. E. Liepins. Punctuated equilibria in genetic search. *Complex Systems*, 5:31–44, 1991.
49. J. Weber. *Dynamics of Neutral Evolution. A case study on RNA secondary structures*. PhD thesis, Biologisch-Pharmazeutische Fakultät der Friedrich Schiller-Universität Jena, 1996. http://www.tbi.univie.ac.at/papers/PhD_theses.html.
50. S. Wright. Character change, speciation, and the higher taxa. *Evolution*, 36:427–43, 1982.
51. J. M. Yeomans. *Statistical Mechanics of Phase Transitions*. Clarendon Press, Oxford, 1992.

Genetic Programming: Biologically Inspired Computation That Creatively Solves Non-trivial Problems

John R. Koza, Forrest H. Bennett III, David Andre, and Martin A. Keane

Abstract. This paper describes a biologically inspired domain-independent technique, called genetic programming, that automatically creates computer programs to solve problems. Starting with a primordial ooze of thousands of randomly created computer programs, genetic programming progressively breeds a population of computer programs over a series of generations using the Darwinian principle of natural selection, recombination (crossover), mutation, gene duplication, gene deletion, and certain mechanisms of developmental biology. The technique is illustrated by its application to a non-trivial problem involving the automatic synthesis (design) of a lowpass filter circuit. The evolved results are competitive with human-produced solutions to the problem. In fact, four of the automatically created circuits exhibit human-level creativity and inventiveness, as evidenced by the fact that they correspond to four inventions that were patented between 1917 and 1936.

1 Introduction

One of the central challenges of computer science is to get a computer to solve a problem without explicitly programming it. In particular, it would be desirable to have a problem-independent system whose input is a high-level statement of a problem's requirements and whose output is a working computer program that solves the given problem. Paraphrasing Arthur Samuel (1959), the challenge is:

> How can computers be made to do what needs to be done, without being told exactly how to do it?

As Samuel also explained (Samuel 1983),

> The aim [is] ... to get machines to exhibit behavior, which if done by humans, would be assumed to involve the use of intelligence.

Three questions then arise:

- Can computer programs be automatically created?
- Can automatically created programs be competitive with human-produced programs?
- Can the automatic process exhibit creativity and inventiveness?

This paper provides an affirmative answer to all three questions.

Section 2 describes genetic programming. Section 3 presents a problem involving the automatic synthesis (design) of an analog electrical circuit, namely a lowpass filter. Section 4 details the circuit-constructing functions used in applying genetic programming to the problem of analog circuit synthesis. Section 5 presents the preparatory steps required for applying genetic programming to the lowpass filter problem. Section 6 shows the results.

2   Background on Genetic Programming

Genetic programming is a biologically inspired, domain-independent method that automatically creates a computer program from a high-level statement of a problem's requirements.

John Holland's pioneering book *Adaptation in Natural and Artificial Systems* (1975) described a domain-independent algorithm, called the genetic algorithm, based on an evolutionary process involving natural selection, recombination, and mutation. In the most commonly used form of the genetic algorithm, each point in the search space of the given problem is encoded into a fixed-length string of characters reminiscent of a strand of DNA. The genetic algorithm then conducts a search in the space of fixed-length character strings to find the best (or at least a very good) solution to the problem by genetically breeding a population of character strings over a number of generations. Numerous practical problems can be solved using the genetic algorithm. Recent work in the field of genetic algorithms is described in Goldberg 1989, Michalewicz 1996, Mitchell 1996, Gen and Cheng 1997, and Back 1997.

Genetic programming is an extension of the genetic algorithm in which the population consists of computer programs. The goal of genetic programming is to provide a domain-independent problem-solving method that automatically creates a computer program from a high-level statement of a problem's requirements. Starting with a primordial ooze of thousands of randomly created computer programs, genetic programming progressively breeds a population of computer programs over a series of generations using the Darwinian principle of natural selection, recombination (crossover), mutation, gene duplication, gene deletion, and certain mechanisms of developmental biology. Work on genetic programming is described in Koza 1992; Koza and Rice 1992; Kinnear 1994; Angeline and Kinnear 1996; Koza, et al. 1996; Koza, et al. 1997; Koza, et al. 1998; Banzhaf, et al. 1998; Spector, et al. 1999, and on the World Wide Web at `www.genetic-programming.org`. The computer programs are compositions of functions (e.g., arithmetic operations, conditional operators, problem-specific functions) and terminals (e.g., external inputs, constants, zero-argument functions). The programs may be thought of as trees whose points are labeled with the functions and whose leaves are labeled with the terminals.

Genetic programming breeds computer programs to solve problems by executing the following three steps:

1. Randomly create an initial population of individual computer programs.
2. Iteratively perform the following substeps (called a *generation*) on the population of programs until the termination criterion has been satisfied:

   (a) Assign a fitness value to each individual program in the population using the fitness measure.
   (b) Create a new population of individual programs by applying the following three genetic operations. The genetic operations are applied to one or two individuals in the population selected with a probability based on fitness (with reselection allowed).

       i. *Reproduction*: Reproduce an existing individual by copying it into the new population.
       ii. *Crossover*: Create two new individual programs from two existing parental individuals by genetically recombining subtrees from each program using the crossover operation at randomly chosen crossover points in the parental individuals.
       iii. *Mutation*: Create a new individual from an existing parental individual by randomly mutating one randomly chosen subtree of the parental individual.

3. Designate the individual computer program that is identified by the method of result designation (e.g., the *best-so-far* individual) as the result of the run of genetic programming. This result may represent a solution (or an approximate solution) to the problem.

Genetic programming starts with an initial population (generation 0) of randomly generated computer programs composed of the given primitive functions and terminals. The creation of this initial random population is a blind random search of the space of computer programs.

The computer programs in generation 0 of a run of genetic programming will almost always have exceedingly poor fitness. Nonetheless, some individuals in the population will turn out to be somewhat more fit than others. These differences in performance are then exploited so as to direct the search into promising areas of the search space. The Darwinian principle of reproduction and survival of the fittest and the genetic operation of crossover (augmented by occasional mutation) are used to create a new population of offspring programs from the current population of computer programs.

The reproduction operation involves probabilistically selecting a computer program from the current population of programs on the basis of fitness (i.e., the better the fitness, the more likely the individual is to be selected) and allowing it to survive by copying it into the new population.

The crossover operation creates new offspring computer programs from two parental programs selected probabilistically on the basis of fitness. The parental programs in genetic programming are typically of different sizes and shapes. The offspring programs are composed of subexpressions (subtrees, subprograms) from their parents.

For example, consider the following computer program (presented here as a LISP S-expression):

```
(+ (* 0.234 Z) (- X 0.789)),
```

which one would ordinarily write as

$$0.234Z + x - 0.789.$$

This program takes two inputs (X and Z) and produces a floating point output.
Also, consider a second program:

```
(* (* Z Y) (+ Y (* 0.314 Z))).
```

One crossover point is randomly and independently chosen in each parent. Suppose that the crossover points are the * in the first parent and the + in the second parent. These two crossover fragments are the subexpressions rooted at the crossover points and are underlined in the above two parental computer programs.

The two offspring resulting from crossover are

```
(+ (+ Y (* 0.314 Z)) (- X 0.789))
```

and

```
(* (* Z Y) (* 0.234 Z)).
```

Crossover creates new computer programs using parts of existing parental programs. Because entire subtrees are swapped, the crossover operation produces syntactically and semantically valid programs as offspring regardless of the choice of the two crossover points. The two offspring here are typical of the offspring produced by the crossover operation in that they are different from both of their parents and different from each other in size and shape. Because programs are selected to participate in the crossover operation with a probability based on fitness, crossover allocates future trials to regions of the search space whose programs contain parts of promising programs.

The mutation operation creates an offspring computer program from one parental program selected based on fitness. One mutation point is randomly and independently chosen and the subtree occurring at that point is deleted. Then, a new subtree is grown at that point using the same growth procedure as was originally used to create the initial random population.

For example, consider the following parental program (presented as a LISP S-expression) composed of Boolean functions and terminals:

```
(OR (AND D2 D1 (NOR D0 D1))).
```

Suppose that the AND is randomly chosen as the mutation point (out of the seven points in the program tree). The three-point subtree rooted at the AND corresponds to the underlined portion of the LISP S-expression above. The subtree rooted at the chosen mutation point is deleted. In this example, the subtree consists of the three points (AND D2 D1). A new subtree, such as

```
(AND (NOT D0) (NOT D1)),
```

is randomly grown using the available functions and terminals and inserted in lieu of the subtree (AND D2 D1). The result of the mutation operation is

```
(OR ((AND (NOT D0) (NOT D1)) (NOR D0 D1))).
```

The offspring here is typical of the offspring produced by the mutation operation in that it is different from its parent in size and shape.

After the genetic operations are performed on the current population, the population of offspring (i.e., the new generation) replaces the old population (i.e., the old generation). Each individual in the new population of programs is then measured for fitness, and the process is repeated over many generations.

The dynamic variability of the computer programs that are created during the run is an important feature of genetic programming. It is often difficult and unnatural to try to specify or restrict the size and shape of the eventual solution in advance.

Scalable automated programming requires some hierarchical mechanism to exploit, by reuse and parameterization, the regularities, symmetries, homogeneities, similarities, patterns, and modularities inherent in problem environments. Subroutines provide this mechanism in ordinary computer programs. Automatically defined functions (Koza 1994a, b) implement this mechanism within the context of genetic programming. Automatically defined functions are implemented by establishing a constrained syntactic structure for the individual programs in the population. Each multi-part program in the population contains one (or more) automatically defined functions and one (or more) main result-producing branches. The result-producing branch usually has the ability to call one or more of the automatically defined functions. An automatically defined function may have the ability to refer hierarchically to other already defined, automatically defined functions.

The initial random generation is created so that every individual program in the population consists of automatically defined function(s) and result-producing branch(es) in accordance with the problem's constrained syntactic structure. Since a constrained syntactic structure is involved, crossover and mutation are performed so as to preserve this syntactic structure in all offspring.

Architecture-altering operations enhance genetic programming with automatically defined functions by providing a way to automatically determine the number of such automatically defined functions, the number of arguments that each automatically defined function possesses, and the nature of the hierarchical references, if any, among such automatically defined functions (Koza 1995). These operations include branch duplication, argument duplication, branch creation, argument creation, branch deletion, and argument deletion. The architecture-altering operations are motivated by the naturally occurring mechanism of gene duplication that creates new proteins (and hence new structures and new behaviors in living things) as described by Susumu Ohno in *Evolution by Gene Duplication* (1970). Details are found in Koza, et al. 1999.

Genetic programming has been applied to numerous problems in fields such as system identification, control, classification, design, optimization, and automatic programming.

3   Statement of the Illustrative Problem

Design is a major activity of practicing engineers. The design process entails creation of a complex structure to satisfy user-defined requirements. Since the design process typically entails tradeoffs between competing considerations, the end product of the process is usually a satisfactory and compliant design as opposed to a perfect design. Design is usually viewed as requiring creativity and human intelligence. Consequently, the field of design is a source of challenging problems for automated techniques of machine intelligence. In particular, design problems are useful for determining whether an automated technique can produce results that are competitive with human-produced results.

The design (synthesis) of analog electrical circuits is especially challenging. The design process for analog circuits begins with a high-level description of the circuit's desired behavior and characteristics and entails creation of both the topology and the sizing of a satisfactory circuit. The topology comprises the gross number of components in the circuit, the type of each component (e.g., a capacitor), and a list of all connections between the components. The sizing involves specifying the values (typically numerical) of each of the circuit's components.

Although considerable progress has been made in automating the synthesis of certain categories of purely digital circuits, the synthesis of analog circuits and mixed analog–digital circuits has not proved to be as amenable to automation. There is no previously known general technique for automatically creating an analog circuit from a high-level statement of the design goals of the circuit. Describing "the analog dilemma," O. Aaserud and I. Ring Nielsen (1995) noted

> Analog designers are few and far between. In contrast to digital design, most of the analog circuits are still handcrafted by the experts or so-called 'zahs' of analog design. The design process is

characterized by a combination of experience and intuition and requires a thorough knowledge of the process characteristics and the detailed specifications of the actual product.

Analog circuit design is known to be a knowledge-intensive, multiphase, iterative task, which usually stretches over a significant period of time and is performed by designers with a large portfolio of skills. It is therefore considered by many to be a form of art rather than a science.

This paper focuses on one particular problem of analog circuit synthesis, namely the design of a lowpass filter circuit composed of capacitors and inductors. A simple *filter* is a one-input, one-output electronic circuit that receives a signal as its input and passes the frequency components of the incoming signal that lie in a specified range (called the *passband*) while suppressing the frequency components that lie in all other frequency ranges (the *stopband*). In particular, the goal is to design a lowpass filter that passes all frequencies below 1,000 Hertz (Hz) and suppresses all frequencies above 2,000 Hz.

The approach described in this paper has been applied to many other problems of analog circuit synthesis, including the design of amplifiers, computational circuits, a temperature-sensing circuit, a voltage reference circuit, a time-optimal robot controller circuit, a difficult-to-design asymmetric bandpass filter, a crossover filter, a double passband filter, a bandstop filter, frequency discriminator circuits, and a frequency-measuring circuit (as described in detail in Koza, et al. 1999).

4   Applying Genetic Programming to the Problem

Genetic programming can be applied to the problem of synthesizing circuits if a mapping is established between the program trees (rooted, point-labeled trees — that is, acyclic graphs — with ordered branches) used in genetic programming and the labeled cyclic graphs germane to electrical circuits. The principles of developmental biology provide the motivation for mapping trees into circuits by means of a developmental process that begins with a simple embryo. For circuits, the embryo typically includes fixed wires that connect the inputs and outputs of the particular circuit being designed and certain fixed components (such as source and load resistors). Until these wires are modified, the circuit does not produce interesting output. An electrical circuit is developed by progressively applying the functions in a circuit-constructing program tree to the modifiable wires of the embryo (and, during the developmental process, to new components and modifiable wires).

An electrical circuit is created by executing the functions in a circuit-constructing program tree. The functions are progressively applied in a developmental process to the embryo and its successors until all of the functions in the program tree are executed. That is, the functions in the circuit-constructing program tree progressively use side-effects to change the embryo and its successors until a fully developed circuit eventually emerges. The functions are applied in a breadth-first order.

The functions in the circuit-constructing program trees are divided into five categories: (1) topology-modifying functions that alter the circuit topology, (2) component-creating functions that insert components into the circuit, (3) development-controlling functions that control the development process by which the embryo and its successors are changed into a fully developed circuit, (4) arithmetic-performing functions that appear in subtrees as argument(s) to the component-creating functions and specify the numerical value of the component, and (5) automatically defined functions that appear in the automatically defined functions and potentially enable certain substructures of the circuit to be reused (with parameterization).

Each branch of the program tree is created in accordance with a constrained syntactic structure, and is composed of topology-modifying functions, component-creating functions, development-controlling functions, and terminals. Component-creating functions typically have one arithmetic-performing subtree, while topology-modifying functions and development-controlling functions do not. Component-creating functions and topology-modifying functions are internal points of their branches and possess one or more arguments (construction-continuing subtrees) that continue the developmental process. The syntactic validity of this constrained syntactic structure is preserved using structure-preserving crossover with point typing. For details, see Koza, et al. 1999.

4.1   The Embryonic Circuit

An electrical circuit is created by executing a circuit-constructing program tree that contains various component-creating, topology-modifying, and development-controlling functions. Each tree in the population creates one circuit. The specific embryo used depends on the number of inputs and outputs.

Figure 1 shows a one-input, one-output embryonic (initial) circuit in which VSOURCE is the input signal and VOUT is the output signal (the probe point). The circuit is driven by an incoming alternating circuit source VSOURCE. There is a fixed load resistor RLOAD and a fixed source resistor RSOURCE in the embryo. In addition to the fixed components, there are two modifiable wires, Z0 and Z1. All development originates from these modifiable wires.

4.2   Component-Creating Functions

The component-creating functions insert a component into the developing circuit and assign component value(s) to the component.

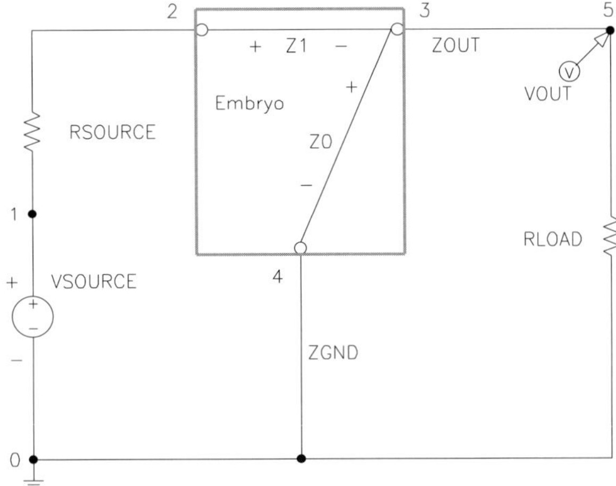

Fig. 1. One-input, one-output embryonic (initial) circuit.

Each component-creating function has a writing head that points to an associated highlighted component in the developing circuit and modifies that component in a specified manner. The construction-continuing subtree of each component-creating function points to a successor function or terminal in the circuit-constructing program tree.

The arithmetic-performing subtree of a component-creating function consists of a composition of arithmetic functions (addition and subtraction) and random constants (in the range −1.000 to +1.000). The arithmetic-performing subtree specifies the numerical value of a component by returning a floating-point value that is interpreted on a logarithmic scale as the value for the component in the range of 10 orders of magnitude (using a unit of measure that is appropriate for the particular type of component).

The two-argument resistor-creating R function causes the highlighted component to be changed into a resistor. The value of the resistor in kΩ is specified by its arithmetic-performing subtree.

Figure 2 shows a modifiable wire Z0 connecting nodes 1 and 2 of a partial circuit containing four capacitors (C2, C3, C4, and C5). Figure 3 shows the result of applying the R function to the modifiable wire Z0 of Fig. 2.

Similarly, the two-argument capacitor-creating C function causes the highlighted component to be changed into a capacitor whose value in μF is specified by its arithmetic-performing subtree. In addition, the two-argument inductor-creating L function causes the highlighted component to be changed into an inductor whose value in μH is specified by its arithmetic-performing subtree.

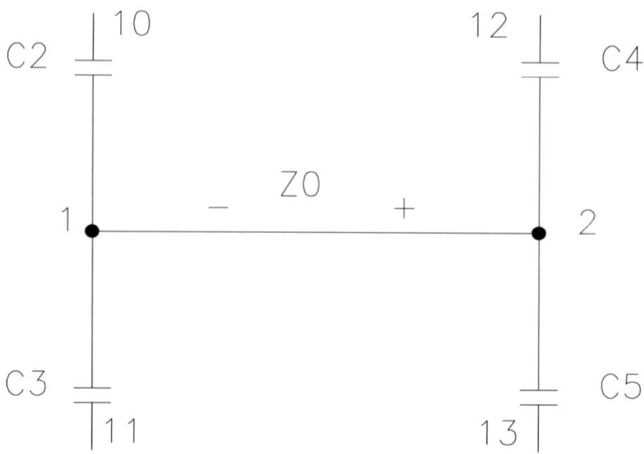

Fig. 2. Modifiable wire Z0.

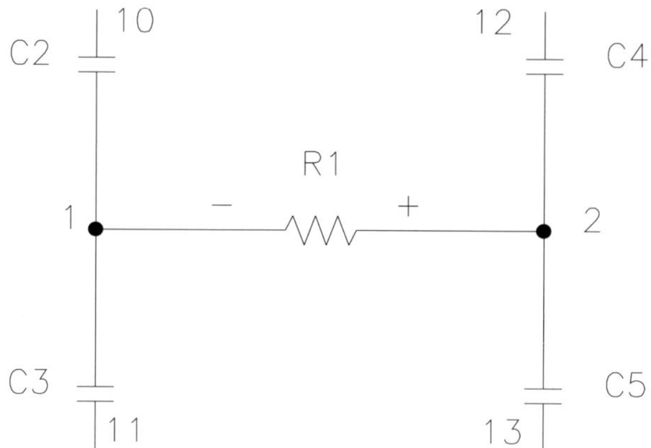

Fig. 3. Result of applying the R function.

4.3   Topology-Modifying Functions

Each topology-modifying function in a program tree points to an associated highlighted component and modifies the topology of the developing circuit.

The three-argument SERIES division function creates a series composition of the modifiable wire or modifiable component with which it is associated, a copy of the modifiable wire or modifiable component with which it is associated, one new modifiable wire (with a writing head), and two new nodes. Figure 4 shows the result of applying the SERIES function to the resistor R1 from Fig. 3. After execution of the SERIES function, resistors R1 and R7 and modifiable wire Z6 remain modifiable. All three are associated with the top-most function in one of the three construction-continuing subtrees of the SERIES function.

The reader is referred to Koza, et al. 1999 for a detailed description of all the circuit-constructing functions mentioned herein.

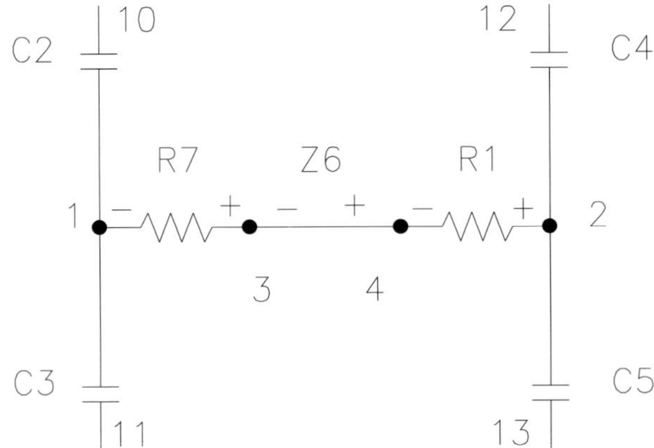

Fig. 4. Result after applying the **SERIES** division function.

The four-argument **PARALLEL0** parallel division function creates a parallel composition consisting of the modifiable wire or modifiable component with which it is associated, a copy of the modifiable wire or modifiable component with which it is associated, two new modifiable wires (each with a writing head), and two new nodes. There are potentially two topologically distinct outcomes of a parallel division. Since we want the outcome of all circuit-constructing functions to be deterministic, there are two members (called **PARALLEL0** and **PARALLEL1**) in the **PARALLEL** family of topology-modifying functions. The two functions operate differently depending on degree and numbering of the preexisting components in the developing circuit. The use of the two functions breaks the symmetry between the potentially distinct outcomes.

The one-argument polarity-reversing **FLIP** function reverses the polarity of the highlighted component.

The two-argument **TWO_GROUND** ("ground") function enables any part of a circuit to be connected to ground. The **TWO_GROUND** function creates a new node and a composition of two modifiable wires and one nonmodifiable wire such that the nonmodifiable wire makes an unconditional connection to ground.

The eight two-argument functions in the **TWO_VIA** family of functions (called **TWO_VIA0**, ..., **TWO_VIA7**) each create a new node and a composition of two modifiable wires and one nonmodifiable wire such that the nonmodifiable wire makes a connection, called a *via*, to a designated one of eight imaginary numbered layers (0 to 7) of an imaginary silicon wafer on which the circuit resides. The **TWO_VIA** functions provide a way to connect distant parts of a circuit.

The zero-argument **SAFE_CUT** function causes the highlighted component to be removed from the circuit provided that the degree of the nodes at both ends of the highlighted component is three (i.e., no dangling components or wires are created).

4.4   Development-Controlling Functions

The one-argument NOOP ("no operation") function has no effect on the modifiable wire or modifiable component with which it is associated; however, it has the effect of delaying activity on the developmental path on which it appears in relation to other developmental paths in the overall circuit-constructing program tree.

The zero-argument END function makes the modifiable wire or modifiable component with which it is associated non-modifiable (thereby ending a particular developmental path).

4.5   Example of the Developmental Process

Figure 5 is an illustrative circuit-constructing program tree shown as a rooted, point-labeled tree with ordered branches. The overall program consists of two main result-producing branches joined by a connective LIST function (labeled 1 in the figure). The first (left) result-producing branch is rooted at the capacitor-creating C function (labeled 2). The second result-producing branch is rooted at the polarity-reversing FLIP function (labeled 3). This figure also contains four occurrences of the inductor-creating L function (at 17, 11, 20, and 12). The figure contains two occurrences of the topology-modifying SERIES function (at 5 and 10). The figure also contains five occurrences of the development-controlling END function (at 15, 25, 27, 31, and 22) and one occurrence of the development-controlling NOOP function (at 6). There is a seven-point arithmetic-performing subtree at 4 under the capacitor-creating C function at 4. Similarly, there is a three-point arithmetic-performing subtree at 19 under the inductor-creating L function at 11. There are also one-point arithmetic-performing subtrees (i.e., constants) at 26, 30, and 21. Additional details can be found in Koza, et al. 1999.

5   Preparatory Steps

Before applying genetic programming to a problem of circuit design, seven major preparatory steps are required: (1) identify the embryonic circuit, (2) determine the architecture of the circuit-constructing program trees, (3) identify the primitive functions of the program trees, (4) identify the terminals of the program trees, (5) create the fitness measure, (6) choose control parameters for the run, and (7) determine the termination criterion and method of result designation.

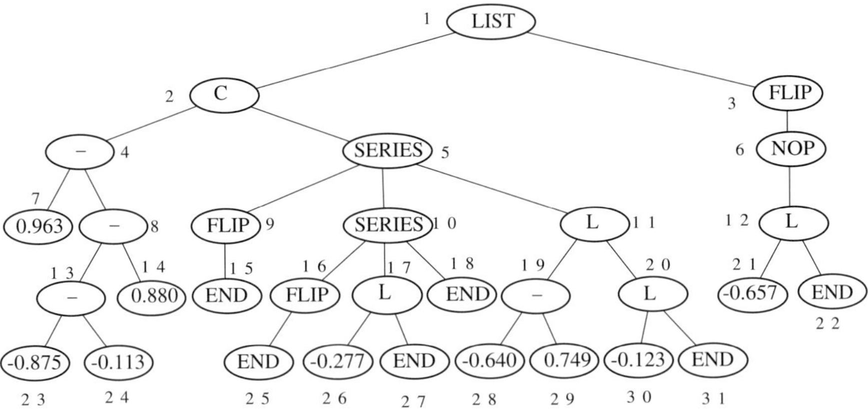

Fig. 5. Illustrative circuit-constructing program tree.

5.1   Embryonic Circuit

The embryonic circuit used on a particular problem depends on the circuit's number of inputs and outputs. A one-input, one-output embryo with two modifiable wires (Fig. 1) was used.

5.2   Program Architecture

Since there is one result-producing branch in the program tree for each modifiable wire in the embryo, the architecture of each circuit-constructing program tree depends on the embryonic circuit. Two result-producing branches were used for the filter problems.

The architecture of each circuit-constructing program tree also depends on the use, if any, of automatically defined functions. Automatically defined functions provide a mechanism enabling certain substructures to be reused and are described in detail in Koza, et al. 1999. Automatically defined functions and architecture-altering operations were not used here.

5.3   Function and Terminal Sets

The function set for each design problem depends on the type of electrical components that are to be used for constructing the circuit.

The function set included two component-creating functions (for inductors and capacitors), topology-modifying functions (for series and parallel divisions and for flipping components), one development-controlling function (NOOP), functions for creating a via to ground, and functions for connecting pairs of points. That is, the function set, $\mathcal{F}_{ccs}$, for each construction-continuing subtree was

$$\mathcal{F}_{ccs} = \{\texttt{L, C, SERIES, PARALLEL0, PARALLEL1, FLIP, NOOP,}$$
$$\texttt{TWO\_GROUND, TWO\_VIA0\_, TWO\_VIA1, TWO\_VIA2, TWO\_VIA3,}$$
$$\texttt{TWO\_VIA4, TWO\_VIA5, TWO\_VIA6, TWO\_VIA7}\}.$$

The terminal set, $\mathcal{T}_{ccs}$, for each construction-continuing subtree was

$$\mathcal{T}_{ccs} = \texttt{END}, \texttt{SAFE\_CUT}.$$

The terminal set, $\mathcal{T}_{aps}$, for each arithmetic-performing subtree consisted of

$$\mathcal{T}_{aps} = \Re,$$

where $\Re$ represents floating-point random constants from -1.0 to $+1.0$. The function set, $\mathcal{F}_{aps}$, for each arithmetic-performing subtree was

$$\mathcal{F}_{aps} = \texttt{+, -}.$$

The terminal and function sets were identical for all result-producing branches for a particular problem.

5.4   Fitness Measure

The evolutionary process is driven by the *fitness measure*. Each individual computer program in the population is executed and then evaluated using the fitness measure. The nature of the fitness measure varies with the problem. The high-level statement of desired circuit behavior is translated into a well-defined measurable quantity that can be used by genetic programming to guide the evolutionary process. The evaluation of each individual circuit-constructing program tree in the population begins with its execution. This execution progressively applies the functions in each program tree to an embryonic circuit, thereby creating a fully developed circuit. A netlist is created that identifies each component of the developed circuit, the nodes to which each component is connected, and the value of each component. The netlist becomes the input to our modified version of the 217,000-line SPICE (Simulation Program with Integrated Circuit Emphasis) simulation program (Quarles, et al. 1994). SPICE then determines the behavior of the circuit. It was necessary to make considerable modifications in SPICE so that it could run as a submodule within the genetic programming system.

The desired lowpass filter has a passband below 1,000 Hz and a stopband above 2,000 Hz. The circuit is driven by an incoming AC voltage source with a 2 V amplitude. In this problem, a voltage in the passband of exactly 1 V and a voltage in the stopband of exactly 0 V is regarded as ideal. The (preferably small) variation within the passband is called the *passband ripple*. Similarly, the incoming signal is never fully reduced to zero in the stopband of an actual filter. The (preferably small) variation within the stopband is called the *stopband ripple*. A voltage in the passband of between 970 mV and 1 V (i.e., a passband ripple of 30 mV or less) and a voltage in the stopband of between 0 V and 1 mV (i.e., a stopband ripple of 1 mV or less) is regarded as acceptable. Any voltage lower than 970 mV in the passband or higher than 1 mV in the stopband is regarded as unacceptable.

Since the high-level statement of behavior for the desired circuit is expressed in terms of frequencies, the voltage VOUT is measured in the frequency domain. SPICE performs an AC small signal analysis and reports the circuit's behavior over five decades (between 1 Hz and 100,000 Hz), with each decade being divided into 20 parts (using a logarithmic scale), so that there are a total of 101 fitness cases.

Fitness is measured in terms of the sum over these cases of the absolute weighted deviation between the actual value of the voltage that is produced by the circuit at the probe point VOUT and the target value for voltage. The smaller the value of fitness, the better. A fitness of zero represents an (unattainable) ideal filter.

Specifically, the standardized fitness is

$$F(t) = \sum_{i=0}^{100} W(d(f_i), f_i) d(f_i),$$

where f_i is the frequency of fitness case i; $d(x)$ is the absolute value of the difference between the target and observed values at frequency x; and $W(y, x)$ is the weighting for difference y at frequency x.

The fitness measure is designed to not penalize ideal values, to slightly penalize every acceptable deviation, and to heavily penalize every unacceptable deviation. Specifically, the procedure for each of the 61 points in the three-decade interval between 1 Hz and 1,000 Hz for the intended passband is as follows:

- If the voltage equals the ideal value of 1.0 V in this interval, the deviation is 0.0.
- If the voltage is between 970 mV and 1 V, the absolute value of the deviation from 1 V is weighted by a factor of 1.0.
- If the voltage is less than 970 mV, the absolute value of the deviation from 1 V is weighted by a factor of 10.0.

The acceptable and unacceptable deviations for each of the 35 points from 2,000 Hz to 100,000 Hz in the intended stopband are similarly weighted (by 1.0 or 10.0) based on the amount of deviation from the ideal voltage of 0 volts and the acceptable deviation of 1 mV.

For each of the five "don't care" points between 1,000 and 2,000 Hz, the deviation is deemed to be zero.

The number of "hits" for this problem (and all other problems herein) is defined as the number of fitness cases that have an acceptable or ideal voltage or lie in the "don't care" band (for a filter).

Many of the random initial circuits and many that are created by the crossover and mutation operations in subsequent generations cannot be simulated by SPICE. These circuits receive a high penalty value of fitness (10^8) and become the worst-of-generation programs for each generation.

5.5   Control Parameters

The population size, M, was 320,000. The probability of crossover was approximately 89%; reproduction 10%; and mutation 1%. Our usual control parameters were used (Koza, et al. 1999, Appendix D).

5.6   Termination Criterion and Results Designation

The maximum number of generations, G, is set to an arbitrary large number (e.g., 501) and the run was manually monitored and manually terminated when the fitness of the best-of-generation individual appeared to have reached a plateau. The best-so-far individual is harvested and designated as the result of the run.

5.7   Implementation on Parallel Computer

The problem was run on a medium-grained parallel Parsytec computer system consisting of 64 80-MHz PowerPC 601 processors arranged in an 8-by-8 toroidal mesh with a host PC Pentium-type computer. The distributed genetic algorithm (Andre and Koza 1996) with unsynchronized generations was used with a population size of $Q = 5,000$ at each of the $D = 64$ demes (semi-isolated subpopulations) for a total population, M, of 320,000. On each generation, four boatloads of emigrants, each consisting of $B = 2\%$ (the migration rate) of the node's subpopulation (selected on the basis of fitness), were dispatched to each of the four adjacent processing nodes.

6   Results

The creation of the initial random population is a blind random search of the search space of the problem. The best circuit from generation 0 has a fitness of 61.7 and scores 52 hits (out of 101).

Figures 6, 7, 8, and 9 show the behavior of the best circuits from generations 0, 10, 15, and 49, respectively, of one run of genetic programming. The horizontal axis represents five decades of frequencies from 1 Hz to 100,000 Hz on a logarithmic scale. The vertical axis represents output voltage on a linear scale. Excluding the fixed source and load resistors of the test fixture of the embryonic circuit, the best-of-generation circuit from generation 0 consists of only a lone 358 nF capacitor that shunts the incoming signal to ground. A good filter cannot be created by a single capacitor. However, even a single capacitor differentially passes higher frequencies to ground and performs a certain amount of filtering. Figure 6 shows that the best circuit from generation 0 bears some resemblance to the desired lowpass filter in that it passes frequencies up to about 70 Hz at nearly a full volt and it almost fully suppresses frequencies near 100,000 Hz. However, its transition region is exceedingly leisurely. Nonetheless, in the valley of the blind, the one-eyed man is king. Moreover, as will be seen momentarily, this modest beginning serves as a building block that will become incorporated in the 100%-compliant lowpass filter that will eventually be evolved.

The evolutionary process produces better and better individuals as the run progresses. For example, the best circuit from generation 10 has inductors in series with the incoming signal as well as a single capacitor shunted to ground. Figure 7 shows that the frequencies up to about 200 Hz are passed at nearly full voltage and that frequencies above 10,000 Hz are almost fully suppressed. Figure 8 shows that the best circuit from generation 15 (with two inductors in series with the incoming signal and three capacitors shunted to ground) comes closer to meeting the requirements of this design problem.

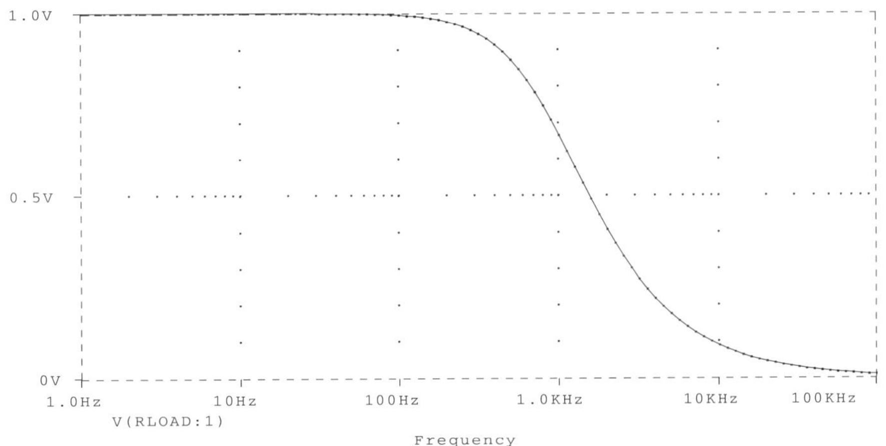

Fig. 6. Frequency domain behavior of the best circuit of generation 0.

6.1   Campbell 1917 Ladder Filter Patent

The best circuit (Fig. 10) of generation 49 from this run is 100% compliant with the problem's design requirements in the sense that it scored 101 hits (out of 101). It has a near-zero fitness of 0.00781 (about five orders of magnitude better than the best circuit of generation 0). As can be seen, this evolved circuit consists of seven inductors (L5, L10, L22, L28, L31, L25, and L13) arranged horizontally across the top of the figure "in series" with the

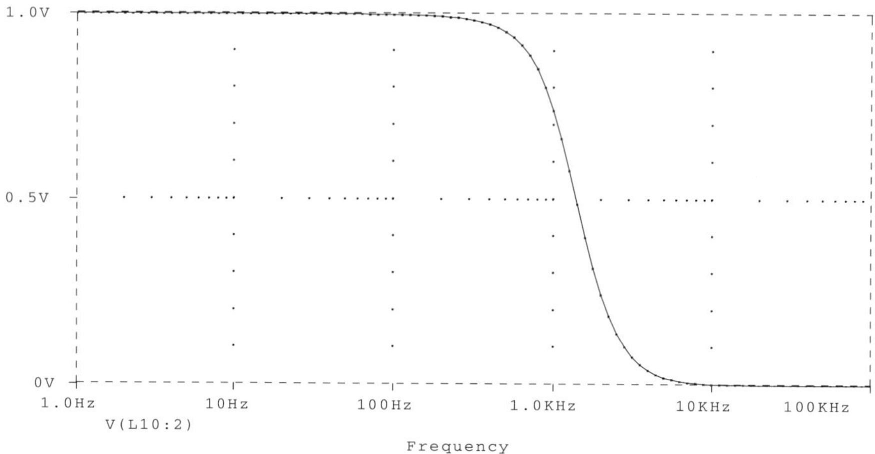

Fig. 7. Frequency domain behavior of the best circuit of generation 10.

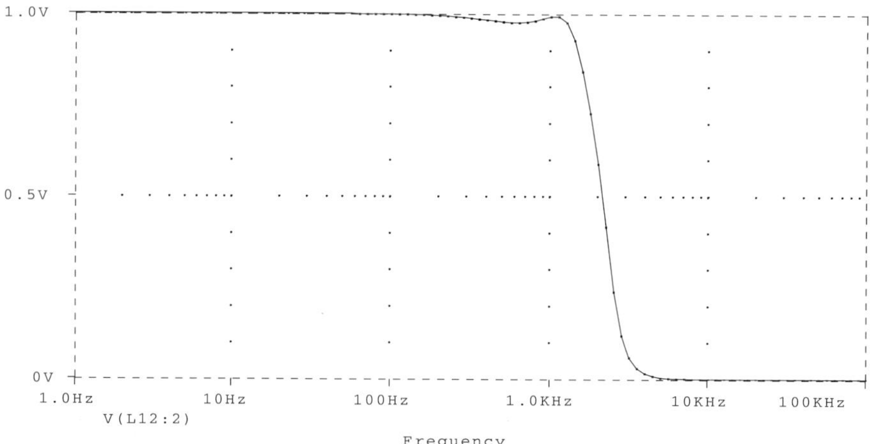

Fig. 8. Frequency domain behavior of the best circuit of generation 15.

incoming signal **VSOURCE** and the source resistor **RSOURCE**. It also contains seven capacitors (C12, C24, C30, C3, C33, C27, and C15) that are each shunted to ground. This circuit is a classical ladder filter with seven rungs (Williams and Taylor 1995).

Figure 9 shows the behavior in the frequency domain of this evolved lowpass filter. As can be seen, the 100%-compliant lowpass filter delivers a voltage of essentially 1 V in the entire passband from 1 Hz to 1,000 Hz and suppresses the voltage of essentially 0 V in the entire stopband starting at 2,000 Hz. There is a sharp drop-off from 1 V to 0 V in the transitional ("don't care") region between 1,000 Hz and 2,000 Hz.

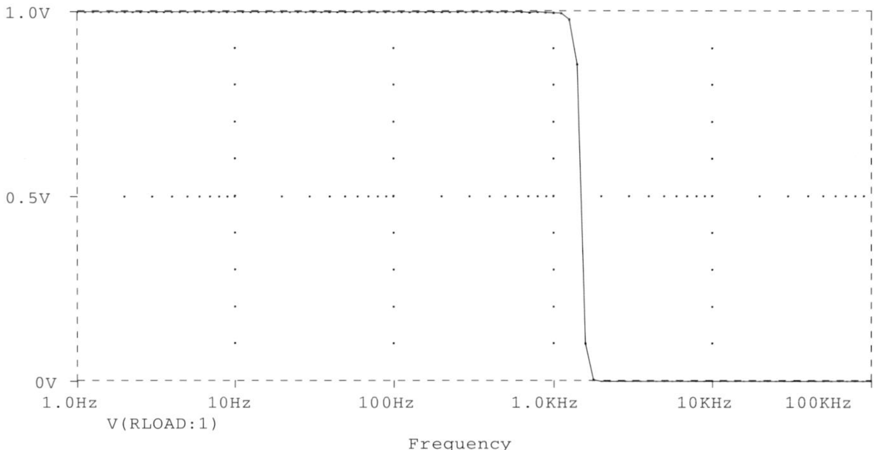

Fig. 9. Frequency domain behavior of 100%-compliant seven-rung ladder circuit from generation 49.

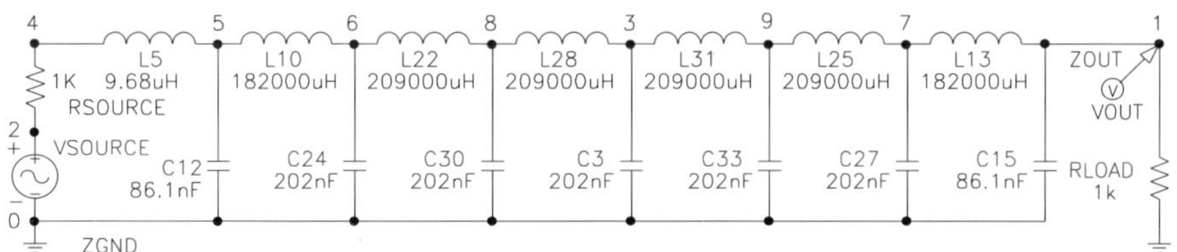

Fig. 10. Evolved seven-rung ladder lowpass filter.

The circuit of Fig. 10 has the recognizable features of the circuit for which George Campbell of American Telephone and Telegraph received U.S. patent 1,227,113 in 1917 (Campbell 1917). Claim 2 of Campbell's patent covered

> An electric wave filter consisting of a connecting line of negligible attenuation composed of a plurality of sections, each section including a capacity element and an inductance element, one of said elements of each section being in series with the line and the other in shunt across the line, said capacity and inductance elements having precomputed values dependent upon the upper limiting frequency and the lower limiting frequency of a range of frequencies it is desired to transmit without attenuation, the values of said capacity and inductance elements being so proportioned that the structure transmits with practically negligible attenuation sinusoidal currents of all frequencies lying between said two limiting frequencies, while attenuating and approximately extinguishing currents of neighboring frequencies lying outside of said limiting frequencies.

An examination of the evolved circuit of Fig. 10 shows that it indeed consists of "a plurality of sections" (specifically, seven). In the figure, "Each section include[s] a capacity element and an inductance element." Specifically, the first of the seven sections consists of inductor L5 and capacitor C12; the second section consists of inductor L10 and capacitor C24; and so forth. Moreover, "one of said elements of each section [is] in series with the line and the other in shunt across the line." Inductor L5 of the first section is indeed "in series with the line" and capacitor C12 is indeed "in shunt across the line." This is also true for the circuit's remaining six sections. Moreover, Fig. 10 herein matches Fig. 7 of Campbell's 1917 patent. In addition, this circuit's 100% compliant behavior in the frequency domain (Fig. 9 herein) confirms the fact that the values of the inductors and capacitors are such that they transmit "with practically negligible attenuation sinusoidal currents" of the passband frequencies "while attenuating and approximately extinguishing currents" of the stopband frequencies.

In short, genetic programming evolved an electrical circuit that infringes on the claims of Campbell's now-expired patent.

In addition to possessing the topology of the Campbell filter, the evolved circuit of Fig. 10 also approximately possesses the numerical values described in Campbell's 1917 patent (Campbell 1917). In fact, this evolved circuit is roughly equivalent to what is now known as a cascade of six identical symmetric π-sections (Johnson 1950). To see this, we modify the evolved circuit of Fig. 10 in four ways.

First, we delete the $9.68\,\mu$H inductor L5 near the upper left corner of the figure. The value of this inductor is more than five orders of magnitude smaller than the value of the other six inductors (L10, L22, L28, L31, L25, and L13) in series across the top of the figure. The behavior of the evolved circuit is not noticeably affected by this deletion for the frequencies of interest in this problem.

Second, we replace each of the five identical 202 nF capacitors (C24, C30, C3, C33, C27) by a composition of two parallel 101 nF capacitors. Since the capacitance of a composition of two parallel capacitors equals the sum of the two individual capacitances, the behavior of the evolved circuit is not changed at all by these substitutions.

Third, we note that the two 86.1 nF capacitors (C12 and C15) at the two ends of the ladder are each approximately equal to the (now) ten 101 nF capacitors. Suppose, for the sake of argument, that these twelve approximately equal capacitors are replaced by twelve equal capacitors with capacitance equal to their average value (98.5 nF). The behavior of the evolved circuit is only slightly changed by these substitutions.

Fourth, we note also that the six non-trivial inductors (L10, L22, L28, L31, L25, and L13) are approximately equal. Suppose, for the sake of argument, that these six approximately equal inductors are replaced by six equal inductors with inductance equal to their average value (200,000 μH). Again, the behavior of the evolved circuit is only slightly changed by these substitutions.

The behavior in the frequency domain of the circuit resulting from the above four changes is almost the same as that of the evolved circuit of Fig. 10. In fact, the modified circuit is 100% compliant (i.e., scores 101 hits). The modified circuit can be viewed as what is now known as a cascade of six identical symmetric π-sections. Each π-section consists of an inductor of inductance L (where L equals 200,000 μH) and two equal capacitors of capacitance $C/2$ (where C equals 197 nF). In each π-section, the two 98.5 nF capacitors constitute the vertical legs of the π and the one 200,000 μH inductor constitutes the horizontal bar across the top of the π. Such π-sections are characterized by two key parameters. The first parameter is the characteristic resistance (impedance) of the π-section. This characteristic resistance should match the circuit's fixed load resistance **RLOAD** (1,000 Ω). The second parameter is the nominal cutoff frequency which separates the filter's passband from its stopband. This second parameter should lie somewhere in the transition region between the end of the passband (1,000 Hz) and the beginning of the stopband (2,000 Hz). The characteristic resistance, R, of each of the π-sections is given by the formula $\sqrt{L/C}$. This formula yields a characteristic resistance, R, of 1,008 Ω. This value is very close to the value of the 1,000 Ω load resistance of this problem. The nominal cutoff frequency, f_c, of each of the π-sections of a lowpass filter is given by the formula $1/(\pi\sqrt{LC})$. This formula yields a nominal cutoff frequency, f_c, of 1,604 Hz (i.e., roughly in the middle of the transition region between the passband and stopband of the desired lowpass filter).

The legal criteria for obtaining a U.S. patent are that the proposed invention be "new" and "useful" and

> the differences between the subject matter sought to be patented and the prior art are such that the subject matter as a whole would [not] have been obvious at the time the invention was made to a person having ordinary skill in the art to which said subject matter pertains (35 *United States Code* 103a).

George Campbell was part of the renowned research team of the American Telephone and Telegraph Corporation. He received a patent for his filter in 1917 because his idea was new in 1917, because it was useful, and because it satisfied the above statutory test for unobviousness. The fact that genetic programming rediscovered an electrical circuit that was unobvious "to a person having ordinary skill in the art" establishes that this evolved result satisfies Arthur Samuel's criterion (Samuel 1983) for artificial intelligence and machine learning; namely,

> The aim [is] ... to get machines to exhibit behavior, which if done by humans, would be assumed to involve the use of intelligence.

6.2   Zobel 1925 "*M*-Derived Half Section" Patent

In another run of this same problem, a 100%-compliant circuit was evolved in generation 34. This evolved circuit is roughly equivalent to what is now known as a cascade of three symmetric T-sections and an *M*-derived half section (Johnson 1950). To see this, we modify this evolved circuit from generation 34 in three ways.

First, we insert wires in lieu of two $0.138\,\mu$H inductors (whose value is about six orders of magnitude smaller than the value of the other inductors in the circuit). The behavior of this slightly modified evolved circuit (Fig. 11) is not noticeably affected by these changes for the frequencies of interest in this problem.

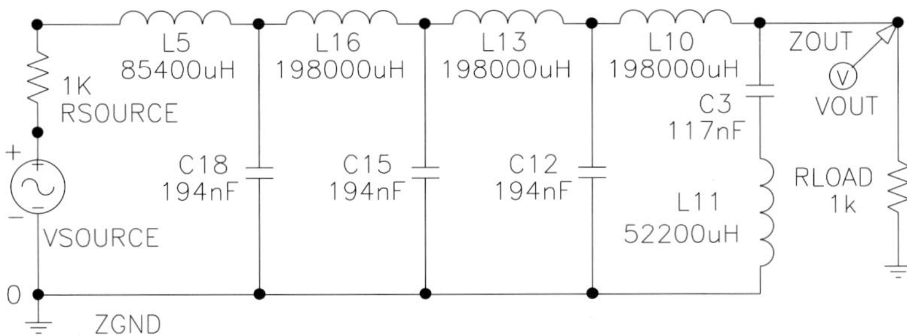

Fig. 11. Slightly modified version of the evolved lowpass filter circuit consisting of three symmetric T-sections and an *M*-derived half section.

Second, we replace each of the three $198,000\,\mu$H inductors in the figure (L16, L13, and L10) with a series composition of two $99,000\,\mu$H inductors. Since the inductance of two inductors in series is equal to the sum of their inductances, this change does not affect the behavior of the circuit at all. The circuit can now be viewed as having one incoming $85,400\,\mu$H inductor (L5) and six $99,000\,\mu$H inductors in series horizontally at the top of the figure.

Third, we note also that the values of the (now) seven inductors in series horizontally across the top of the figure are approximately equal. Suppose, for the sake of argument, that each of these seven approximately equal inductors are replaced by an inductor with inductance equal to their average value ($97,000\,\mu$H). This change does not appreciably affect the behavior of the circuit for the frequencies of interest.

After the above changes, the evolved lowpass filter can be viewed as consisting of a cascade of three identical symmetric T-sections and an *M*-derived half section. In particular, each T-section consists of an incoming inductor of inductance $L/2$ (where L equals $194,000\,\mu$H), a junction point from which a capacitor of capacitance C (where C equals $194\,$nF) is shunted off to ground, and an outgoing inductor of inductance $L/2$. The two inductors are the horizontal arms of the "T." The final half section (so named because it has only one arm of a "T") has one incoming inductor of inductance 2 and a junction point from which a capacitive-inductive shunt (C3 and L11) is connected to ground.

The first three symmetric T-sections are referred to as "constant K" filter sections (Johnson 1950, page 331). Such filter sections are characterized by two key parameters. The characteristic resistance, R, of each of the three T-sections is given by the formula $R = \sqrt{L/C}$. When the inductance, L, is 194,000 μH and the capacitance, C, is 194 nF, then the characteristic resistance, R, is 1,000 Ω according to this formula (i.e., equal to the value of the actual load resistor). The nominal cutoff frequency, f_c, of each of the three T-sections of a lowpass filter is given by the formula $f_c = 1/(\pi\sqrt{LC})$. This formula yields a nominal cutoff frequency, f_c, of 1,641 Hz (which is near the middle of the transition band for the desired pass filter).

In other words, both of the key parameters of the three T-sections are very close to the canonical values of constant K sections designed with the aim of satisfying this problem's design requirements.

The final section of the evolved circuit closely approximates a section now called an M-derived half section. This final section is said to be "derived" because it is derived from the foregoing three identical constant K prototype sections. In the derivation, m is a real constant between 0 and 1. Let m be 0.6 here. In a canonical M-derived half section that is derived from the above constant K prototype section, the value of the capacitor in the vertical shunt of the half section is given by the formula mC (116.4 nF). The actual value of C3 in the evolved circuit is 117 nF. The value of the inductor in the vertical shunt of an M-derived half section is given by the formula $L(1 - m^2)/4m$. This formula yields a value of 51,733. The actual value of L5 in the evolved circuit is 52,200 μH. The frequency, f_∞, where the attenuation first becomes complete, is given by the formula $f_\infty = f_c/\sqrt{1 - m^2}$. This formula yields a value for f_∞ of 2,051 Hz (i.e., near the beginning of the desired stopband).

Taken as a whole, the topology and component values of the evolved circuit are reasonably close to the canonical values for the three identical symmetric T-sections and a final M-derived half section that is designed with the aim of satisfying this problem's design requirements.

Otto Zobel of the American Telephone and Telegraph Company invented the idea of adding an M-derived half section to one or more constant K sections. As Zobel (1925) explains in U.S. patent 1,538,964,

The principal object of my invention is to provide a new and improved network for the purpose of transmitting electric currents having their frequency within a certain range and attenuating currents of frequency within a different range. . . . Another object of my invention is to provide a wave-filter with recurrent sections not all of which are alike, and having certain advantages over a wave-filter with all its sections alike.

The advantage of Zobel's approach is a "sharper transition" in the frequency domain behavior of the filter. Claim 1 of Zobel's 1925 patent covers

> A wave-filter having one or more half-sections of a certain kind and one or more other half-sections that are M-types thereof, M being different from unity.

Claim 2 covers

> A wave-filter having its sections and half-sections so related that they comprise different M-types of a common prototype, M having several values for respectively different sections and half-sections.

Claim 3 goes on to cover

> A wave-filter having one or more half-sections of a certain kind and one or more half-sections introduced from a different wave-filter having the same characteristic and the same critical frequencies and a different attenuation characteristic outside the free transmitting range.

Viewed as a whole, the evolved circuit here infringes the claims of Zobel's 1925 patent.

6.3   Johnson 1926 "Bridged T" Patent

In another run, a 100% compliant recognizable "bridged T" arrangement was evolved. The bridged T filter topology was invented and patented by Kenneth S. Johnson of the Western Electric Company in 1926 (Johnson 1926). As U.S. patent 1,611,916 (Johnson 1926) states

> In accordance with the invention, a section of an artificial line, such as a wave filter, comprises in general four impedance paths, three of which are arranged in the form of a T network with the fourth path bridged across the transverse arms of the T. The impedances of this network, which for convenience, will be referred to as a bridged T network, bear a definite relationship to a network of the series shunt type, the characteristics of which are well known.
>
> In the forms of the invention described herein, the arms of the bridged T network consist of substantially pure reactances. Its most useful forms are found to be wave filter networks in which there is a substantially infinite attenuation at a frequency within the band to be suppressed and the network may be designed so that this frequency is very near the cut-off frequency of the filter, thus producing a very sharp separation between the transmitted and suppressed bands.

Claim 1 of patent 1,611,916 covers

An electrical network comprising a pair of input terminals and a pair of output terminals, an impedance path connected directly between an input terminal and an output terminal, a pair of impedance paths having a common terminal and having their other terminals connected respectively to the terminals of said first path, and a fourth impedance path having one terminal connected to said common terminal and having connections from its other terminal to the remaining input terminal and output terminal, each of said paths containing a substantial amount of reactance, the impedances of said network having such values that said network is the equivalent of a series-shunt network having desired transmission characteristics.

The bridged T of Fig. 12 involves L14, C3, C15, and L11. In particular, L14 is the "impedance path connected directly between an input terminal and an output terminal" that is referred to later as "the first path." The junction of C3, C15, and L11 is the "common terminal." C3 and C15 are the "pair of impedance paths having a common terminal and having their other terminals connected respectively to the terminals of said first path." L11 is the "fourth impedance path having one terminal connected to said common terminal and having connections from its other terminal to the remaining input terminal and output terminal" (namely, the input and output terminals of the section that are both grounded).

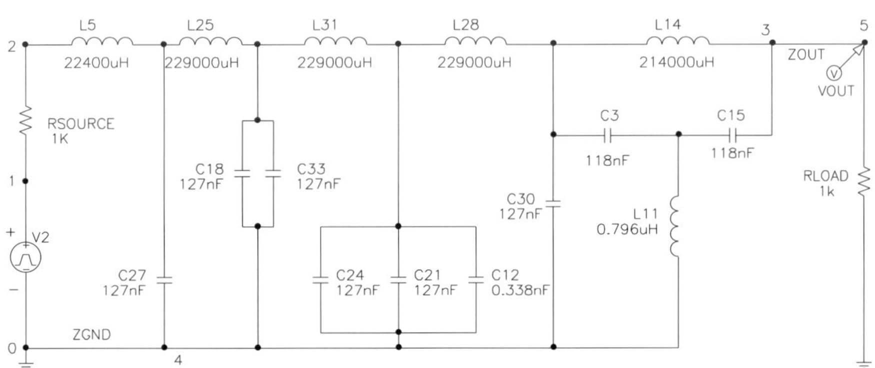

Fig. 12. "Bridged T" circuit from generation 64.

6.4   Cauer 1934–1936 Elliptic Patents

In a run of this same problem using automatically defined functions (described in Koza, et al. 1999), a 100% compliant circuit emerged in generation 31. After all of the pairs and triplets of series inductors in the evolved circuit are consolidated (as shown in Fig. 13), it can be seen that the circuit has the equivalent of six inductors horizontally across the top of the circuit and five vertical shunts. Each vertical shunt consists of an inductor and a capacitor.

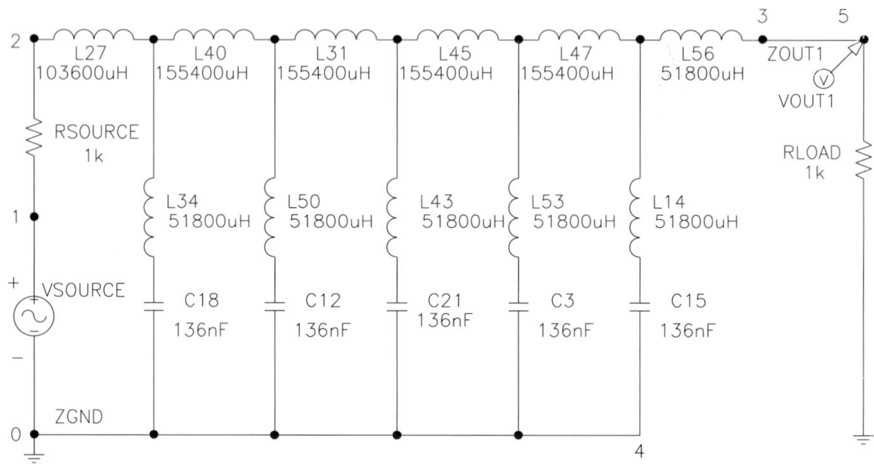

Fig. 13. Evolved Cauer (elliptic) filter topology from generation 31.

This circuit has the recognizable elliptic topology that was invented and patented by Wilhelm Cauer (1934, 1935, 1936). The Cauer filter was a significant advance (both theoretically and commercially) over the earlier filter designs of Campbell, Zobel, Johnson, Butterworth, and Chebychev. For example, for one commercially important set of specifications for telephones, a fifth-order elliptic filter matches the behavior of a 17th-order Butterworth filter or an eighth-order Chebychev filter. The fifth-order elliptic filter has one less component than the eighth-order Chebychev filter. As Van Valkenburg (1982, page 379) relates in connection with the history of the elliptic filter:

> Cauer first used his new theory in solving a filter problem for the German telephone industry. His new design achieved specifications with one less inductor than had ever been done before. The world first learned of the Cauer method not through scholarly publication but through a patent disclosure, which eventually reached the Bell Laboratories. Legend has it that the entire Mathematics Department of Bell Laboratories spent the next two weeks at the New York Public library studying elliptic functions. Cauer had studied mathematics under Hilbert at Goettingen, and so elliptic functions and their applications were familiar to him.

Genetic programming did not, of course, study mathematics under Hilbert or anybody else. Instead, the elliptic topology invented and patented by Cauer emerged from this run of genetic programming as a natural consequence of the problem's fitness measure and natural selection — not because the run was primed with domain knowledge about elliptic functions or filters or electrical circuitry. Genetic programming opportunistically *reinvented* the elliptic topology because necessity (fitness) is the mother of invention.

7 The Illogical Nature of Creativity and Evolution

Many computer scientists and mathematicians unquestioningly assume that every problem-solving technique must be logically sound, logically consistent, deterministic, and parsimonious. Accordingly, most conventional methods of artificial intelligence and machine learning are constructed so as to possess these characteristics. However, in spite of this strong predisposition by computer scientists and mathematicians, the features of logic do not govern two of the most important types of complex problem-solving processes, namely the invention process performed by creative humans and the evolutionary process occurring in nature.

A new idea that can be logically deduced from facts that are known in a field, using transformations that are known in a field, is not considered to be an invention. There must be what the patent law refers to as an "illogical step" (i.e., an unjustified step) to distinguish a putative invention from that which is readily deducible from that which is already known. Humans supply the critical ingredient of "illogic" to the invention process. Interestingly, everyday usage parallels the patent law concerning inventiveness: People who mechanically apply existing facts in well-known ways are summarily dismissed as being uncreative. Logical thinking is unquestionably useful for many purposes. It usually plays an important role in setting the stage for an invention. But, at the end of the day, logical thinking is the antithesis of invention and creativity.

Recalling his invention in 1927 of the negative feedback amplifier, Harold S. Black of Bell Laboratories (1977) said,

> Then came the morning of Tuesday, August 2, 1927, when the concept of the negative feedback amplifier came to me in a flash while I was crossing the Hudson River on the Lackawanna Ferry, on my way to work. For more than 50 years, I have pondered how and why the idea came, and I can't say any more today than I could that morning. All I know is that after several years of hard work on the problem, I suddenly realized that if I fed the amplifier output back to the input, in reverse phase, and kept the device from oscillating (singing, as we called it then), I would have exactly what I wanted: a means of canceling out the distortion of the output. I opened my morning newspaper and on a page of *The New York Times* I sketched a simple canonical diagram of a negative feedback amplifier plus the equations for the amplification with feedback.

Of course, inventors are not oblivious to logic and knowledge. They do not thrash around using blind random search. Black did not try to construct the negative feedback amplifier from neon bulbs or doorbells. Instead, "several years of hard work on the problem" set the stage and brought his thinking into the proximity of a solution. Then, at the critical moment, Black made his "illogical" leap. This unjustified leap constituted the invention.

The design of complex entities by the evolutionary process in nature is another important type of problem-solving that is not governed by logic. In nature, solutions to design problems are discovered by the probabilistic process of evolution and natural selection. There is nothing logical about this process. Indeed, inconsistent and contradictory alternatives abound. In fact, such genetic diversity is necessary for the evolutionary process to succeed. Significantly, the solutions evolved by evolution and natural selection almost always differ from those created by conventional methods of artificial intelligence and machine learning in one very important respect. Evolved solutions are not brittle; they are usually able to grapple with the perpetual novelty of real environments.

Similarly, genetic programming is not guided by the inference methods of formal logic in its search for a computer program to solve a given problem. When the goal is the automatic creation of computer programs, all of our experience has led us to conclude that the nonlogical approach used in the invention process and in natural evolution is far more fruitful than the logic-driven and knowledge-based principles of conventional artificial intelligence and machine learning. In short, "logic considered harmful."

8   Conclusion

We illustrated genetic programming by applying it to a non-trivial problem, namely the synthesis of a design for a lowpass filter circuit. The results were competitive with human-produced solutions to the problem. The results exhibited creativity and inventiveness and correspond to four inventions that were patented between 1917 and 1936.

References

1. Aaserud, O. and Nielsen, I. Ring. 1995. Trends in current analog design: a panel debate. *Analog Integrated Circuits and Signal Processing.* 7(1), 5–9.
2. Andre, David and Koza, John R. 1996. Parallel genetic programming: a scalable implementation using the transputer architecture. In Angeline and Kinnear. 1996.
3. Angeline, Peter J. and Kinnear, Kenneth E. Jr. (editors). 1996. *Advances in Genetic Programming 2.* Cambridge, MA: MIT Press.
4. Bäck, Thomas (editor). 1997. *Genetic Algorithms: Proceedings of the Seventh International Conference.* San Francisco, CA: Morgan Kaufmann.

5. Banzhaf, Wolfgang; Nordin, Peter; Keller, Robert E.; and Francone, Frank D. 1998a. *Genetic Programming — An Introduction.* San Francisco, CA: Morgan Kaufmann.

6. Banzhaf, Wolfgang; Poli, Riccardo; Schoenauer, Marc; and Fogarty, Terence C. 1998b. *Genetic Programming: First European Workshop. EuroGP'98. Paris, France, April 1998 Proceedings.* Lecture Notes in Computer Science. Volume 1391. Heidelberg: Springer-Verlag.

7. Black, Harold S. 1977. Inventing the negative feedback amplifier. *IEEE Spectrum.* December, pp. 55–60.

8. Campbell, George A. 1917. *Electric Wave Filter.* U.S. Patent 1,227,113. Filed July 15, 1915. Issued May 22, 1917.

9. Cauer, Wilhelm. 1934. *Artificial Network.* U.S. Patent 1,958,742. Filed June 8, 1928 in Germany. Filed December 1, 1930 in United States. Issued May 15, 1934.

10. Cauer, Wilhelm. 1935. *Electric Wave Filter.* U.S. Patent 1,989,545. Filed June 8, 1928 in Germany. Filed December 6, 1930 in United States. Issued January 29, 1935.

11. Cauer, Wilhelm. 1936. *Unsymmetrical Electric Wave Filter.* Filed November 10, 1932 in Germany. Filed November 23, 1933 in United States. Issued July 21, 1936.

12. Gen, Mitsuo and Cheng, Runwei. 1997. *Genetic Algorithms and Engineering Design.* New York, NY: John Wiley and Sons.

13. Goldberg, David E. 1989a. *Genetic Algorithms in Search, Optimization, and Machine Learning.* Reading, MA: Addison-Wesley.

14. Holland, John H. 1975. *Adaptation in Natural and Artificial Systems.* Ann Arbor, MI: University of Michigan Press.

15. Johnson, Kenneth S. 1926. *Electric-Wave Transmission.* U.S. Patent 1,611,916. Filed March 9, 1923. Issued December 28, 1926.

16. Johnson, Walter C. 1950. *Transmission Lines and Networks.* New York, NY: McGraw-Hill.

17. Kinnear, Kenneth E. Jr. (editor). 1994. *Advances in Genetic Programming.* Cambridge, MA: MIT Press.

18. Koza, John R. 1992. *Genetic Programming: On the Programming of Computers by Means of Natural Selection.* Cambridge, MA: MIT Press.

19. Koza, John R. 1994a. *Genetic Programming II: Automatic Discovery of Reusable Programs.* Cambridge, MA: MIT Press.

20. Koza, John R. 1994b. *Genetic Programming II Videotape: The Next Generation.* Cambridge, MA: MIT Press.

21. Koza, John R. 1995. Evolving the architecture of a multi-part program in genetic programming using architecture-altering operations. In McDonnell, John R., Reynolds, Robert G., and Fogel, David B. (editors). 1995. *Evolutionary Programming IV: Proceedings of the Fourth Annual Conference on Evolutionary Programming.* Cambridge, MA: MIT Press. pp. 695–717.

22. Koza, John R.; Banzhaf, Wolfgang; Chellapilla, Kumar; Deb, Kalyanmoy; Dorigo, Marco; Fogel, David B.; Garzon, Max H.; Goldberg, David E.; Iba, Hitoshi; and Riolo, Rick L. (editors). *Genetic Programming 1998: Proceedings of the Third Annual Conference, July 22–25, 1998, University of Wisconsin, Madison, Wisconsin.* San Francisco, CA: Morgan Kaufmann.

23. Koza, John R.; Bennett III, Forrest H; Andre, David; and Keane, Martin A. 1999. *Genetic Programming III: Darwinian Invention and Problem Solving.* San Francisco, CA: Morgan Kaufmann.

24. Koza, John R.; Deb, Kalyanmoy; Dorigo, Marco; Fogel, David B.; Garzon, Max; Iba, Hitoshi; and Riolo, Rick L. (editors). 1997. *Genetic Programming 1997: Proceedings of the Second Annual Conference.* San Francisco, CA: Morgan Kaufmann.

25. Koza, John R.; Goldberg, David E.; Fogel, David B.; and Riolo, Rick L. (editors). 1996. *Genetic Programming 1996: Proceedings of the First Annual Conference.* Cambridge, MA: MIT Press.

26. Koza, John R., and Rice, James P. 1992. *Genetic Programming: The Movie.* Cambridge, MA: MIT Press.

27. Michalewicz, Zbigniew. 1996. *Genetic Algorithms + Data Structures = Evolution Programs,* 3rd edition. Springer-Verlag.

28. Mitchell, Melanie. 1996. *An Introduction to Genetic Algorithms.* Cambridge, MA: MIT Press.

29. Ohno, Susumu. 1970. *Evolution by Gene Duplication.* New York, NY: Springer-Verlag.

30. Quarles, Thomas; Newton, A. R.; Pederson, D. O.; and Sangiovanni-Vincentelli, A. 1994. *SPICE 3 Version 3F5 User's Manual.* Department of Electrical Engineering and Computer Science, University of California, Berkeley, CA. March 1994.

31. Samuel, Arthur L. 1959. Some studies in machine learning using the game of checkers. *IBM Journal of Research and Development.* 3(3): 210–229.

32. Samuel, Arthur L. 1983. AI: Where it has been and where it is going. *Proceedings of the Eighth International Joint Conference on Artificial Intelligence.* Los Altos, CA: Morgan Kaufmann, pp. 1152–1157.

33. Spector, Lee; Langdon, William B.; O'Reilly, Una-May; and Angeline, Peter (editors). 1999. *Advances in Genetic Programming 3.* Cambridge, MA: MIT Press.

34. Van Valkenburg, M. E. 1982. *Analog Filter Design.* Fort Worth, TX: Harcourt Brace Jovanovich.

35. Williams, Arthur B. and Taylor, Fred J. 1995. *Electronic Filter Design Handbook,* 3rd Edition. New York, NY: McGraw-Hill.

36. Zobel, Otto Julius. 1925. *Wave filter.* Filed January 15, 1921. U.S. Patent 1,538,964 Issued May 26, 1925.

Is Ours the Best of All Possible Codes?

Stephen J. Freeland

Abstract. Although evidence is accumulating that the genetic code arose through stereochemical interactions between individual amino acids and oligonucleotides, its subsequent evolution remains contentious. On the one hand, the present structure of the code may be an end-point of natural selection such that codon assignments are organised to minimise the deleterious phenotypic effects of genetic error. On the other hand, the structure of the code may simply reflect its history, whereby novel amino acids were synthesised as the by-products of metabolism and subsequently incorporated into the code by capturing a subset of the codons previously assigned to their biosynthetic precursors. Both processes could potentially produce a code in which similar amino acids are assigned to similar codons. Here I argue for the plausibility of the adaptive ('error minimisation') interpretation of code structure and present quantitative evidence for this model. I further demonstrate that this evidence cannot be explained away as an artifact of a biosynthetic model of code evolution.

1 Introduction

The genetic code represents an interface between genotype (nucleic acids) and phenotype (proteins). Although this functional dichotomy of biomolecules is not absolute [15], this merely emphasizes that the arrival of proteins within the primordial biosphere, and hence the arrival of the genetic code, was probably an adaptive progression from a metabolically complex, exclusively RNA world.

Beyond this, the origin and evolution of the genetic code remain obscure: whilst the assignments of individual codons to individual amino acids have been known for three decades, we still do not know why the canonical genetic code takes the form that it does. Why, for instance, does AUU encode isoleucine rather than some other amino acid? Why are some amino acids assigned more codons than others? And why do similar amino acids tend to cluster together? In the accompanying chapter, Knight describes three general classes of explanation that have been proposed to explain the code's structure, giving a detailed consideration to the idea that codon assignments originated through simple chemical interactions between oligonucleotides and individual amino acids. In this chapter I present evidence that the code is organized to minimize the detrimental effects of genetic error, such that it represents an optimal interface for translating genotype into phenotype. I further demonstrate that these results cannot be dismissed as an artifact of code coevolution, which might allocate biosynthetically related amino acids to similar codons for historical reasons.

1.1   The Plausibility of an Optimal Genetic Code

An optimal code is one in which natural selection has organised codon assignments to minimise the average change in amino acid meaning when genetic errors (such as point mutation or mistranslation) change codon identity by a single nucleotide. Such a code would be advantageous in that it would maximize the production of functional proteins in the context of the background noise associated with metabolic processes.

For natural selection to produce an optimal genetic code, codon assignments must be capable of variation over an evolutionary timescale: why should we consider the code capable of adaptive evolution? The simple answer is that the last 20 years have produced a large body of empirical evidence to show that the 'universal' genetic code is a misnomer. Variations in codon assignment have now been described for a phylogenetically diverse group of species, for organelle genomes, prokaryotic genomes and eukaryotic nuclear genomes [20] (Fig. 1).

The sheer diversity of organisms involved precludes any reasonable argument other than that the genetic code can and does vary between populations over evolutionary time, but how is such variation realized? Two general mechanisms have been proposed to explain how individual codons may change their amino acid meaning: 'codon capture' [10,11,20–23] and 'codon ambiguity' [26,27,39]. Codon capture envisages a four step process: (i) through any combination of chance, mutational bias or selection, a given codon disappears entirely from the genome; (ii) without selective constraints, the corresponding tRNA is inactivated through point mutation; (iii) changes in mutational bias, selection pressures or chance cause the lost codon to re-emerge; and (iv) a new tRNA becomes assigned to the codon through gene duplication of an existing tRNA followed by an appropriate point mutation of the codon. Codon ambiguity envisages a simpler three step process: (i) a point mutation within a pre-existing tRNA weakens its specificity to its assigned codon, permitting it to bind to (and therefore translate) an alternative codon; (ii) subsequent changes in codon usage exploit this ambiguity; and (iii) further mutations in the tRNA favour and eventually fix the new (changed) amino acid meaning.

These two hypothesized mechanisms differ only in degree. The selective effect of a particular codon ambiguity is inversely proportional to its frequency within the genome. Empirical evidence now exists to support both mechanisms. Codon ambiguity is supported by the finding that CUG can be read as either Serine or Leucine within *Candida* [25,29], and experimental manipulation has demonstrated that tRNA duplication followed by mutation can lead to the reassignment of a tRNA from one isoaccepting group to another (an important step in codon capture) [24]. It thus seems likely that both mechanisms contribute to code variation, at least in extant taxa. In addition, individual codon reassignments iteratively applied to random initial code structures demonstrate that such 'code shuffling' works at a very general

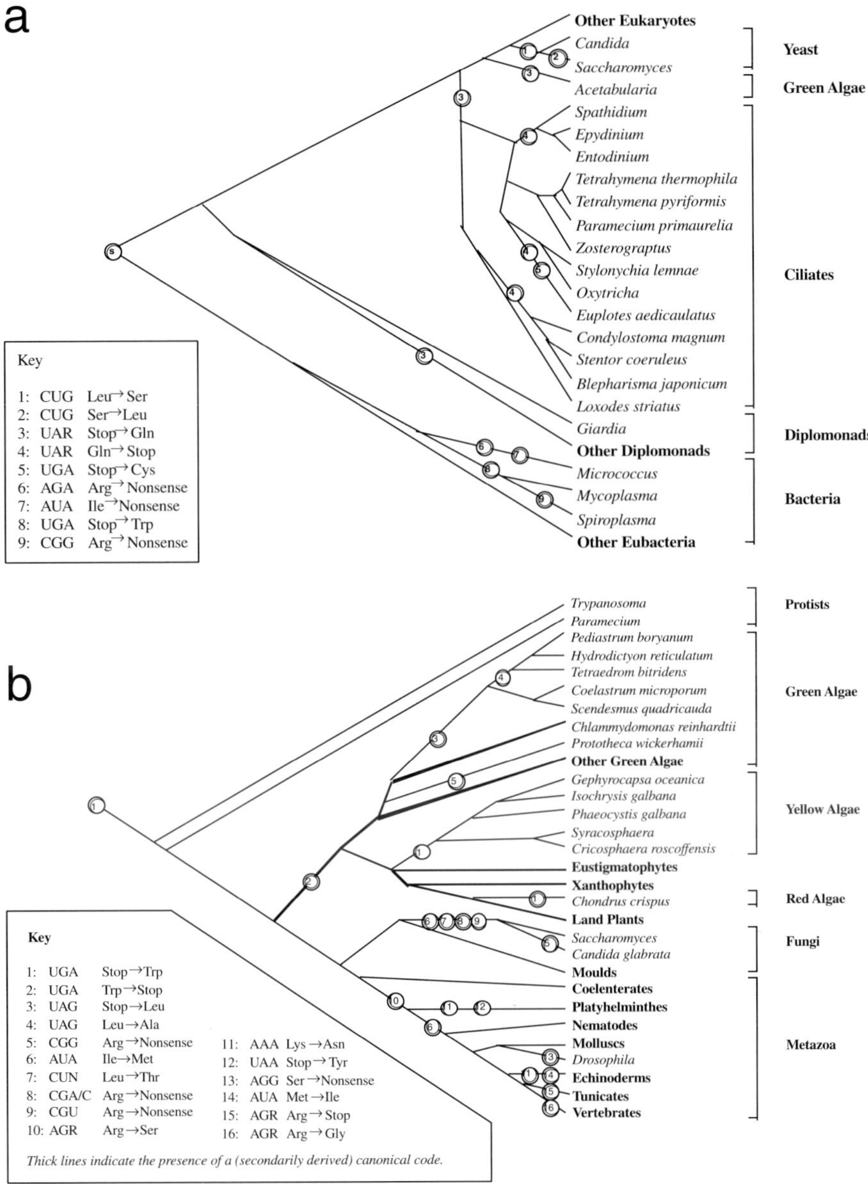

Fig. 1. Naturally occurring variants of the canonical genetic code; (a) mitochondrial variants; (b) 'nuclear' variants (including changes effective within bacterial genomes) *from R. Knight, unpublished*

level (assuming wobble rules remain invariant) to enable gradual transition between any two code structures [30]. In other words, individual codon reassignments, iteratively applied, unfreeze the code to the extent of allowing significant optimization.

Although extant code variation is of limited scope, both in terms of the proportion of species involved and in terms of the extent of codon reassignment, this does not necessarily weaken the case for significant adaptive code evolution prior to the Last Universal Ancestor (LUA). First, if the canonical code used by the LUA is a (near) optimal solution for error minimization, then further codon reassignments would mostly be deleterious: the paucity of extant code variation may reflect the early fixation of a highly adaptive configuration. Second, Crick's frozen accident hypothesis implicitly states that code malleability is inversely proportional to genome size. The LUA's predecessors likely possessed smaller genomes. The very fact that fully functioning organisms such as *Candida* can undergo codon reassignment suggests that the LUA's simpler ancestors, competing in a less sophisticated biotic environment, possessed relatively malleable codes. Particularly intriguing in this context is the proposal that the very idea of a universal ancestor might be misleading [34–36]. When rRNA phylogenies are compared to other deeply rooted phylogenies, fundamental inconsistencies suggest that early life may have been dominated by a very different evolutionary dynamic. Specifically, Woese proposes that lateral transmission was the primary evolutionary mechanism, implying a selective advantage to the evolution of a consensus code whereby unrelated organisms could 'share' genes. Under this scenario, it is easy to envisage an optimal code structure acting as an attractor for this consensus code.

1.2 Apparent Evidence for an Optimized Genetic Code, and Alternative Interpretations

Almost as soon as the canonical genetic code was fully deciphered, various researchers noted that similar codons tend to be assigned to physiochemically similar amino acids [6,8,28,33,40]. Although this certainly meets the expectations of an optimal code, and may be interpreted as the end point of selection for error minimization, other interpretations are possible. For example, an alternative model of code evolution suggests that the canonical genetic code evolved from a simpler ancestral form, which encoded fewer amino acids with greater redundancy [38]. During the course of early evolution, novel amino acids were synthesized as the by-products of metabolism. New amino acids that were found to be advantageous would have been incorporated into the code by capturing a subset of the codons previously used by their biosynthetic pre-cursors. This explanation for code structure could plausibly explain the apparent 'adaptive' features of the genetic code as a simple artifact: biosynthetically related amino acids that share similar physiochemical properties might be assigned to similar codons by historical constraints.

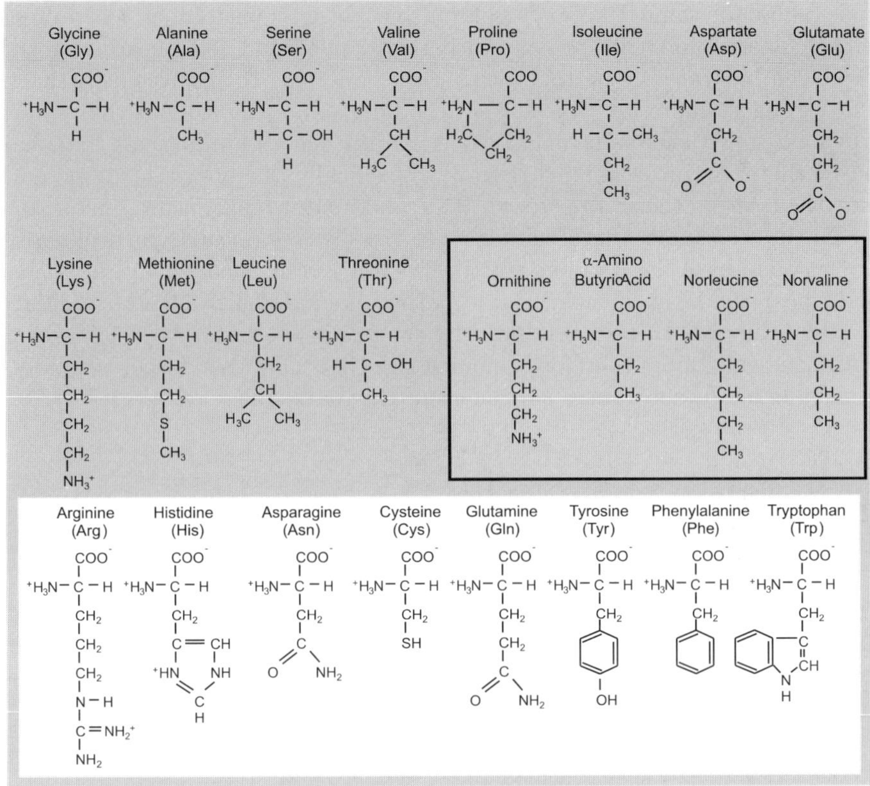

Fig. 2. Amino acid structures; gray background indicates those found either in the products of pre-biotic simulation ('spark experiments' [17]) or in the Murchison meteorite [14]. Boxed amino acids are a subset of those found in the Murchison meteorite which are not found within the genetic code.

In fact, significant evidence supports the idea that the code has undergone expansion from a simpler, primordial state. For example, only a subset of the 20 amino acids currently encoded have been produced in prebiotic simulations (the 'spark experiments' [17]), and this same subset was found within the Murchison meteorite [14] (Fig. 2). But does this process of expansion explain extant codon assignments? Several analyses have proposed patterns of biosynthetic relatedness that seem to link related amino acids to similar codons. The first major contribution in this field [14] was flawed because it used hypothetical patterns of biosynthetic relatedness which bear little resemblance to the pathways of real organisms. This was followed by a very detailed description of perceived networks of biosynthetic relatedness within the code [38], which grouped all amino acids into 'precursor/product' relationships matching cognate codon clusters. Once again, however, the analysis was subsequently shown to be flawed [1] in that many randomly generated

codes show the same degree of apparent biosynthetic patterning: the problem
is that most amino acids are biosynthetically related at some level, such that
perceived biosynthetic patterns within the code may themselves be artifacts.
The strongest claim for a biosynthetic subdivision of the code is that codons
which begin with the same nucleotide generally encode amino acids that share
biosynthetic pathways [18,32] (Fig. 3). Specifically, the aromatic amino acids
and their biosynthetic precursors (the 'shikimate' family), are encoded by
UNN; the 'glutamate' family are encoded by CNN and the 'aspartate' family
are encoded by ANN. Codons starting with a guanine (i.e., GNN) all appear
at or near the head of biosynthetic pathways, and specify likely candidates
for the 'primordial' code, produced by the prebiotic world. Any claim for an
optimized code must therefore demonstrate that 'adaptive' features of codon
assignments are not a mere side effect of this pattern.

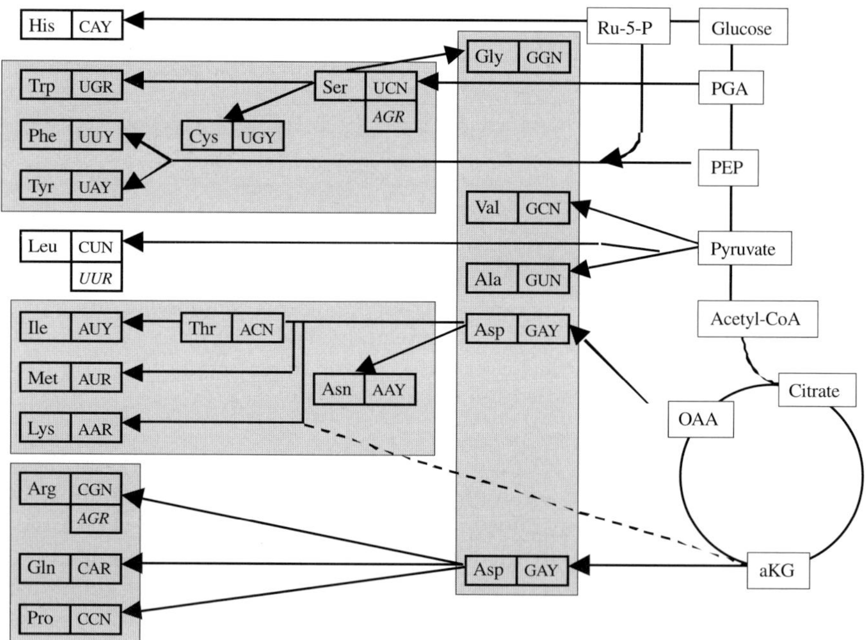

Fig. 3. Biosynthetic pathways and codon assignments for the 20 amino acids coded
for by the canonical genetic code, adapted from [32]. Grey shaded areas highlight
codon assignments which suggest a relationship between codon families and amino
acid meanings. In addition to the four specific nucleotide abbreviations (U, C, A
and G), I use N to refer to 'any nucleotide', Y to refer to 'any pyrimidine' (U or C)
and R to refer to 'any purine' (A or G).

2   Methods

Assessing the extent to which the canonical code minimizes the effects of genetic error is a two-step process: first, the structure of the canonical code is quantified into a single 'error value' which reflects the average effect of a point-mutation. Second, a large random sample of plausible alternative code structures is created, the error value of each sample member is calculated, and the canonical code is compared with the resulting distribution.

2.1   Calculating a Code Error Value

Each codon is connected by single nucleotide substitution to 9 alternatives (3 replacements are possible at each of the 3 codon positions). Of the 576 possible single base errors, 50 involve changes to or from a termination codon and cannot be quantified. Of the remaining 526 changes, some are synonymous (resulting in no change in amino acid meaning), and some result in a different amino acid 'meaning'. Synonymous changes are given a value of 0 (indicating no difference between intended and actual meaning), non-synonymous changes are given a value corresponding to some quantified measure of amino acid similarity. For the purpose of this paper, I use the physiochemical property 'polar requirement' to quantify amino acid similarity. This measure of hydrophobicity was empirically determined in the 1960's [37] and multivariate analyses of code structure indicate its probable importance in this context [4,9,31]. Given this measure, the error value of a genetic code may be calculated quite simply as the total modular difference in amino acid polar requirement resulting from all single nucleotide changes to all codons of the code, divided by the number of changes.

However, much research indicates that the relative frequency of errors, at least in terms of point mutation, varies according to the nucleotide identities involved. In particular, the 4 letters of the genetic alphabet fall into 2 distinct groups: the single ring pyrimidines (U and C) and the double ring purines (A and G). Point mutations that swap one member of a group for another (transition mutations) occur more frequently than those that swap a member of one group for a member of the other group (transversion mutations). The simple measure of code efficiency described above is thus given more sophistication and biological relevance by subdividing individual codon changes into transitions and transversions, and weighting the differences caused by transitions more heavily than those caused by transversions. By incorporating the weighting factor into the divisor (the number of changes in a code), the simple average polarity difference is transformed into a weighted average polarity difference.

Finally, the overall error value of a code may be partitioned into three separate components representing the weighted mean polarity difference resulting from all possible single nucleotide changes in the first, second and third codon positions of the codons.

2.2 Generating a Sample of Plausible Codes

Such code error values only take on significance when compared with a distribution of plausible code configurations. To do this, I defined a set of possible code configurations, generated a sample of 1 million possible codes according to this definition, and measured the proportion of these random variant codes with lower error value than the canonical code (i.e., the proportion which minimize the effects of genetic error better than the canonical code).

At least two assumptions restrict the set of possible genetic codes ('variants') against which the canonical code may be compared: (i) all variants must comprise the 64 codons NNN (where N is one of the four nucleotides U, C, A or G) and (ii) all variants must encode the 20 amino acids and translation termination signal used by the entirety of known life. Without these assumptions, the definition of an 'optimal' genetic code configuration becomes trivial and biologically meaningless. For example, an optimal variant would encode 1 amino acid or comprise 1 codon. For the following analysis, I assumed two further restrictions on possible code structure: that all variants maintain the pattern of redundancy, and the number and position of 'stop' codons of the canonical genetic code. In the absence of restrictions on redundancy patterns, an optimal code would comprise 44 synonyms for the amino acid with the similarity value closest to the average of the 20 amino acids, and 1 codon assigned to each of the other meanings [1]. This solution is biologically unrealistic; for example, such a code would require larger amounts of more highly specific translation machinery (e.g., tRNA species and associated aminoacyl synthetases) than does the canonical code, and this would probably entail a fitness cost. Although redundancy patterns vary over evolutionary time (most naturally occurring variant genetic codes exhibit different patterns of redundancy; Fig. 1), current theory does not describe a biologically realistic restriction of this feature, so invariance represents the most realistic condition available. Similarly, the number and position of stop codons have also varied (the reassignment of the UGA 'stop' codon to Tryptophan within animal and fungal mitochondria is an obvious example), but because current theory does not define plausible limits to the scope of such variation, the undefined error value associated with changes to and from stop codons once again means that invariance remains the best approach.

The inclusion of all four assumptions reduces the set of possible codes to 20! ($= 2.4 \times 10^{18}$) configurations (Fig. 4a), which is henceforth referred to as the 'unrestricted' set (though as noted above, it probably represents an underestimate of the actual number of possible code configurations).

In order to test for the possible confounding effect of biosynthetic code expansion, additional assumptions may be made which restrict the set of possible codes to reflect plausible historical constraints on amino acid assignments. If amino acids from the same biosynthetic pathway are only allowed to take codon assignments of their biosynthetic relatives, then the set of possible codes is limited to $5!^4$ ($= 2 \times 10^9$) configurations (Fig. 4b), henceforth referred

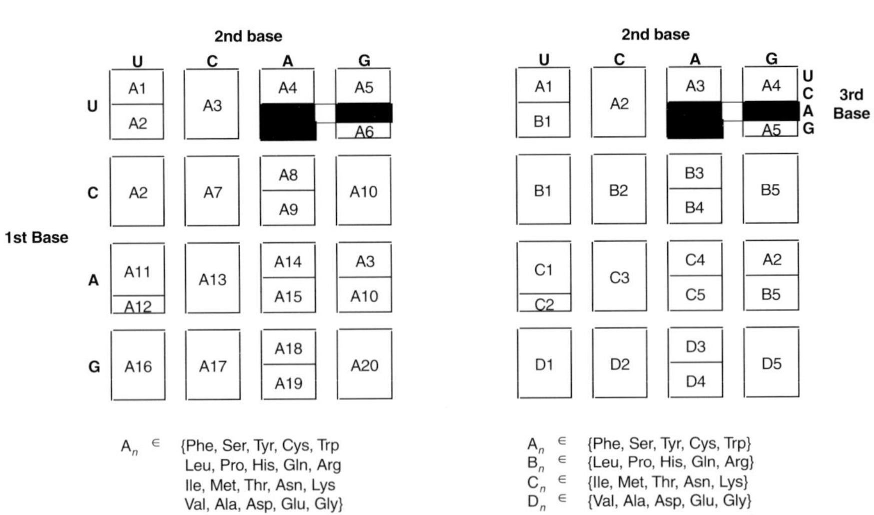

$A_n \in$ {Phe, Ser, Tyr, Cys, Trp
Leu, Pro, His, Gln, Arg
Ile, Met, Thr, Asn, Lys
Val, Ala, Asp, Glu, Gly}

$A_n \in$ {Phe, Ser, Tyr, Cys, Trp}
$B_n \in$ {Leu, Pro, His, Gln, Arg}
$C_n \in$ {Ile, Met, Thr, Asn, Lys}
$D_n \in$ {Val, Ala, Asp, Glu, Gly}

Fig. 4. Definitions of the set of possible codes. (a) The unrestricted set: each synonymous codon block takes the assignment of one of the 20 amino acids; no restrictions are placed on this mapping (b) The restricted set: codon assignments consistent with the Taylor/Coates model of historical constraint. Codon assignments are divided into 4 groups (A to D), each containing 5 members (1 to 5). Codons assignments are allowed to vary randomly within members of a group but not between. Thus, for example, amino acids Phe, Ser, Tyr, Cys and Trp are randomly assigned one synonymous block each of elements A1 to A5. In both sets of possible codes, the arrangement and size of synonymous codon blocks remains constant, as do the assignments of 'TER' codons UAR and UGA.

to as the 'restricted' set. These assumptions of biosynthetic restriction thus reduce the set of possible codes to such an extent that the distributions of error values for the restricted and unrestricted sets of codes may be regarded as effectively independent [6]. Given the unproven possibility of historical restrictions, analyses were undertaken for both the unrestricted and restricted code sets.

2.3   Basic Comparison

Within a sample of one million randomly generated unrestricted code variants, only 1 in ten thousand out-perform the canonical genetic code using the simple (unweighted) measure of code efficiency (Fig. 5). In other words, the estimated probability that chance alone would produce a code structure as or more efficient than the one chosen by nature is 0.0001. Furthermore, this result is not changed qualitatively when the sample is drawn from the restricted set of codes (Table 1): the canonical code still appears remarkably optimized if we accept plausible biosynthetic restrictions.

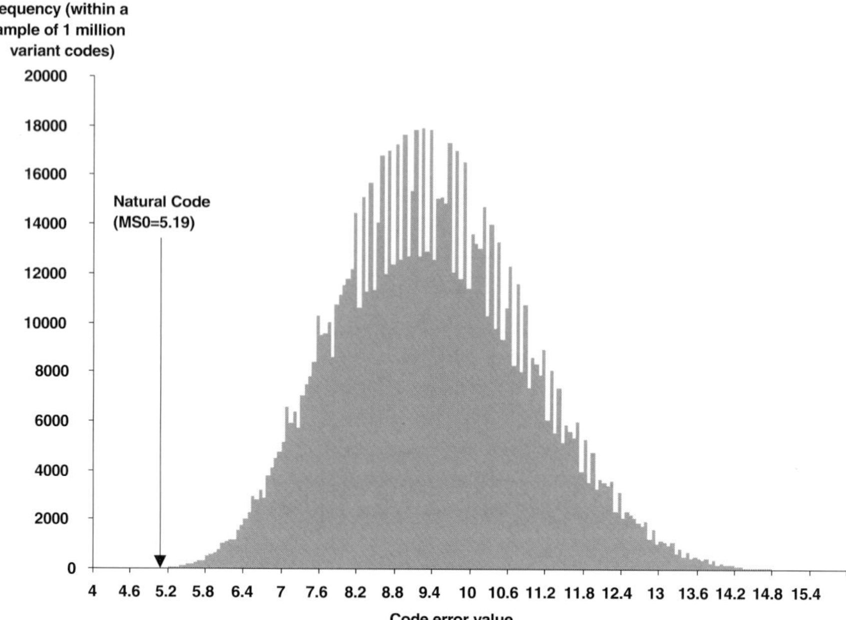

Fig. 5. Frequency distributions for the code error values obtained from one million randomly generated variant codes. In each plot, the X axis gives the range of error values encountered, and the Y axis gives the frequency with which codes of that error value were encountered within the sample. In addition, an arrow indicates the error value of the canonical code: the cumulative frequency to the left of this arrow thus indicates the number of more conservative codes found among the random variants, and is used to estimate the probability of evolving a code as efficienct as the canonical code by chance alone.

Interestingly, the perceived overall efficiency of the canonical code is not uniform across different codon base positions (Table 1). Codon position 3 appears most efficient (with a proportion of 0.00009 better codes in the sample of one million variants), followed by codon position 1 (a proportion of 0.003 better codes) followed by codon position 2 (a proportion of 0.2216 better codes). In other words, there is strong evidence to suggest that codon assignments are arranged to minimize the effects of errors at the first and third codon positions, but the average effect of mistakes occurring at the second codon position is no smaller than would be expected from a randomly arranged code. Once again, this pattern remains qualitatively unchanged whether codes are drawn from the restricted or unrestricted set of possible configurations.

2.4   Introducing a Transition/Transversion Bias

When the code's efficiency is measured in a more sophisticated manner, the apparent adaptive arrangement of codon assignments becomes even more

Measure		Unconstrained Model (n=1,000,000)		Constrained Model (n=1,000,000)		t$\tilde{\text{O}}$ (p)*
Mean	all	9.41	(1.51)	8.87	(0.95)	302.9 (<<0.01)
and Std. Dev.	base 1	12.04	(2.80)	12.38	(2.49)	90.81 (<<0.01)
	base 2	12.63	(2.60)	11.24	(0.99)	499.0 (<<0.01)
	base 3	3.59	(1.50)	3.02	(1.18)	298.9 (<<0.01)
Proportion of	all	0.0001		0.0003		
better	base 1	0.0030		0.0006		
codes found	base 2	0.2216		0.2466		
	base 3	0.0001		0.0006		

Table 1. Descriptive statistics for the distributions of error values formed by a sample of one million variant codes drawn from the constrained and unconstrained sets of possible codes. * The probabilities that the two samples are drawn from the same population (i.e., that the biosynthetic restriction rules make no difference to the mean error value for random codes), were calculated using the t' test (for difference of means with unequal sample variance) [7].

pronounced. Although the precise transition bias occurring in genomes varies according to the system under consideration, it is generally thought to range between 2 and 10 [3,13,19]. Within this region of parameter space, the overall efficiency of the code is consistently better for the restricted set of codes (Fig. 6). For the unrestricted set of codes, optimal efficiency is achieved at a transition bias of about 3. Once again, the effect of an increasing transition bias varies greatly across different codon positions: by far the greatest improvement in perceived code efficiency is seen at codon position 2, where a transition bias of 3 effects a five-fold increase (Fig. 7).

3   Discussion

The genetic code can and does vary between populations over evolutionary time, and such variation was probably more common amongst the predecessors of the LUA than in extant lineages. It is intuitive that a code which minimizes the phenotypic effects of genetic error will be more fit than one which does not do so (particularly if the LUA's ancestors were more error prone), and it is obvious that in the absence of mutation, the structure of the genetic code within a particular lineage is a heritable characteristic. In short, the genetic code meets the prerequisites of natural selection: variation in form, concomitant variation in fitness of carriers, and heritability.

When the canonical code is compared against a large random sample of possible alternatives, it outperforms the vast majority. Furthermore, this

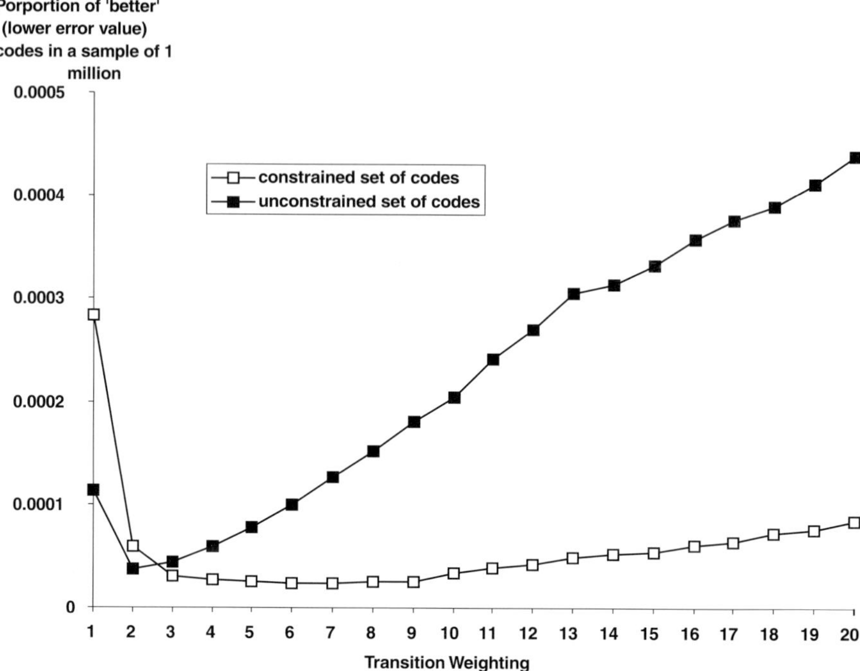

Fig. 6. The overall efficiency of the genetic code at a range of transition/transversion weightings, for the restricted and unrestricted sets of possible codes

apparent optimality increases when the efficiency metric is improved by including appropriate biological weighting reflecting the higher frequency of transition errors over transversions. This behavior is not found in the few theoretical codes which appear more efficient when transitions and transversions are considered equally likely.

These characteristics are exactly what would be expected if natural selection had steered primordial code evolution towards an optimal structure for error minimization. This interpretation is further strengthened by the observation that these 'adaptive' results remain qualitatively unchanged when the sample of alternative codes is restricted to reflect plausible historical constraints, according to which amino acids from the same metabolic pathways share similar codons. It is entirely plausible that the present catalogue of 20 amino acids arose from a simpler primordial code, but it appears that the process of code expansion did not produce a superficially adaptive arrangement of codon assignments as an artifact.

Taken as a whole, this evidence strongly suggests that natural selection for error minimization played an important role in code evolution, but how does this relate to other explanations for code structure? Recent evidence has shown that Arginine shows the greatest stereochemical affinity for those

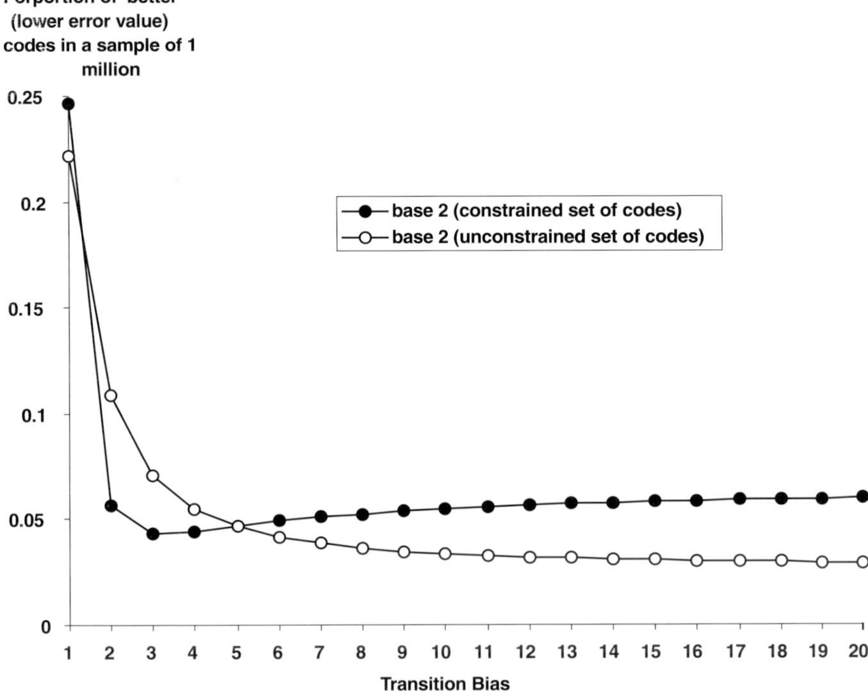

Fig. 7. The perceived efficiency of the second codon base position of the genetic code over a range of transition weightings, for the restricted and unrestricted sets of possible codes

codons to which it is assigned within the canonical code [11] Further evidence suggests that this relationship may hold true for isoleucine [15]. At least two explanations are possible. If only a few amino acids show such affinities for their present codons, then the code probably originated through stereochemical interactions, but was later reshuffled by natural selection for error minimization. Alternatively, if most amino acids interact preferentially with their present codons, then the present catalogue of amino acids may have been 'chosen' as those that satisfied criteria of both stereochemistry and error minimization. This highlights the importance of investigating the stereochemical affinities of the remaining 18 amino acids, and those of amino acids likely formed under prebiotic conditions but not found in the present code (Fig. 2). Only in the light of this information can we progress towards a comprehensive explanation for code evolution.

References

1. Ardell, D.H. (1998). On error minimisation in a sequential origin of the genetic code. *J. Mol. Evol.* 47:1–13.

2. Amirnovin, R. (1997). An analysis of the metabolic theory of the origin of the genetic code. *J. Mol. Evol.* 44:473–476.

3. Collins, D.W. (1994). Rates of transition and transversion in coding sequences since the human-rodent divergence. *Genomics* 20:386–396.

4. DiGiulio, M. (1989). Some aspects of the organisation and evolution of the genetic code. *J. Mol. Evol.* 29:191–201.

5. Dillon, L.S. (1973). The origins of the genetic code. *The Botanical Review* 39:301–345.

6. Epstein, C.J. (1966). Role of the amino acid 'code' and of selection for conformation in the evolution of proteins. *Nature* 210:25–28.

7. Freeland, S.J. & Hurst, L.D. (1998). Load minimisation of the code: history does not explain the pattern. *Proc. Roy. Soc. Lond. B* 265:2111–2119.

8. Goldberg, A.L. & Wittes, R.E. (1966). Genetic code: aspects of organisation. *Science* 153:420–424.

9. Haig, D. & Hurst, L.D. (1991). A quantitative measure of error minimisation within the genetic code. *J. Mol. Evol.* 33:412–417.

10. Jukes, T.H. & Osawa, S. (1996). CUG codons in *Candida* spp. *J. Mol. Evol.* 42:321–322.

11. Jukes, T.H. & Osawa, S. (1997). Further comments on codon reassignment. *J. Mol. Evol.* 45:1–3.

12. Knight, R.D. & Landweber, L.F. (1998). Rhyme or reason: RNA-arginine interactions and the genetic code. *Chemistry and Biology* 5:R215–R220.

13. Kumar, S. (1996). Patterns of nucleotide substitution in mitchondrial protein-coding genes of vertebrates. *Genetics* 143:537–548.

14. Kvenvolden, K.A., Lawless, J.G. et al. (1971). Non-protein amino acids in the Murchison meteorite. *Proc. Natl. Acad. Sci. USA* 68:486–490.

15. Landweber, L.F., Simon, P.J. & Wagner, T.A. (1998). Ribozyme design and early evolution. *BioScience* 48:94–103.

16. Majerfield, I. & Yarus, M. (1998). Isoleucine: RNA sites with essential coding sequences. *RNA* 4:471–478.

17. Miller, S.L. (1987). Which organic compounds could have occurred on pre-biotic earth? *Cold Spring Harbour Symp. Quant. Biol.* 52:17–27.

18. Miseta, A. (1989). The role of protein associated amino acid precursor molecules in the organisation of genetic codons. *Physiol. Chem. Phys. Med. NMR* 21:237–242.

19. Moriyama, E.N. & Powell, J.R. (1997). Synonymous substitution rates in Drosophila: mitochondrial versus nuclear genes. *J. Mol. Evol.* 45:378–391.

20. Osawa, S. (1995). *The evolution of the genetic code.* Oxford, Oxford University Press.

21. Osawa, S. & Jukes, T.H. (1988). Evolution of the genetic code as affeected by anticodon content. *Trends Genet.* 4:191–198.

22. Osawa, S. & Jukes, T.H. (1989). Codon reassignment (codon capture) in evolution. *J. Mol. Evol.* 21:271–278.

23. Osawa, S., Jukes, T.H., Watanabe, K. & Muto, A. (1992). Recent evidence for the evolution of the genetic code. *Microbiological Rev.* 56:229–264.

24. Saks, M.E., Sampson, J.R. & Abelson, J. (1998). Evolution of a transfer RNA gene through point mutation in the anti-codon. *Science* 279:1665–1670.

25. Santos, M.A.S., Veda T., Watanabe, K. & Tuite, M.F. (1997). The non-standard genetic code of Candida spp.: an evolving genetic code or a novel mechanism for adaptation? *Molecular Microbiology* 26:423–431.

26. Schultz, D.W. & Yarus, M. (1994). Transfer RNA mutation and the malleability of the genetic code. *J. Mol. Biol.* 235:1377–1380.

27. Schultz, D.W. & Yarus, M. (1996). On malleability in the genetic code. *J. Mol. Evol.* 42:597–601.

28. Sonneborn, T.M. (1965). Degeneracy of the Genetic Code: Extent, Nature, and Genetic Implications. In *Evolving Genes and Proteins*, V. Bryson and H.J. Vogel, eds. Academic Press, New York, pp. 377–297.

29. Suzuki, T., Ueda, T. & Watanabe, K. (1997). The 'polysemous' codon — a codon with multiple amino acid assignment caused by dual specificity of tRNA identity. *EMBO Journal* 16:1122–1134.

30. Szathmary, E. (1991). Codon swapping as a possible evolutionary mechanism. *J. Mol. Evol.* 32:178–182.

31. Szathmary, E. & Zintzaras, E. (1992). A statistical test of hypotheses on the organization and origin of the genetic code. *J. Mol. Evol.* 35:185–189.

32. Taylor, F.J.R. & Coates, D. (1989). The code within the codons. *BioSystems* 22:177–187.

33. Woese, C.R. (1965). Order in the genetic code. *Proc. Natl. Acad. Sci. USA* 54:71–75.

34. Woese, C.R. (1998a). The universal ancestor. *Proc. Natl. Acad. Sci. USA* 95:6854–6859.

35. Woese, C.R. (1998b). The universal ancestor (vol. 95, pg. 6854, 1998). *Proc. Natl. Acad. Sci. USA* 95: 9710.

36. Woese, C.R. (1998c). Default taxonomy: Ernst Mayr's view of the microbial world. *Proc. Natl. Acad. Sci. USA* 95:11043–11046.

37. Woese, C.R., Dugre, D.H., Dugre, S.A., Kondo, M. & Saxinger, W.C. (1966). On the fundamental nature and evolution of the genetic code. *Cold Spring harbour Symp. Quant. Biol.* 31:723–736.

38. Wong, J.T-F. (1975). A co-evolution theory of the genetic code. *Proc. Natl. Acad. Sci. USA* 72:1909–1912.

39. Yarus, M. & Schultz, D.W. (1997). Response. *J. Mol. Evol.* 45:1–8.

40. Zuckerkandl, E. & Pauling, L. (1965). Evolutionary Divergence and Convergence in Proteins. In *Evolving Genes and Proteins*, V. Bryson and H.J. Vogel, eds. Academic Press, New York.

The Impact of Message Mutation
on the Fitness of a Genetic Code

Guy Sella and David H. Ardell

Abstract. The standard genetic code (SGC) is organized in such a way that similar codons encode similar amino acids. One of the earliest explanations for this was that the SGC is the result of natural selection to reduce the fitness cost, or "load," from mutations in and mistranslation of protein-coding genes. However, it was later argued on both empirical and conceptual grounds that the SGC could not have evolved to reduce load. We claim that the empirical evidence has been misinterpreted and review how the pattern of amino acid similarities in the SGC are consistent with the "load minimization hypothesis" or "LM hypothesis." We then present a model which addresses a second classical objection to the load minimization hypothesis: that selection for load minimization must be indirect or weak because it acts across generations. In this model, individual fitness is determined by a protein distribution resulting from the translation of its genetic message using a genetic code. Amino acids contribute independently and multiplicatively to the fitness of the protein distribution, which is defined relative to a fixed target protein. We show that in mutation-selection balance a fitness can be associated with a population of individuals with the same genetic code, and illustrate that structure-preserving codes that assign similar codons to similar amino acids confer higher fitness. We also show that in mutation-selection balance the total message of any individual behaves like a population of sites. That is, the usage of codons in a message in almost any individual reflects the codon frequencies across the population for a site of a given type, and consequently the fitness of almost all individuals is equal to the population fitness associated with their genetic code. We thereby establish that selection for load minimization in genetic codes acts at the level of an individual in a single generation. Comparing the fitness of two genetic codes, one more structure-preserving than the other, we find that the more structure-preserving code is associated with lower load and consequently higher fitness, despite the fact that the equilibrium fraction of mutant codons in its mutation-selection balance is larger; these results are explained and conjectured to hold in general. We conclude with comments on the origin of the genetic code from the perspective of a model of this kind, including its shortcomings and advantages over other hypotheses as a comprehensive explanation for the origin of the SGC.

Introduction

A *genetic code* is a specific mapping between words of RNA called *codons*, which are three nucleotides long, and functional units in proteins called *amino acids*, of which there are twenty. There are four types of nucleotides, distinguished by, and denoted by, their *bases*, written $A, C, G,$ and U. The three positions of nucleotides in codons are referred to as *codon positions*. The nearly universal mapping between codons and amino acids experimentally deduced by Nirenberg and his contemporaries is now called the *standard genetic code* (SGC). Out of the 64 codons in the Standard Genetic Code, 61 code for the twenty amino acids while the other 3 are *stop codons* denoting the end of a message.

Shortly after biochemical methods were used to decipher the genetic code of the bacterium *Escherichia coli* [1], it was recognized that almost every organism, organelle and virus exploits the same mapping, despite the potentially enormous number of equivalent genetic codes. Sonneborn, Zuckerkandl and Pauling, and others were prompted to ask whether the SGC became nearly-universal by virtue of its superiority to other genetic codes, such that it became fixed in the early history of life through the action of natural selection. In support of this position many noticed that the organization of the SGC is nonrandom in that physicochemically similar amino acids tend to be assigned to codons that differ by a single base. The SGC, it was argued, may have been selectively superior to other codes because it corrects for errors in the translation and genetic transmission of protein-coding information (*message errors*) [1–5]. We call this the *error-correction hypothesis*, in which no distinction is made as to the types of message errors thought to have influenced the origin of the SGC. It should be noted that by "error correction" we refer specifically to effects of the organizational *pattern* of amino acids in the SGC rather than to error-correcting *processes* such as proofreading or repair.

Soon after the general error-correction hypothesis was put forward, at least three objections were raised to the notion that the SGC evolved to correct for the effects of mutations on protein function, or for any errors at all. These objections are both empirical and conceptual in nature. The empirical objection is that the pattern of amino acids in the SGC appears to be inconsistent with evolution of the SGC to correct for the effects of mutations. In 1965, Carl Woese connected two observations: that translational errors occur more frequently in the first and third positions of codons [6], and that amino acids encoded by codons that differ only in the first or third positions tend to be similar or identical [7]. Woese and later authors used these observations to argue that the genetic code evolved to correct for the deleterious effects of translational misreading (which he called the *error-minimization hypothesis*, or *EM hypothesis*) but not of message mutations (which he called the *load minimization hypothesis*, or *LM hypothesis*). Because mutations occur invariantly in all three codon positions, Woese and later authors argued, a null expectation for the SGC under the LM hypothesis is that it should have evolved to be equally conservative of amino acid properties in all three codon positions [8–12]. As we review in the next section, this null expectation is incorrect. It does not take into account that any selective influence of mutations on genetic codes must be filtered through translational error. Because translational error varies with codon position, so should the selective influence of mutation on the evolution of amino acid assignments in the SGC [13].

Three conceptual objections were raised to the notion that the SGC evolved by natural selection to correct for any type of errors at all. First, changing an existing code appears likely to render preexisting messages meaningless, which implies a devastating loss in fitness. Second, a code that expresses many amino acids is useful even if it is not error-correcting; therefore, selection for adding amino acids should dominate over selection for error correction. As an extreme position, Crick argued that genetic codes should not be able to evolve at all, but remain a "frozen accident" [14]. To these objections Jungck added that the space of possible codes is too large to have been effectively searched by natural selection, especially in the face of the freezing arguments of Crick [15]. Yet the observed non-randomness of amino acid assignments in the SGC still had to be explained. Crick, as well as Woese [8,16], sought to explain these patterns in the SGC with the notion that certain codons or anticodons have a specific stereochemical affinity for amino acids, and that related codons are associated chemically with related amino acids. For a comprehensive review of this direction see [17] in this volume.

Until recent times, the great extent and intricacy of the potentially error-correcting structure in the SGC had not come to light. For example, the SGC allocates more similar amino acids to the most frequent type of message mutations in the codon position that has the highest translational fidelity [13,18]. Recent quantitative studies indicate that the SGC may be in the upper millionth of error-correcting codes with respect to a certain class of randomized genetic codes [18]. Finally, the pattern of amino acids in the SGC suggests that they did not become fixed in the code all at once. Rather, amino acid assignments show evidence of having been fixed separately in disjoint parts of the Code [13], as was suggested by Eigen and Schuster [19] and others. Furthermore, the assignment of amino acids in the hypothetically oldest part of the SGC appears to have been influenced by the putatively G/C-biased mutation and base composition patterns in ancient genomes, again exactly in the codon position with the highest translational fidelity [13]. These observations suggest that the theoretical basis for selection on genetic codes to be mutation-correcting, that is, the LM hypothesis, should be reassessed.

A step in this direction was taken in 1989, when Figureau [20] applied the quasispecies formulation [19] to compare the fitnesses of two populations with different genetic codes in the face of message mutations. In this treatment, an individual is a composite of a message of a single codon and a simple genetic code which translates that codon. Figureau compares the growth rate of two asexual, non-interbreeding, and infinite-sized populations of code-message individuals. One population has a load-minimizing code, that is, a code with codons likely to mutate to one another encoding similar amino acids, while the other has a less load-minimizing code. Figureau illustrates, using an example, that a population with a load-minimizing code has a higher growth rate. This treatment establishes that selection acting at the level of two or more quasispecies with different genetic codes can result in fixation of the more load-minimizing code, provided that these populations can sustain themselves long enough to compete as quasispecies in a common ecological niche.

In the next two Secs. we address a different question arising from Figureau's result: is the advantage of a load-minimizing code realized by an individual in a single generation, or is it necessary to rely on selection of lineages in order to obtain a fitness advantage for a load-minimizing code? From kin selection arguments in population genetics [21] we know that lineage selection plays a significant role in the evolution of a variety of traits. But here we exemplify that it is unnecessary to invoke arguments based on lineages for having effective selection for LM, because a load-minimizing genetic code improves the fitness of a single individual within a single generation. To do this, we extend Figureau's result for one codon, or site, to longer messages of multiple sites, assuming multiplicative fitness interactions across sites. In developing this formalism we introduce the concept of a message as a population of sites, wherein we find that, in mutation-selection balance, codon frequencies in homologous sites across the population are realized as the codon usage within the messages of almost all individuals.

Evidence that the SGC Evolved to Correct Message Mutations

Mutation has a nonuniform structure that differentiates two subsets of nucleotides: $Y = \{U, C\}$, called *pYrimidines*, and $R = \{A, G\}$, called *puRines*. The rates of mutations within these sets, called *transitions*, are generally greater than those of mutations between them, called *transversions*. The manifold causes for the transition biases in mutation are well understood [22], and because they have, in part, an intrinsic chemical basis, it probably existed in some form in the era or eras in which the SGC originated. One type of transition may have even been more frequent during the early history of life [13].

Because transitions were the more frequent type of mutation at the time that the SGC evolved, the hypothesis that selection on codes is load-minimizing might predict that the SGC would evolve to allocate more similar amino acids to codons that differ from each other by transitions. The degree to which a code conserves amino acid chemical properties can be estimated quantitatively using methods such as those elaborated in [13] and in [18]. In order to make such a measurement, a *chemical distance* on amino acids must be defined using either measurable chemical properties of the amino acids or statistical measures taken from the patterns of amino acid alignments in homologous protein sequences. Using a chemical distance on amino acids based on direct chemical measurements, the first codon position in the SGC was shown to be one to two orders of magnitude more conservative in amino acid chemical distance than the second codon position [8–11]. This difference was attributed to the higher rate of translational misreading in the first codon position. However, separate analysis of transitions and transversions in the SGC revealed underlying structure [13,18]. Table 1 shows a representative analysis from data in [13]. Transitions are an order of magnitude more conservative in chemical distance than transversions in the second codon position, while the two types of mutation are approximately equally conservative in the first codon position.

		Position I		Position II	
Distance	Statistic	Transitions	Transversions	Transitions	Transversions
$D = \|\Delta PR\|^2$	P	5.6×10^{-3}	6.2×10^{-3}	3.2×10^{-2}	6.5×10^{-1}
$D = D_B$	P	2×10^{-4}	1×10^{-4}	4.2×10^{-2}	2.8×10^{-1}

Table 1. The proportion P of randomized genetic codes at least as conservative in two types of amino acid distances as the SGC when analyzed by codon position and type of error (transition or transversion). The $|\Delta PR|^2$ distance is the squared difference in a measurable chemical index on amino acids called polar requirement, and D_B is an empirical distance from amino acid evolutionary substitution data (see [13] for details). Overall, the first position is about two orders of magnitude more conservative than the second position. Transitions are an order-of-magnitude more conservative in position II than transversions in that position, while they are about equally conservative in position I.

If mutation affects the three codon positions equally, why should the SGC have evolved to correct for the effects of mutation in one codon position more strongly than in any other? And why should the characteristic signal of mutation, that is transition/transversion bias, be reflected in the least chemically conservative position in the SGC? Ardell [13] conjectured that, despite the homogeneity of mutation in different codon positions, the pattern of amino acids in the SGC evolved to correct for the effects of mutation differently in different codon positions because the rates of translational error vary by codon position. The argument goes as follows: one must postulate that there is some population variation in genetic codes. Codes contribute to the fitness of individuals through translation of their messages. In order for mutations in messages to influence the component of fitness of individuals that depends on translation, those mutations must be translated. Now suppose that a certain class of mutations (transitions) are more frequent than another class. Also suppose that translational error occurs in one codon position (the first) at a much higher rate than in another (the second), and that the distribution of translational errors in each position is uniform. This is an approximation to the general trend as reviewed in [6], and implies that translational misreading is not itself transition-biased. In such a scenario, one must compare the translational consequences of mutations in those parts of messages that correspond to the first codon position ("first-position mutations") to those that occur in the second codon position ("second-position mutations"). Although the distribution or spectrum of first-position and second-position mutations is equally transition-biased, the higher rate of translational error in the first position will mask any specific spectrum of first-position mutations. To an approximation, if the noise level is high enough, any first-position mutation spectrum with a given code will yield the same translation distribution (or spectrum) of amino acids, all other things being equal; this implies that the transition-biased pattern of first position mutations is irrelevant to the pattern of amino acids produced. This in turn implies that selection will not be able to distinguish between genetic codes that preferentially correct the transition-biased mutation spectrum in the first codon position. In the second codon position however, the relatively high fidelity of translation can transmit the transition-biased mutation pattern to the proteins being produced, such that selection may act to minimize the load specifically associated with transitions in this position. Thus, the LM and EM hypotheses are not mutually exclusive. The above argument assumes that mutations and translational errors occur independently by codon position, which for translational misreading is reasonably substantiated [6]. Quantitative evaluation of these conjectures awaits a theory of how genetic codes evolve.

The pattern shown in Table 1 could also be consistent with a transition bias in translational misreading in position two, as reviewed in [18]. However, there is other evidence in the pattern of amino acids in the SGC suggesting that mutation played a role in shaping its evolution. For example, there is good independent evidence that the earliest genomes were biased in their composition towards being rich in the nucleotides G and C, reviewed in [13]. This G/C *bias* in early genomes reflects a mutational or selective bias that does not appear to have a translational counterpart [23]. Furthermore, from empirical models of prebiotic chemistry we know that certain amino acids are likely to have been encoded earlier than other amino acids (also reviewed in [13]). These "early" amino acids are encoded by the GNN and ANY codons, where "N" means "any base" and "Y" means "pyrimidine" (see above). A consequence of the G/C bias in early genomes may have been a high rate of effective mutation between C and G and a low rate of effective mutation between A and U. For details of this argument consult [13]. Suppose that the G/C bias existed in genetic messages at the earliest stages in the origin of the SGC, when the "early" amino acids were being incorporated into the code, and that there was selection on codes to minimize load. By the generalized error-correction hypothesis, which accounts for both mutation and translational error, one expects that in the GNN and ANY codons, the most similar amino acids will be allocated to codons that differ by G or C in the second codon position, while the most dissimilar amino acids will be encoded by codons that differ by A or U in the second codon position. Fig. 1 confirms the existence of this pattern in the assignments of amino acids in the dimension of the second codon position in six different codon contexts, from data given in [13].

This concludes our brief review of evidence that the SGC was selected to correct for message mutation. In the following section we lay the foundations of a formalism to address the question of how load minimization may have affected the evolution of the SGC.

A Model for the Association of Individual Fitness with a Genetic Code

Basic Definitions

We define a genetic code as a deterministic mapping between a set C of N codons and a set A of M amino acids:

$$\mathbf{c} : C \to A. \tag{1}$$

We denote by $c(i) \in A$ the amino acid corresponding to codon i in code $\mathbf{c}$.

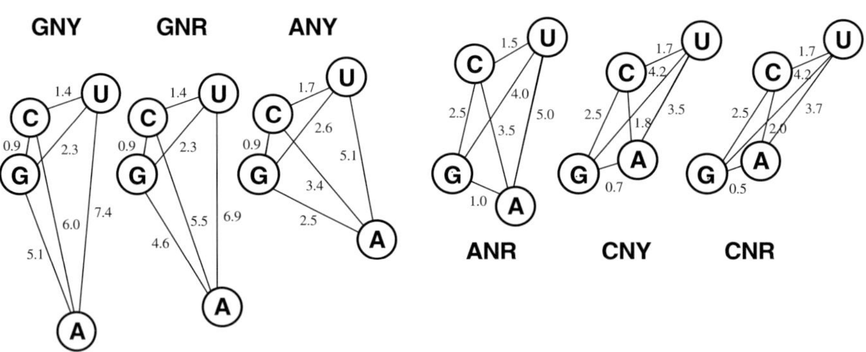

Fig. 1. The similarity of amino acids after errors in the second codon position, in six different codon contexts. Drawn distances approximate the data taken from [13]. The similarity of amino acids is measured by absolute value difference in polar requirement, the square root of the measure used to create Table 1. The GNY, GNR and ANY codons map to the putatively oldest amino acids in the code, and the pattern of amino acid polar requirement after errors in the second codon position is consistent with a primordial G/C bias in mutation. In the ANR, CNY and CNR codons, this pattern is consistent with a transition bias in mutation. For more detail about these data consult the text or [13].

A structure in the amino acid space A reflects the functional manifestation of a "chemical distance" between amino acids through the average effect that replacement of one amino acid by another in proteins has on fitness. We shall use the term *site* to mean both the specific location in a protein occupied by an amino acid and its corresponding locus occupied by a codon in a genetic message. A *site type* is defined by the assumption that the fitness contribution of any amino acid in a site of certain type is uniquely defined in an average sense independently of its genetic or biochemical context. A *site of type* α is a site in a protein or a genetic message where $\alpha \in A$ is the uniquely most fit amino acid. This uniquely most fit amino acid is called the *target* amino acid. In this manuscript we assume that there are only as many site types as there are amino acids, and that each site type has a distinct amino acid target. The fitness of an amino acid $\beta \in A$ in a site $\alpha \in A$ is described by the matrix

$$\mathbf{w} = \{w(\beta|\alpha)\}_{\alpha,\beta=1}^{M}, \tag{2}$$

where $w(\alpha|\alpha) = 1$ and $1 \geq w(\beta|\alpha) > 0$ for all $\beta \neq \alpha \in A$. We assume that the messages of all individuals in a population may be aligned so that homologous sites correspond to the same target amino acids.

A structure in the codon space C is defined here strictly through mutation, but could be generalized to include translational misreading. We assume that mutations occur between codons at any site in any individual according to the matrix

$$\boldsymbol{\mu} = \{\mu(j|i)\}_{i,j=1}^{N}, \tag{3}$$

where $\mu(j|i)$ is the probability per generation of codon i mutating to codon j.

We introduce the notion of a *structure-preserving* genetic code, which is a property of a genetic code as a mapping between codon space and amino space, such that codons which are close in codon space map to amino acids that are close in amino acid space. One measure of structure preservation in a genetic code $\mathbf{c}$ is:

$$\langle w \rangle_{\mathbf{c}} = \sum_{i=1}^{N} \sum_{j \neq i}^{N} \mu(j|i) w(c(j)|c(i)). \tag{4}$$

The greater $\langle w \rangle_{\mathbf{c}}$ is, the more that close codons (large $\mu(j|i)$) are associated by the code $\mathbf{c}$ with close amino acids (large $w(c(j)|c(i))$), and the more structure-preserving the code. This intuitive measure of structure preservation in a genetic code is related to previously-used measures [11,13,18]. In this paper we show that one consequence of a structure-preserving genetic code is that it minimizes or reduces the load due to mutation. Furthermore, the load due to mutation is lower with a structure-preserving code even though the equilibrium frequency of mutant codons is higher. Secondly, we show that a growth rate λ, which corresponds to individual fitness, may be associated with a genetic code, and that this growth rate increases with the degree to which a code minimizes load, as a result of being structure-preserving. In a later section, we demonstrate this result by comparing two different genetic codes to show how the intuitive measure of Eq. (4) correlates with larger growth rates. Ultimately, the growth rate λ may be the best measure of the degree to which a code is load-minimizing.

The Quasispecies Model for Individuals with a Single Site

We first consider a population dynamic model in which an individual is defined by its genetic code and a message consisting of one codon. The fitness of an individual is determined by the site α to which the single codon corresponds. An infinite population with a fixed genetic code $\mathbf{c}$ may be described at time t by the distribution of codons in different individuals. We denote by $\vec{u}^{\alpha}(t) = (u_1^{\alpha}(t), \ldots, u_N^{\alpha}(t))$ the vector of frequencies of different codons in the population at time t, where $u_i^{\alpha}(t)$ is the frequency of individuals with codon i in their message at time t, assuming a single site of type α. The equation describing the dynamic of $\vec{u}^{\alpha}(t)$ is then:

$$\vec{u}^{\alpha}(t+1) = \frac{1}{\bar{w}(t)} \mathbf{Q}^{\alpha} \vec{u}^{\alpha}(t) \tag{5}$$

where $\mathbf{Q}^{\alpha}$ is the iteration matrix reflecting the application of selection followed by mutation:

$$\mathbf{Q}^{\alpha} \equiv \begin{pmatrix} \mu(1|1) & \mu(1|2) & \ldots & \mu(1|N) \\ \mu(2|1) & \mu(2|2) & \ldots & \mu(2|N) \\ \vdots & \vdots & \ddots & \vdots \\ \mu(N|1) & \mu(N|2) & \ldots & \mu(N|N) \end{pmatrix} \begin{pmatrix} w(c(1)|\alpha) & 0 & \ldots & 0 \\ 0 & w(c(2)|\alpha) & \ldots & 0 \\ \vdots & \vdots & \ddots & \vdots \\ 0 & & \ldots & 0 \ w(c(N)|\alpha) \end{pmatrix},$$

$$\tag{6}$$

and $\bar{w}(t)$ is the average fitness of the population at time t:

$$\bar{w}(t) \equiv \sum_{i=1}^{N} w(c(i)|\alpha) u_i^{\alpha}(t). \tag{7}$$

The equilibrium codon frequencies at site α, which may also be called the mutation-selection balance or quasispecies for this site, is given as a solution to the eigensystem:

$$\lambda_{\alpha} \hat{\vec{u}}^{\alpha} = \mathbf{Q}^{\alpha} \hat{\vec{u}}^{\alpha}. \tag{8}$$

In this eigensystem the matrix is non-negative and irreducible. Thus, we know from the Perron–Frobenius theorem [24] that there exists a unique largest eigenvalue λ_{α} to which corresponds the only positive eigenvector $\hat{\vec{u}}^{\alpha}$, the equilibrium codon frequency vector at this site α.

The eigenvalue λ_{α}, which depends on the genetic code $\mathbf{c}$, is the (unnormalized) *fitness* or *growth rate* associated with that code in a site α. Given two subpopulations with different codes sharing common resources (but not exchanging messages), the growth rate associated with the two codes will determine which subpopulation will fix, and at what relative rate. Using a derivation of the same nature, but with a flow reactor formalism, Figureau illustrated that with a message of length 1, the more structure-preserving genetic code will have a higher growth rate [20]. We shall illustrate a similar result in a later section after we extend the derivation of this section to a message of any length.

The Quasispecies Model for Individuals with any Message Length

We now extend the formalism of the previous section to treat messages of any length L. We consider a message of length L as a vector of codons $\vec{m} = (m_1, \ldots, m_L)$, $m_i \in C$. To the messages corresponds a *target protein* which is a vector of site types $\vec{t} = (t_1, \ldots, t_L)$ where $t_i \in A$. This target protein may be considered as the concatenation of all the proteins, which would optimize the contribution of messages to fitness through translation by a genetic code.

In order to treat longer messages, we must define how amino acids interact in proteins to determine individual fitness. As before, we assume that all sites in all proteins encoded by an individual genome may be partitioned into a fairly small number of site types, taken here to be equal to the number of amino acids. Furthermore, we assume that the contributions to fitness of individual amino acids are independent of context, and are given by the matrix $\mathbf{w}$ in Eq. (2). Loci in messages may be collected together into equivalence classes according to the site types in the target protein to which they correspond.

We refer to the set of sites in a single message corresponding to the same site type α, as the *sites of class* α. We denote by l_α the *size of site class* α in the target protein or in messages, with $\sum_{\alpha=1}^{M} l_\alpha = L$. To the messages corresponds a *target protein* which is a vector of site types $\vec{t} = (t_1, \ldots, t_L)$ where $t_i \in A$. This target protein may be considered as the concatenation of all the encoded proteins which, if optimally made, would maximize the translational contribution of messages to individual fitness.

The vector of fractions of codons used in the sites of class α in a single message will be referred to as the *codon usage* in that class. When we consider the fractions of codons used in a homologous site across the population, we will refer to them as the *codon frequencies* corresponding to that site. The fitness contributions of amino acids at different sites combine multiplicatively to determine individual fitness. Assuming multiplicative epistasis across sites, the fitness of an individual with message $\vec{m}$ and code $\mathbf{c}$ is then:

$$w(\vec{m}|\mathbf{c}) = \prod_{l=1}^{L} w(c(m_l)|t_l). \tag{9}$$

In analogy to the previous section, an infinite population of individuals with messages of length L, and a fixed genetic code $\mathbf{c}$, may be described at any time t by the frequency of each of the N^L possible messages. Let $P(\vec{m}, t)$ be the fraction of individuals with message $\vec{m}$ at time t. Then

$$P(\vec{m}, t+1) = \frac{1}{\bar{w}(t)} \sum_{\vec{m}'} Q(\vec{m}|\vec{m}') P(\vec{m}', t), \tag{10}$$

where $Q(\vec{m}|\vec{m}')$ is proportional to the frequency of message $\vec{m}$ in one generation generated by a message $\vec{m}'$ in the previous generation, and is given by

$$Q(\vec{m}|\vec{m}') = \left(\prod_{i=1}^{L} \mu(m_i|m_i') \right) w(\vec{m}') = \prod_{i=1}^{L} \mu(m_i|m_i') w(m_i'|t_i), \tag{11}$$

and $\bar{w}(t)$ is the population average fitness at time t given by

$$\bar{w}(t) \equiv \sum_{\vec{m}'} w(\vec{m}') P(\vec{m}', t). \tag{12}$$

The equilibrium message frequencies $\hat{P}(\vec{m})$ satisfy the equilibrium eigensystem corresponding to Eq. (10):

$$\lambda \hat{P}(\vec{m}) = \sum_{\vec{m}'} Q(\vec{m}|\vec{m}') \hat{P}(\vec{m}'). \tag{13}$$

Applying once again the Perron–Frobenius theorem, there exists a unique largest eigenvalue λ to which corresponds the unique positive eigenvector $\{\hat{P}(\vec{m})\}_{\vec{m}}$ (normalized to one), describing the quasispecies distribution at equilibrium.

In the Appendix we show that at the equilibrium given by Eq. (13):

1. The message distribution is such that the codon distributions in all sites are independent of one another. Moreover, the codon frequency in any site of type α is given by Eq. (8). Thus, the probability $\hat{P}(\vec{m}|\mathbf{c})$ of message $\vec{m}$ given code $\mathbf{c}$ is:

$$\hat{P}(\vec{m}|\mathbf{c}) = \prod_{l=1}^{L} \hat{u}_{m_l}^{t_l}. \tag{14}$$

2. The growth rate corresponding to the mutation-selection balance on messages is given by:

$$\lambda = \prod_{\alpha=1}^{M} (\lambda_\alpha)^{l_\alpha}, \tag{15}$$

where λ is a growth rate in the sense described in the previous section.

An Individual Message is a Population of Sites

In the last section we established that any genetic code is associated with a growth rate which may determine the outcome of competition between genetically isolated populations with different genetic codes. In this section we show that, assuming that messages contain a large number of sites of each type, i.e., that $l_\alpha \gg 1$ for all α, the growth rate of the quasispecies derived in Eq. (15) is equal to the fitness of almost any individual in the population. Moreover, for all site types, the codon usage in a site class in almost all individuals is equal to the codon frequency in any homologous site of that type across the population. Because the quasispecies distribution of Eq. (13) is instantiated through codon usage in almost all individuals, the individual message may considered as a population of sites.

According to the results of the last section, at equilibrium any message may be regarded as the composite of M independent, identically distributed samples of codons, where each of the M samples is the codon usage in sites of class α, which is of size l_α. Let n_i^α be the number of times codon i appears in site class α in a given message. From the results of the previous section we know that n_i^α is a binomial random variable

$$n_i^\alpha \sim B(l_\alpha, \hat{u}_i^\alpha). \tag{16}$$

Hence, for the usage in site class α, $\frac{n_i^\alpha}{l_\alpha}$, we get

$$E\left(\frac{n_i^\alpha}{l_\alpha}\right) = \hat{u}_i^\alpha, \tag{17}$$

and

$$\sigma\left(\frac{n_i^\alpha}{l_\alpha}\right) = \sqrt{\frac{\hat{u}_i^\alpha(1 - \hat{u}_i^\alpha)}{l_\alpha}}. \tag{18}$$

From Eq. (18), we conclude that as l_α increases, the usage of codon i in site class, α $\frac{n_i^\alpha}{l_\alpha}$ will converge quickly to the equilibrium codon frequency $\hat{u}_i^\alpha$ in any of the homologous sites across the population, for nearly all messages. This is a manifestation of the central limit theorem applied to a message as a population of sites.

We illustrate the notion of a message as a population of sites in fig. 2. Note that the approximate equality of codon usage in sites of a given class across any one message and of codon frequencies in a homologous site of that type across a population, derives from the multiplicative epistasis model in equilibrium. With different forms of site interaction, the codon usage in messages will still reflect the frequency in the population, although it will not be strictly identical to the codon frequencies in homologous sites.

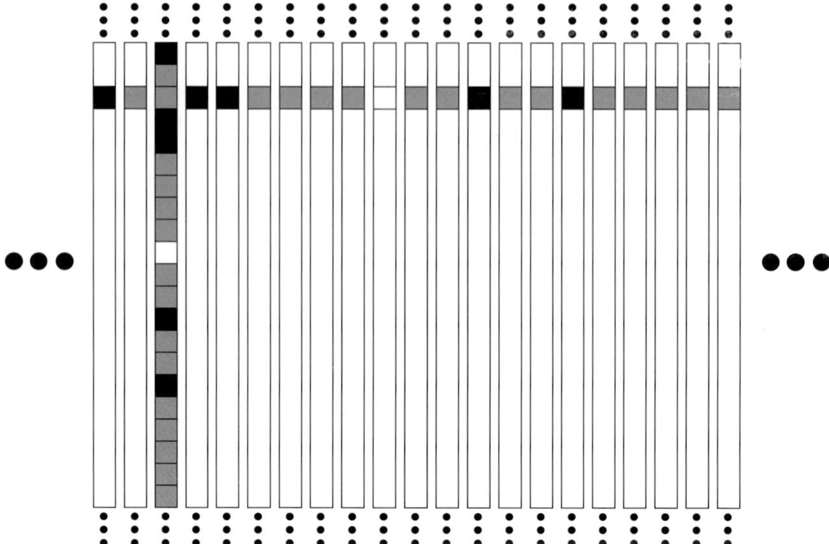

Fig. 2. An heuristic illustration of the fact that codon usage within a site class is approximately equal to the codon frequencies in one homologous site of the same type across the population. Different individuals in a population are aligned vertically with homologous sites in different messages arrayed by row. The sites illustrated in the third individual from the left belong to the same site class.

An Example of Higher Fitness in a Structure-Preserving Genetic Code

In order to illustrate that structure-preserving genetic codes have higher fitness, we present an example where both the codon space C and the amino acid space A have the topology of a ring, and $N = M = 5$. These codon and amino acid spaces are depicted in fig. 3. In codon space each codon may mutate to its two neighbors with probability μ, giving a mutation matrix of the form:

$$\boldsymbol{\mu} = \begin{pmatrix} 1-2\mu & \mu & 0 & 0 & \mu \\ \mu & 1-2\mu & \mu & 0 & 0 \\ 0 & \mu & 1-2\mu & \mu & 0 \\ 0 & 0 & \mu & 1-2\mu & \mu \\ \mu & 0 & 0 & \mu & 1-2\mu \end{pmatrix}. \tag{19}$$

The structure on amino acid space is that of a ring of circumference 1. The abstract chemical distance between amino acids is measured in units of $d = \frac{1}{5}$. The fitness matrix $\mathbf{w} = \{w(\beta|\alpha)\}_{\alpha,\beta=1}^{M}$ is derived from the distance with the following relation:

$$w(\beta|\alpha) = \phi^{d(\beta,\alpha)} \tag{20}$$

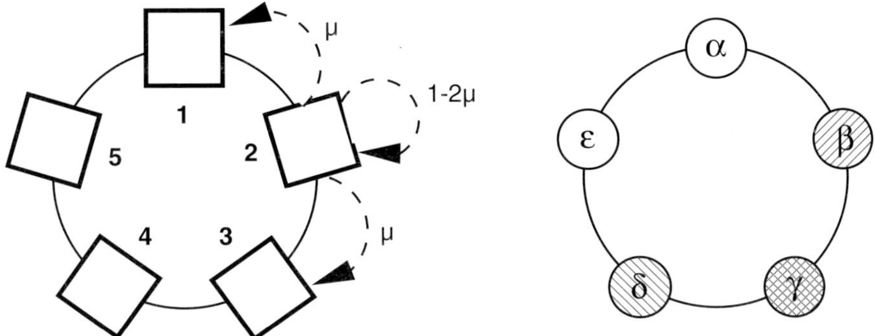

Fig. 3. The codon space and amino acid space used in the example of this section. Both have a ring-like topology. Codons mutate to their neighbors on the ring with mutation rate μ. Amino acids are arrayed evenly in an abstract chemical metric. See text for further details.

where $1 > \phi > 0$ provides a scale for the effect of chemical distance on fitness. The corresponding fitness matrix is then:

$$\mathbf{w} = \begin{pmatrix} 1 & \phi^d & \phi^{2d} & \phi^{2d} & \phi^d \\ \phi^d & 1 & \phi^d & \phi^{2d} & \phi^{2d} \\ \phi^{2d} & \phi^d & 1 & \phi^d & \phi^{2d} \\ \phi^{2d} & \phi^{2d} & \phi^d & 1 & \phi^d \\ \phi^d & \phi^{2d} & \phi^{2d} & \phi^d & 1 \end{pmatrix}. \tag{21}$$

Next, we consider the two genetic codes diagrammed in fig. 4. Code A is maximally structure-preserving while code B is not. Using the measure $\langle w \rangle_{\mathbf{c}}$ we described in Eq. (4):

$$\langle w \rangle_A = 0.964 < 0.980 = \langle w \rangle_B. \tag{22}$$

In order to calculate the growth rates associated with both codes, we begin by calculating the growth rates associated with a site of type α. For that we write the recursion matrices associated with site type α for both codes:

$$\mathbf{Q}_A^\alpha \equiv \begin{pmatrix} 1-2\mu & \mu & 0 & 0 & \mu \\ \mu & 1-2\mu & \mu & 0 & 0 \\ 0 & \mu & 1-2\mu & \mu & 0 \\ 0 & 0 & \mu & 1-2\mu & \mu \\ \mu & 0 & 0 & \mu & 1-2\mu \end{pmatrix} \begin{pmatrix} w(\alpha|\alpha) & 0 & 0 & 0 & 0 \\ 0 & w(\beta|\alpha) & 0 & 0 & 0 \\ 0 & 0 & w(\gamma|\alpha) & 0 & 0 \\ 0 & 0 & 0 & w(\delta|\alpha) & 0 \\ 0 & 0 & 0 & 0 & w(\epsilon|\alpha) \end{pmatrix}, \tag{23}$$

$$\mathbf{Q}_B^\alpha \equiv \begin{pmatrix} 1-2\mu & \mu & 0 & 0 & \mu \\ \mu & 1-2\mu & \mu & 0 & 0 \\ 0 & \mu & 1-2\mu & \mu & 0 \\ 0 & 0 & \mu & 1-2\mu & \mu \\ \mu & 0 & 0 & \mu & 1-2\mu \end{pmatrix} \begin{pmatrix} w(\alpha|\alpha) & 0 & 0 & 0 & 0 \\ 0 & w(\delta|\alpha) & 0 & 0 & 0 \\ 0 & 0 & w(\beta|\alpha) & 0 & 0 \\ 0 & 0 & 0 & w(\epsilon|\alpha) & 0 \\ 0 & 0 & 0 & 0 & w(\gamma|\alpha) \end{pmatrix}. \tag{24}$$

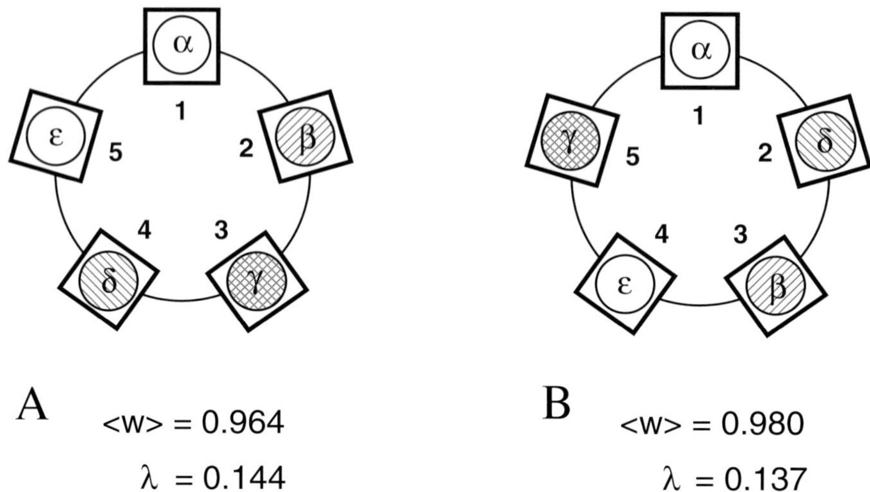

Fig. 4. A comparison between two genetic codes "A" and "B," with codons and amino acids as in fig. 3. Code "A" is more structure-preserving, and an individual with this code has higher individual fitness. The fitnesses were calculated for $l_\alpha = l_\beta = l_\gamma = l_\delta = l_\epsilon = 20$.

Solving the corresponding eigensystems 8 with $\mu = 0.1$ and $\phi = 0.8$, we get the following growth rates and usage vectors:

$$\lambda_\alpha^A = 0.980787 \quad \hat{\vec{u}}_A^\alpha = (0.9058, 0.0460, 0.0011, 0.0011, 0.0460), \quad (25)$$

$$\lambda_\alpha^B = 0.980337 \quad \hat{\vec{u}}_B^\alpha = (0.9446, 0.0268, 0.0009, 0.0009, 0.0268). \quad (26)$$

The two codes were chosen so that the growth rates associated with all the site types are identical, while their corresponding usage vectors are identical up to a permutation in the site indices. Therefore, the growth rates associated with any message of length L for both codes is:

$$\lambda^A = (\lambda_\alpha^A)^L \quad \lambda^B = (\lambda_\alpha^B)^L, \quad (27)$$

where $\lambda^A > \lambda^B$ for any L. In fig. 4 we calculated these growth rates for $l_\alpha = l_\beta = l_\gamma = l_\delta = l_\epsilon = 20$. It is important to note in this example that the usage of mutant codons is *greater* with the more structure-preserving code, even though the more structure-preserving code has higher fitness. Mutants which come from the wild-type codons (encoding the target amino acids of site classes) at equilibrium encode amino acids which are more similar to their targets in a structure-preserving code. Consequently, in a more structure-preserving code, natural selection is not as strong in focusing the usage distribution around the wild-type codon; this results in more diffuse codon usages. In balance, however, the fitness of the more structure-preserving code is higher, because the combined load of the mutant codons is smaller. We believe this result will hold in general.

We see here, that this more diffuse pattern of codon usage associated with the more structure-preserving genetic code is results in higher individual fitness.

l_α	$\sigma(\frac{n_1^\alpha}{l_\alpha})$	$\sigma(\frac{n_2^\alpha}{l_\alpha}) = \sigma(\frac{n_5^\alpha}{l_\alpha})$	$\sigma(\frac{n_3^\alpha}{l_\alpha}) = \sigma(\frac{n_4^\alpha}{l_\alpha})$
10	0.0924	0.0662	0.0105
100	0.0292	0.0209	0.00331
1000	0.00924	0.00662	0.00105

Table 2. Standard error of codon usage across individuals with genetic code "A" from fig. 4 in different numbers of sites l_α of a site type α in messages.

The notion of the population of sites for codes is illustrated in Table 2. The standard deviations of the codon usage in sites of any class α of different sizes l_α are shown for an individual message with the same genetic code A as in fig. 4, using Eq. (18) with the usages from Eq. (25). The mean codon usage is equal to that calculated in Eq. (8). As the the class size l_α increases, almost all messages accurately reflect the codon frequencies in the population at equilibrium.

Discussion

We have shown that a genetic code which is "structure-preserving" is also load-minimizing, and that this load-minimizing property of a genetic code may lead to higher individual fitness. First we defined the "structure-preserving" property of a genetic code as a static property of its mapping between codons and amino acids. A structure-preserving code maps codons which are mutational neighbors to amino acids which are physicochemically related. Then, following Figureau, we showed that the equilibrium message distribution for a message of a single site depends on the genetic code, and restated his result that the quasispecies with a structure-preserving code has a higher quasispecies growth rate. Then we extended the model to give the fitness associated with genetic codes of individuals with messages of any length. We showed that both the codon frequencies at equilibrium corresponding to a given code and the growth rate associated with it will be reflected in a single individual in a single generation — thereby establishing that selection for LM may act effectively on single individuals in a single generation.

The form of codon usage as well as the effect of mutations on fitness depend on the organization of the genetic code. The codon usage within a site class will be dominated by the codon or codons that encode the target amino acid of that site class. This will also increase the usage in that site class of the codons which are mutational neighbors of those dominant, or wild-type, codons. If the code is structure-preserving, the neighboring mutant codons will encode amino acids more similar to the target, so that selection to concentrate usage at the wild-type codon or codons will be weaker. For example, we saw in the last section that the more structure-preserving genetic code resulted in a more diffuse equilibrium codon distribution within every site class. This means that the more structure-preserving genetic code actually populates its sites with more erroneous codons than a less structure-preserving genetic code. This tendency operates to lower the fitness of a more structure preserving genetic code. On the other hand, the amino acid meaning of these mutant codons is more closely related to that of the target amino acid in a more structure-preserving genetic code. This tendency operates to increase the fitness of a more structure-preserving code. In balance, the combination of these effects is reflected in the growth rate associated with a code, which is the most appropriate measure of its load-minimizing quality. In the example of the last section, the more structure-preserving genetic code had a lower load and consequently higher fitness. We conjecture that this will be true in almost all reasonable examples.

In developing the notion of a population of sites, we assumed that different sites in the set of proteins of an individual are divisible into a relatively small number of equivalence classes of site types, called site classes, encompassing all sites in the target protein of an individual. All sites within a site class contribute identically to fitness depending on the amino acid that occupies that site, and this contribution to fitness is independent of the identity of amino acid occupying any other site in the protein. Also, single codons are assumed to mutate independently of one another. Several studies of codon and amino acid usage suggest that these assumptions are generally invalid [26–29]. On the other hand, it is clear that analysis of amino acid substitution data at sites corresponding to single amino acids has proven sufficient for a variety of predictive and descriptive applications of protein evolution, such as structure-prediction, alignment of homologous proteins, phylogeny estimation, multivariate analysis of amino acid physicochemical properties, and so on. These practical observations support our premise that at least in an average sense, sites consisting of single amino acids are meaningful units for selection, and that sites in a given individual may be grouped into a fairly small number of equivalence classes covering most sites. Our assumption of a multiplicative fitness scheme across sites was convenient for analysis. The exact equality of codon usage within site classes in an individual with the codon frequencies at homologous sites of that site type in a population follows from this assumption. However, even when the interaction among sites takes a different form, and the strict equality doesn't hold, when many sites affect fitness in a similar way (depending on the type of amino acid which occupies them) then a message may be treated as a population of sites. And when an individual message may be treated as a population of sites, a genetic code that is load-minimizing will increase individual fitness even when our specific assumptions regarding fitness do not hold.

How do the formulations in this paper bear on the question of the evolution of an error-correcting genetic code? Let us begin with what we view to be the shortcomings of explaining the informatic structure in the SGC by only specific stereochemical associations of amino acids with their codons or anticodons. As Crick argued in 1968, and has since been further developed [17], the error-correcting structure of the SGC might be explained by the fact that related codons could have specific stereochemical affinity with physicochemically related amino acids. This premise, which may be called "continuity of stereochemical association," although not unreasonable, has not yet been tested (see discussion in [17]). There are as yet no aptamers reported for a majority of the canonical amino acids. Of those reported, none show evidence that similar codon or anticodon motifs are found for related amino acids. Even if such a relationship were to be demonstrated, it would have to be extremely specific to explain the SGC without incorporating the evolutionary advantage of structure-preservation demonstrated here. The reason for this is that as noted in the introduction, the error-correcting structure in the SGC is multidimensional. One dimension corresponds to a relation between mutational distances and the distance among amino acids and may in principle be accounted for by a continuity of stereochemical associations. Substructures in the mutational error-correcting structure are associated with transition/transversion bias and historically different patterns of mutation, and these seem to be more difficult to explain by the continuity of stereochemical association. The overall association of conservation in different codon positions with rates of translational misreading seems completely orthogonal to these patterns. To us it seems extremely unlikely that continuity of stereochemical association can account for these independent dimensions of error-correcting structure in the SGC.

Selection upon variation in genetic codes, where individual fitness is associated with the error-correcting quality of a code, can provide an explanatory mechanism for the informatic structure of the SGC. However, the plausibility of such a mechanism suffers in the face of Crick's freezing argument when the quantitative quality of the code is considered. A simplistic explanation of the error-correcting structure in the SGC by the action of natural selection for a code which is in the upper 10^{-5}th – 10^{-6}th of genetic codes, would require a population with 10^5-10^6 competing genetic codes. Given Crick's freezing arguments, this degree of variability in a single population seems highly unlikely. However, there is good evidence that the SGC may have evolved in stages [13], in which case the required variability at any given time may be reduced. Although the result on the selectability of load-minimizing codes presented here seems encouraging, it appears that it also is insufficient to explain the high degree of error-correcting structure in the SGC.

The two hypotheses discussed above for the origin of the SGC, continuity of stereochemical association and evolution by natural selection, are not mutually exclusive. The model developed in this paper may eventually be extended to synthesize a single explanation for the origin of the Standard Genetic Code.

Acknowledgements. Research supported by NIH grants to M.W. Feldman — contribution number 12 from the Center for Computational Genetics and Biological Modeling. We thank Lauren Ancel, Laura Landweber, Aaron Hirsh and Marcus Feldman for critical readings of the manuscript.

References

1. Nirenberg M., Jones O., Leder P., Clark B., Sly W., and Pestka S. (1963) Cold Spring Harbor Symp. Quant. Biol. 28:549-558
2. Sonneborn T. (1965) in Bryson V. Vogel H (eds) Evolving genes and proteins. Academic Press, NY pp 377-397
3. Zuckerkandl E. and L. Pauling (1965) Evolving Genes and Proteins, Academic Press, NY pp. 97
4. Goldberg A.L. and Wittes R. Science 153:420-424
5. Epstein C. (1966) Nature 210(5031):25-28
6. Parker J. (1989) Microbiol. Rev. 53(3):273-298
7. Woese C. (1965) Proc. Natl. Acad. Sci. USA, 54, 1546-1552
8. Woese C., Dugre D., Dugre S., Kondo M., and Saxinger W. (1966) Cold Spring Harbor Symp. Quant. Biol. 31:723-736
9. Alff-Steinberger C. (1969) Proc. Natl. Acad. Sci. USA 64:584-591
10. Swanson R. (1984) Bull. Math. Biol. 46(2):187-203
11. Haig D. and Hurst LD. (1991) J. Mol. Evol. 33:412-417
12. Goldman N. (1993) J. Mol. Evol. 37:662-664
13. Ardell, D.H. (1998) J. Mol. Evol. 47:1-13
14. Crick F.H.C. (1968) J. Mol. Biol. 38:367-379
15. Juncgk J. (1978) J. Mol. Evol. 11:211-224
16. Woese C. (1965) Proc. Natl. Acad. Sci. USA, 54, 1546-1552
17. Knight R.D. Genetic code evolution in the RNA world and beyond. In *Evolution as Computation*, this volume.
18. Freeland S. and Hurst L.D. (1998) J. Mol. Evol., 47(3):238-248
19. Eigen M. and Schuster P. (1979) The Hypercycle: a Principle of Natural Self-Organization. Springer, Berlin.
20. Figureau A. (1989) Orig. Life Evol. Biosph. 19:57-67
21. Hamilton W.D. (1964) J. Theor. Biol. 7:1-52
22. Lewin B. (1997) Genes VI. Oxford University Press, Oxford, England.
23. Grosjean H., de Henau S., and Crothers D. (1978) Proc. Natl. Acad. Sci. USA. 75(2):610-614
24. Wilkinson J.H., The Algebraic Eigenvalue Problem, Oxford UP,1965.
25. Hofbauer J. and Sigmund K. (1988) The Theory of Evolution and Dynamical Systems. Cambridge, England.
26. Bulmer M. (1988) J. Evol. Biol. 1:15-26
27. Maynard-Smith J. and N.H. Smith (1996) Genetics 142:1033-1036
28. Berg O.G. and P.J.N. Silva (1997) Nucl. Acids Res. 25(7):1397-1404
29. Pollock D.D. and Taylor W.R. (1997) Prot. Eng. 10(6):647-657

Appendix: Solution of the Eigensystem Equation for Messages of Any Length

We know that eigensystem of equation 13 has a unique solution with a positive eigenvector. Therefore, it is enough to find a positive defined eigenvector and its corresponding eigenvalue that solve the eigensystem, and they will necessarily be the unique suitable solution for the mutation-selection balance. We shall assume that:

$$\hat{P}(\vec{m}) = \prod_{i=1}^{L} \hat{u}_{m_i}^{t_i} \tag{28}$$

is the sought solution, which is, by definition, positively defined.

In order to check this is indeed a solution we substitute Eq. (28) into the eigensystem 13:

$$\lambda \prod_{i=1}^{L} \hat{u}_{m_i}^{t_i} = \sum_{\vec{m}'} Q(\vec{m}|\vec{m}') \prod_{j=1}^{L} \hat{u}_{m_i'}^{t_i} \tag{29}$$

$$= \sum_{\vec{m}'} (\prod_{i=1}^{L} \mu(m_i|m_i') w(c(m_i')|t_i)) \prod_{j=1}^{L} \hat{u}_{m_i'}^{t_i} \tag{30}$$

$$= \sum_{\vec{m}'} \prod_{i=1}^{L} \mu(m_i|m_i') w(c(m_i')|t_i) \hat{u}_{m_i'}^{t_i} \tag{31}$$

$$= \prod_{i=1}^{L} (\sum_{j=1}^{N} \mu(m_i|j) w(c(j)|t_i) \hat{u}_{j}^{t_i}) \tag{32}$$

$$= \prod_{i=1}^{L} (\sum_{j=1}^{N} \mathbf{Q}^{t_i}(m_i|j) \hat{u}_{j}^{t_i}) \tag{33}$$

where the right hand side in Eq. (33) is exactly a product of right hand side terms in Eq. (8), hence:

$$\lambda \prod_{i=1}^{L} \hat{u}_{m_i}^{t_i} = \prod_{i=1}^{L} (\lambda_{t_i} \hat{u}_{m_i}^{t_i}). \tag{34}$$

We conclude that $\lambda = \prod_{i=1}^{L} \lambda_{t_i} = \prod_{\alpha=1}^{M} (\lambda_\alpha)^{l_\alpha}$ and 28 is indeed the sought solution.

Genetic Code Evolution in the RNA World and Beyond

Robin D. Knight

Abstract. Although the translation apparatus presumably arose in an RNA world, subsequent modifications obscure its origins. The genetic code, fixed in the Last Universal Ancestor may contain clues about the types of chemical interaction that led to early correspondences between RNA and protein. The extent to which contemporary translation reflects these primordial influences depends on the processes that have shaped the genetic code since its inception: stereochemical interaction between amino acids and RNA, historical constraints ensuring continuity between successive codes, and optimization to minimize the effects of errors caused by translation and mutation. This chapter explains how these processes, typically presented as mutually antagonistic, may actually be viewed as complementary on different timescales, and I suggest how the "first" codons could have been established in the context of an RNA world.

1   Introduction

In modern organisms, the genetic code links inheritance and development. By establishing a mapping function between RNA and protein molecules, the genetic code allows stable inheritance of the phenotypic variation on which selection acts. Before the genetic code evolved, primitive organisms must have used either of two simpler strategies: (a) inheritance of metabolic states rather than of physical carriers of information (limited replicators) or (b) restriction of phenotypes to nucleic acids and their reactions (unlimited replicators [1]). The former type of inheritance limits the transmission of variability, while the latter limits the range.

The stage at which the genetic code developed should affect its properties in predictable ways. Theories of the origin of the genetic code can be broadly divided into "early" theories, which imply that the genetic code developed before macromolecules were common, and "late" theories, which imply that it developed after macromolecules were widely available. Early development of the genetic code implies greater reliance on simple stereochemical interactions, since complex catalytic biopolymers would have been unavailable. For instance, the genetic code could have been established at the start of life, when most "metabolites" were actually synthesized by abiotic processes. If so, there must have been some stereochemical mechanism allowing specific pairing between the limited repertoire of amino acids and nucleic acids available at the time; furthermore, the original genetic code would have been restricted to those amino acids that can be synthesized under plausible prebiotic conditions [2,3]. Alternatively, the genetic code may have developed in the context of a chemoton [4], an autocatalytic chemical system composed primarily of small molecules. If so, the initial choice of amino acids would not be restricted to those with prebiotically plausible syntheses, but the code could still only be established by simple stereochemical relationships between amino acids and oligonucleotides. Late theories, in contrast, are necessarily statistical because macromolecules acting as adaptors could enhance any arbitrary pairing between codons and amino acids. Thus, the genetic code might be a "frozen accident" [5], persisting because any change would be deleterious. This serves as a null model, as it predicts (a) that there should be no particular similarity between amino acids that have related codons, (b) that amino acids should have no particular affinity for their codons or for other short RNA motifs, and (c) that the present genetic code should not be particularly adaptive relative to other possible codes. Other late theories question these assumptions (reviewed in [6]).

Stereochemical theories of the code's origin [7,8] postulate some chemical interaction between specific amino acids and short RNA motifs. Since translation presupposes the presence of RNA as a messenger, it seems plausible that the genetic code developed directly in the RNA world [9], a time at which metabolism relied exclusively on RNA as catalysts. If so, (a) RNA catalysts (ribozymes) must be capable of catalyzing amino acid biosynthesis and peptide condensation, and (b) RNA must be able both to discriminate between and bind to amino acids [10,11]. If RNA molecules were the original adaptors specifying particular amino acids, any stereochemical relationship that tended to promote specific amino acid-oligonucleotide interactions would influence the developing genetic code. Thus amino acids should tend to bind to their corresponding codons, or to some simple transform thereof.

Coevolution theories [12–14] suggest that the code has been successively refined as more amino acids become available, perhaps through an expanded repertoire of reactions catalyzed either by ribozymes or early peptides. Thus the genetic code may have developed by progressive refinements of the translation apparatus [12], with each generation of proteins providing greater accuracy by discriminating more finely between related amino acids, some of which may only have become available after early protein catalysts became active [13]. Alternatively, metabolic "descendants" of the original amino acid set may have co-opted codons from their metabolic precursors [13,15–18]. Such coevolution models imply that related amino acids should have similar codons, although the block of codons assigned to each group could be arbitrary (and hence uninfluenced by simple stereochemical constraints).

Adaptationist theories [19,20] question the assumption that the actual code (typically the "universal" code found in the Last Universal Ancestor) is no better than randomly generated codes. Recent evidence suggests that the actual code is at or near a global optimum for reducing the errors caused by accidental substitution of one amino acid for another due to translation error or point mutation [21–24]. These results indicate either that the code underwent a period of strong selection for error minimization or that other processes (such as stereochemical constraints) had the effect of ensuring that similar amino acids were assigned to similar codons.

Although these models have typically been presented as competitors, they may actually represent processes that have acted in synergy. In this paper, I outline the evidence for the various models, focusing primarily on the type of stereochemical interaction that could have led to the initial establishment of a primordial code (see accompanying papers by Freeland, Ardell and Sella for more detail on adaptive and coevolutionary models). I conclude with a timeline showing the periods in code evolution at which each mechanism probably exerted greatest influence on the present "universal" genetic code.

2   Chemical Evolution

Of the 20 amino acids in the contemporary genetic code, only twelve have been extracted from plausible prebiotic syntheses [2,3,25–27] or from extraterrestrial sources such as the Murchison meteorite [28,29] (Fig. 1). At least some amino acids must have been added to the code after the evolution of sophisticated macromolecular catalysts: since the genetic code may have evolved entirely in an RNA world in which such catalysts were available, prebiotic availability did not necessarily influence the set of amino acids that were eventually incorporated into proteins.

Prebiotically available amino acids do seem to be concentrated in certain parts of the genetic code table (Fig. 2). GNN codons all specify amino acids found both in spark tube experiments and in the Murchison meteorite: because these amino acids also tend to be precursors to other, more complex amino acids, it is possible that the initial code was a GNN code [16]; G,C based codes using primarily Val, Ala, Asp/Glu, and Gly have been suggested by several authors [5,14,30,31]. The association of prebiotic amino acids with second position pyrimidines is less compelling, but still striking. The fact that the 40% of amino acids not found in prebiotic syntheses occupy only about 30% of the code table may or may not be significant.

One difficulty with the suggestion that the code developed in a simple chemical environment is the fact that many amino acids that are produced at high frequency under such situations are not found in the code at all. α-Amino-n-butyric acid and norvaline are synthesized in spark-tube experiments at higher concentration than any amino acid other than Gly and Ala, and a total of 15 α-amino and imino acids not in the code occur at higher frequency than does Lys, which is in the code [25,26]. In particular, the absence of norvaline, norleucine, and α-Amino-n-butyric acid suggests that prebiotic abundance was not the sole criterion for incorporation of the first amino acids [27].

Experiments seeking association between free amino acids and nucleotide bases under simple chemical conditions have not strongly supported the idea

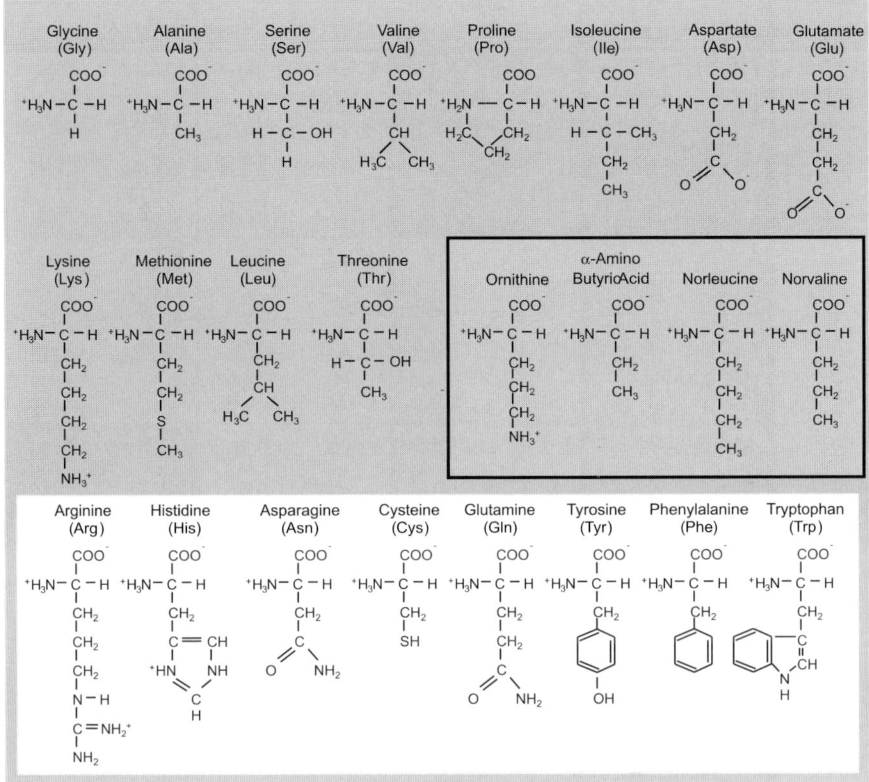

Fig. 1. Amino acid structures. Gray background indicates those found either in the products of prebiotic simulation ("spark experiments": see [27] and references therein) or in the Murchison meteorite [28,29]. Boxed amino acids are a subset of those found in the Murchison satellite which are not found within the genetic code. *In vitro* selection can determine which RNA triplets associate most strongly with each amino acid, and may reveal why certain prebiotic amino acids are absent from the code.

that such interactions led to the present genetic code. For instance, chromatographic separation of amino acids and nucleotide monophosphates showed that alanine (GCN) comigrates with cytosine monophosphate, and glycine (GGN) comigrates with guanosine monophosphate on silica, but such relationships did not hold for other amino acid-nucleotide pairs or on other surfaces [32]. Similarly, partitioning of amino acids and nucleotides between aqueous and organic phases (as in a primordial oil slick) might have associated AAA codons with Lys and UUU codons with Phe [33] if any of these molecules had existed prebiotically. One chromatographic property of amino acids, the "polar requirement" (measured as the ratio of the log relative mobility to the log mole fraction water in a water-pyridine mixture)

	U	C	A	G
U	UUU Phe UUC Phe UUA Leu UUG Leu	UCU Ser UCC Ser UCA Ser UCG Ser	UAU Tyr UAC Tyr UAA TER UAG TER	UGU Cys UGC Cys UGA TER UGG Trp
C	CUU Leu CUC Leu CUA Leu CUG Leu	CCU Pro CCC Pro CCA Pro CCG Pro	CAU His CAC His CAA Gln CAG Gln	CGU Arg CGC Arg CGA Arg CGG Arg
A	AUU Ile AUC Ile AUA Ile AUG Met	ACU Thr ACC Thr ACA Thr ACG Thr	AAU Asn AAC Asn AAA Lys AAG Lys	AGU Ser AGC Ser AGA Arg AGG Arg
G	GUU Val GUC Val GUA Val GUG Val	GCU Ala GCC Ala GCA Ala GCG Ala	GAU Asp GAC Asp GAA Glu GAG Glu	GGU Gly GGC Gly GGA Gly GGG Gly

Fig. 2. The "Universal" Genetic Code. Shading indicates abundance in prebiotic syntheses and in the Murchison meteorite [27]; shading is proportional to log (abundance) in spark tubes. White amino acids were not found in either source. Dark borders indicate amino acids not found on the Murchison meteorite but found in spark-tube experiments. Amino acids with G at the first position tend to be common in prebiotic syntheses.

varies regularly with second-position base such that amino acids with U in the second position of their codon are hydrophobic while those with A are hydrophilic; those with C are intermediate, and those with G are mixed; and codons that share a doublet have almost identical polar requirements even if not otherwise related (e.g., His and Gln, Cys and Trp) [34,35]. However, this type of association can be explained by several other models, notably selection and ambiguity reduction. Relative hydrophobicity of the homocodonic amino acids (Phe UUU, Pro CCC, Lys AAA, Gly GGG) and the four nucleotides (or the corresponding dinucleotide monophosphates) in an ammonium acetate/ammonium sulfate system showed an anticodonic association [36], as did a multivariate analysis of the properties of dinucleoside monophosphates and amino acids focusing on hydrophobicity [37]. Thus, chromatography data tend to support the idea that amino acids and their *anticodons* have similar hydrophobicities.

While chromatography looks for common partitioning of particular amino acids and nucleotides, it is also possible to test for direct interactions between these molecules. Mononucleotides interact nonspecifically and charge-dependently with polyamino acid chains, as measured by the change in turbidity of the solution [38]. Affinity chromatography testing retardation of the four nucleotide monophosphates by each of nine amino acids (Gly, Lys, Pro, Met, Arg, His, Phe, Trp, Tyr) immobilized by their carboxyl groups showed no association between binding strength and codon or anticodon assignments [39]. NMR studies of interactions between free amino acids and poly(A) are also "not easily reconcilable with the genetic code" [40]; nor are the selective interactions between amino acids and mono-, di-, and trinucleotides [41]. Although imidazole-activated amino acids esterize with $2'$-OH groups of RNA homopolymers with high specificity, both amino acids tested (Phe and Gly) much preferred poly(U) over any other polynucleotide [42]. The dissociation constants (K_D) of AMP complexes with the methyl esters of amino acids show strong selectivity, ranging nearly an order of magnitude from Trp (120 mM) to Ser (850 mM), and show a strong negative correlation between the association constant $(1/K_D)$ and amino acid hydrophobicity [43]. However, neither Trp (UGG) nor Ser (CUN, AGY) have unusually many or few A residues in their codons and anticodons, while the amino acids that do (Lys AAR, Phe UUY) have intermediate dissociation constants (320 and 196 mM, respectively).

Amino acids can also affect the reactivity of specific nucleotides, and vice versa; however, the relationship of these interactions to the code is as eqivocal as the data from the other chemical approaches. Condensation of dipeptides of the form Gly–X in the presence of AMP, CMP, poly(A), and poly(U) was mainly enhanced by the anticodonic nucleotides in those cases where a pattern was apparent [44]. Although different amino acids differ in their ability to stabilize poly(A)–poly(U) and poly(I)–poly(C) double helices [45], the order is the same for nearly every amino acid and so is unlikely to have contributed to the establishment of the genetic code. Finally, D-ribose adenosine biases esters with L-Phe but not D-Phe towards the $3'$-OH (the pattern is reversed with L-ribose adenosine), indicating that single nucleotides can stereoselectively aminoacylate themselves [46].

Two comprehensive reviews of these and other data [47,48] concluded that the weight of evidence favored specific association between free amino acids and their anticodons, rather than their codons or other related motifs. This inference relied on two major assumptions: (a) that the genetic code arose directly from a prebiotic environment containing the exact twenty amino acids found in present proteins, and (b) that similar hydrophobicity measurements imply that amino acids and anticodons would associate together in such a way that the relationship would be preserved in the present genetic code. The first assumption is unlikely because, as discussed earlier in this section, the set of amino acids in proteins only partially overlaps the set of amino acids probably found on the early earth. Although no data exist for the interaction of prebiotic but nonproteinaceous amino acids with nucleotides (or nucleotide derivatives), it is unlikely that measures such as hydrophobicity will separate norvaline and norleucine from Val, Leu, and Ile. The second assumption is also doubtful, especially since the hydrophobicities of the amino acids span a much greater range than those of the bases. None of the evidence to date compels the conclusion that the genetic code arose directly from interactions between its constituent monomers in a simple chemical milieu.

3   Stereochemical Models of Codon Assignment

If the genetic code arose in the context of an already complex metabolism, the range of possible evolutionary mechanisms is greater than that available in a simple metabolism. Macromolecular adaptors that bring together unrelated smaller molecules (for instance, the aminoacyl-tRNA synthetases that link amino acids with tRNAs in modern translation) could potentially impose any arbitrary correspondence between codons and amino acids. Later evolution of these adaptors, such as alterations in specificity, could easily obscure the relationships that led to the initial pairings. All stereochemical models of code evolution assume some form of the "codon correspondence hypothesis": for each amino acid, there must be a unique set of base triplets for which it has greatest affinity and which are found at RNA sites that bind that amino acid.

The codon-correspondence hypothesis is compatible with establishment of the genetic code either before or after the RNA world. A direct association between trinucleotides and their cognate amino acids would imply that this part of the genetic code could have been established prior to the evolution of complex RNA catalysts, since trinucleotides would likely be randomly synthesized before the directed synthesis of longer oligonucleotides. This might be the case if, for instance, a primitive hypercyclic metabolism relied on trinucleotides to sequester undesirable amino acids, or to act as an amino acid transport mechanism that was able to affect the production of short peptides or the composition of longer ones. An association between trinucleotides in the context of a folded RNA tertiary structure and their cognate amino acids would imply an origin of the genetic code in the RNA world, since this would be the earliest point in evolution at which long RNA molecules became available. This might be the case if amino acids were originally used as coenzymes for ribozymes [49], or to stabilize RNA double helices [45] or label tRNA-like genomic tags [50,51]. In any case, a strong correspondence between amino acid binding sites and codons would imply that this mechanism largely determined the form of the genetic code. More plausibly, the pattern might hold for only certain amino acids. The interpretation of such a result would depend on the particular amino acids showing code-dependent interactions with RNA. If prebiotically available amino acids such as Leu, Ile, Val, Ala, Asp/Glu, and Gly tend to associate with a superset of their present codons (or anticodons), whereas other amino acids show no such association, the most likely scenario would be that direct chemical interactions led to a primordial code based on these amino acids, which was elaborated as further amino acids were made available by metabolism. Similarly, if complex and/or unstable amino acids such as Trp, His, Arg, Gln, and Asn show codonic associations whereas prebiotic amino acids do not, then it would be more probable that these later amino acids displaced earlier amino acids from those codons for which they had greatest affinity. In contrast, a mixture of prebiotic and non-prebiotic amino acids associated with their present codons would imply that these amino acids were added to the code first (or last), and that the code as a whole arose after all amino acids were available.

If there is no association between any trinucleotide and its cognate amino acid, there are several alternative explanations. First, such associations might not exist. This would imply that the genetic code evolved from an already complex metabolism, in which specific macromolecular adaptors bound amino acids through complex and arbitrary interactions. The diversity of RNA molecules that bind arginine (Section 4) shows that efforts to recreate a single, primordial adaptor for each amino acid would be futile. Second, specific associations might exist but be entirely different from the actual codon assignments. This would imply either (a) that these interactions played no role in establishing the genetic code, perhaps because more complex and specific binders were available at its inception; (b) that these interactions established the original genetic code by direct templating [11], but the transfer to the present adaptor system caused the original codon assignments to be lost or transformed by some complex function; or (c) that these interactions established the original genetic code, but a long process of codon swapping erased the original assignments as a result of optimization or drift.

It should be noted that in modern organisms, there is no direct interaction between codon (or anticodon) and amino acid. The specific coding between codon and amino acid takes place in a two-step process. In the first step, a specific enzyme (an aminoacyl-tRNA synthetase) simultaneously recognizes the correct amino acid and the correct tRNA, pairing them. This synthetase recognizes the anticodons of some, but not all, tRNAs; thus, the correct charging of certain amino acids does not depend at all on the presence of the appropriate anticodon. In the second step, the tRNA anticodon pairs with the mRNA codon in the ribosome. This takes place largely by Watson-Crick pairing, although many bases in the tRNA are modified. At no stage does the amino acid ever explicitly pair with the codon. However, this indirect system may be quite different from the primordial condition; primitive translation may have relied on direct nucleotide-amino acid pairing.

Molecular model-building has provided an embarrassment of possible schemes for such pairing. The most important criterion is that of continuity: the transition from a primordial coding scheme to the present coding scheme must not destroy the utility of the information already stored at the time of the transition. Thus, the primordial codons with which pairing occurred must either be the actual codons, or a related variation [34]. Consequently, interactions have been proposed between amino acids and nearly every such simple transform: codons [8], anticodons [7,52,53], codons read 3' to 5' instead of 5' to 3' [54,55], a complex of four nucleotides or "C4N" formed by the three 5' nucleotides of tRNA (assumed to be the anticodon) with the fourth nucleotide from the 3' end [56], or a double-stranded complex of the codon and anticodon [57,58]. Until recently, none of these models had any convincing empirical support. However, sequence and structural analysis of aptamers to amino acids may reveal a role for codon-amino acid interactions in establishing the present code.

The apparent universality of the genetic code once provided the strongest support for stereochemical theories, because it suggested that the actual code was the only possible code. However, the known variations in the present code do not disprove the stereochemical theories. All deviations from the canonical code are recently derived compared to the last universal ancestor: the deepest branching is probably in the diplomonads [59], and most are much more recent. The processes leading to recent variation in the code are probably quite different from those that established the initial code, and rely more on the details of tRNA gene mutation than on simple stereochemical interactions between amino acids and trinucleotides. However, a different type of stereochemical constraint does appear to explain the degeneracy of even contemporary codes: the GC content of codon doublets affects whether or not the "family box" formed by variation at the third position is split between multiple amino acids. Codons in which the doublet (the first two bases) is composed solely of G and C form four-codon boxes, while those in which the doublet is composed solely of A and U form split boxes (either two two-codon boxes or one three-codon box and one one-codon box). This pattern might arise because all-GC doublets bind sufficiently strongly to their cognate anticodons that the third base is irrelevant, while all-AU doublets bind weakly enough to allow discrimination. Mixed doublets form a four-codon box if the second base is a pyrimidine, but form split boxes if the second base is a purine [60,61]. These rules hold true even for all known variant codes, with the minor exceptions that CUN is split between Ser and Leu in *Candida*, and that the CGN box is sometimes split between arginine and nonsense codons (Fig. 3). This consistency may imply that the degeneracy in the code is still influenced by chemical considerations.

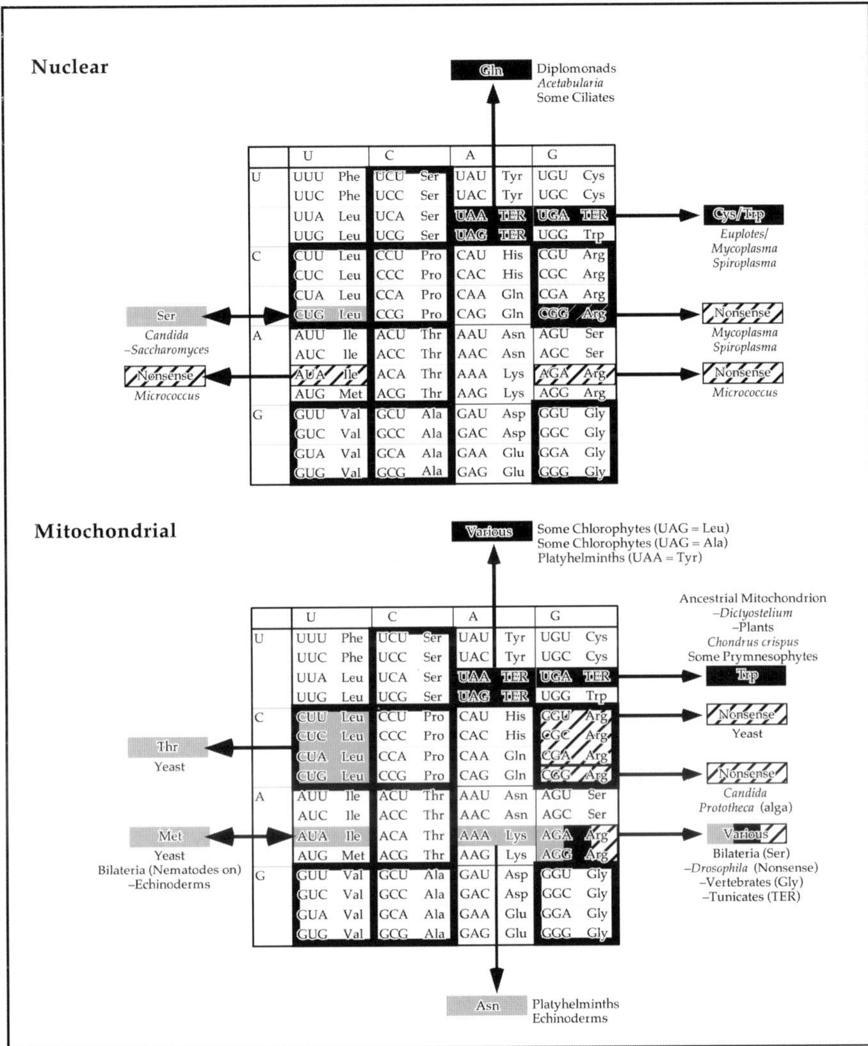

Fig. 3. Naturally occurring variants of the canonical genetic code. (a) "Nuclear" variants (including changes effective within bacterial genomes), synthesized from [83,84,59]; (b) mitochondrial variants, synthesized from [83,85,86]; yeasts from http://www.ncbi.nlm.nih.gov. Grey indicates missense changes, hatching indicates nonsense changes, and black indicates changes in termination codons. The variation in the code provides a base for natural selection to act on. Dark borders indicate codon classes that are family boxes under the rules proposed in [60]: these family boxes are never split among amino acids (except for a CUG Ser/Leu ambiguity in certain yeasts), although sometimes certain members of a family box are unassigned.

4   Amino Acid Aptamers

The most direct test of RNA-amino acid interactions is to find the RNA sequences with greatest affinity for each amino acid. The technique of *in vitro* selection permits a direct test of the codon correspondence hypothesis [62]: this procedure searches a large space of possible sequences for optimal or near-optimal "solutions" to particular binding problems, isolating nucleic acid molecules that bind a particular target by selective amplification over several generations [63–66]. Aptamers (RNA ligands) now exist for a variety of amino acids, including tryptophan [67], valine [68], phenylalanine [69], citrulline [70], and isoleucine [71]. Of these, however, the tryptophan aptamer is dependent on the chromatography support for binding, and the phenylalanine aptamer is not specific.

Interactions between arginine and RNA have received most attention: several laboratories have selected arginine aptamers [70,72–75], and the solution structure of one of these has been deciphered by NMR [76]. The interaction between the HIV Tat protein and TAR RNA takes place at an arginine-rich motif that can be emulated by free arginine in solution [77], which has stimulated interest in artificially selected arginine binders. Arginine also interacts with the guanosine-binding site in the group I intron via its codons [78].

Each stereochemical hypothesis suggests that particular short RNA motifs will be found at amino acid binding sites. Analysis of the published aptamers shows that arginine binding sites significantly overrepresent all six arginine codons, but not anticodons or other motifs (Table 1): in fact, the set of arginine codons in the "universal" genetic code shows a higher association with arginine binding sites than does *any* other set of six codons composed of a family box and a doublet [62]. The published isoleucine aptamers also have the appropriate codons at their binding sites [71] (although there are insufficient data for statistical analysis).

		Codon	nt in Codon	nt not in Codon	G	P
Arg Aptamers		Binding	23	9		
		Not Binding	83	190	20.2	3.4×10^{-6}
Others		Binding	17	42		
		Not Binding	29	82	0.14	0.35

Table 1: Arginine codon/binding site frequencies for arginine and non-arginine aptamers (from [62]). Tests for association between codons and binding sites were directional. The number of nucleotides involved in arginine codons need not be a multiple of three, because some codons overlap.

Although the available data allow only tentative conclusions about the role of direct templating in determining the present genetic code [11], they rule out some hypotheses. Because arginine, a metabolically complex amino

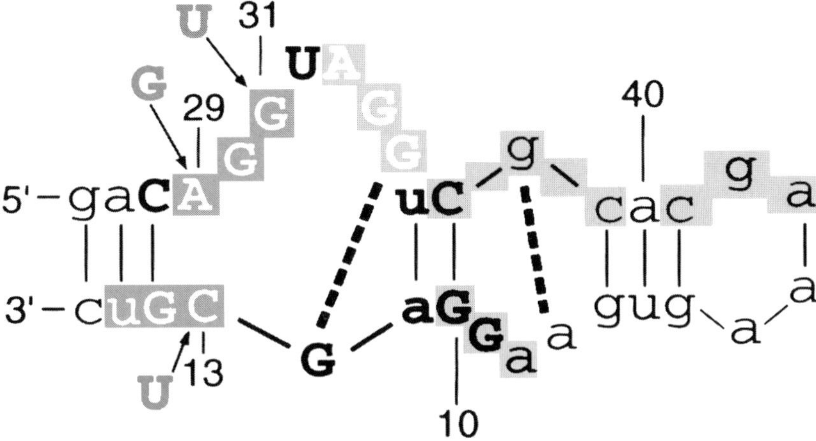

Fig. 4. Arginine aptamer structure, adapted from [76]. Bases in capital letters are directly involved in arginine binding; bases boxed in gray participate in arginine codons (CGN or AGN) in some reading frame. The three bases with arrows indicate differences between the original citrulline aptamer and the final arginine aptamer: note that in the arginine codon these three bases contribute to two new arginine codons that bind directly to arginine [62].

acid not found under prebiotic conditions, shows an association with its codons, either arginine annexed those codons for which it had highest affinity from another amino acid in an earlier code, or arginine was available at the time the code was established. However, this refutes the hypothesis that only prebiotic amino acids will show codonic associations. Because isoleucine, a prebiotically plausible amino acid, also seems to show affinity for its codons, it seems likely that the genetic code became established after a variety of amino acids were available, and that each amino acid in this initial set acquired those codons to which it was most strongly attracted.

The mode of binding also reveals clues about the types of interactions that could have been important in establishing the initial genetic code. The available arginine aptamer NMR structure [76] (Fig. 4) shows that binding sites are dispersed throughout the molecules, and that many arginine codons contribute to the arginine binding site. Consequently, it is unlikely that a single codon can ever directly attract an amino acid; rather, binding sites must be made up of many codons in a large RNA molecule. This indicates an RNA world origin for the genetic code, because long RNA molecules would not have been available until an RNA-based metabolism capable of synthesizing and polymerizing nucleotides was already present.

If the amino acid binds directly to the codon, why do tRNAs instead contain anticodons? One possibility is that tRNAs are extensions of primordial adaptor molecules that transported amino acids to the ribozymes that required them as coenzymes. An amino acid aptamer could evolve into an aminoacylating ribozyme [11], covalently attaching the amino acid to the terminal 2′ or 3′ hydroxyl of itself or other RNA molecules (the same way that amino acids are presently attached to tRNAs by protein-based synthetases). Once the amino acid was covalently attached, the aptamer domain would be free to base-pair with other RNA molecules. If the amino acid binding region overrepresented codons, then RNA molecules overrepresenting anticodons would statistically be more likely to associate by base pairing, and would therefore be more likely to become aminoacylated if the ribozyme had aminoacyltransferase activity in trans. The evolution of short RNA molecules able to act as carriers for specific amino acids would be useful for delivering these early coenzymes to particular ribozyme targets, and these carriers would preferentially contain anticodons corresponding to the amino acids. Elaboration into a modern tRNA-like structure might occur to promote stability or to allow interaction with other RNA components [79]. Later, incipient translation systems would be able to take advantage of this existing amino acid delivery system.

5   The RNA world as the Milieu of Code Evolution?

Translation presents a problem of origins: many critical components required for translation, such as aminoacyl-tRNA synthetases, release factors, and much of the ribosome, are themselves made of protein. Consequently, translation is a "chicken or egg" situation: protein synthesis itself is required to make the machinery that makes the proteins. This difficulty is surmounted by the RNA world hypothesis [9], which suggests that RNA originally acted as both a genetic material and a catalyst, roles subsequently usurped by DNA and protein. Because the genetic code has a non-random structure, it may contain clues that hint at the chemical milieu in which it evolved.

There are two main pathways that could have produced the present genetic code (or its predecessor) within the context of an RNA world. The first possibility is based on peptide-specific translation mechanisms: instead of a generalized translation apparatus, RNA catalysts could have produced peptides residue by residue, much as some short peptides are produced by specific enzymes today. A general translation system, once it evolved, would displace these original peptide-specific pathways as a more efficient solution to the problem of making the diversity of enzymes required for protein-based life. The second possibility is that amino acids, and later peptides, acted as cofactors for certain ribozymes [49]. As the importance of peptide synthesis increased, the RNA components would increasingly have been displaced by the protein parts of the hybrid catalysts. Eventually, only the protein and a few essential nucleotide cofactors as molecular fossils would remain. In either of these scenarios, the initial coding system must have been established by

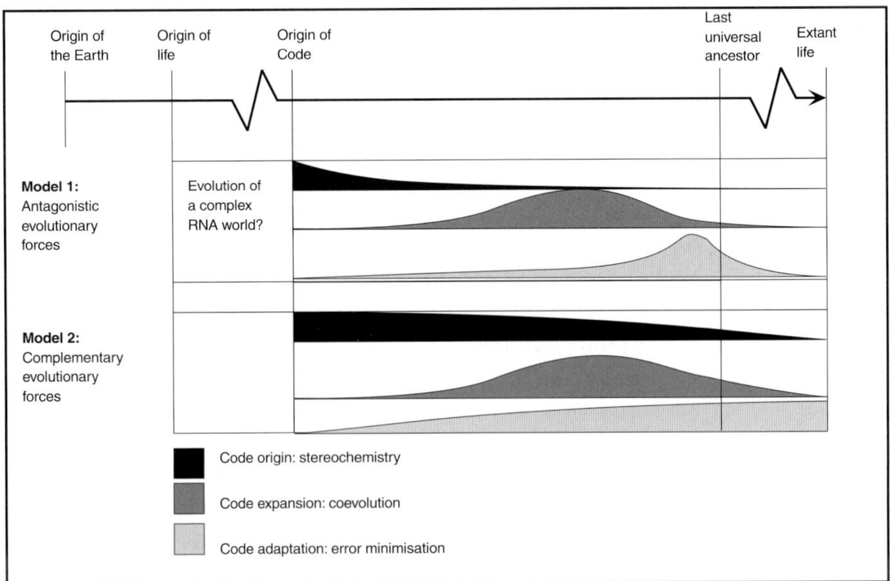

Fig. 5. Three facets of code evolution. The genetic code probably originated with stereochemical interactions, then underwent a period of expansion whereby new amino acids were incorporated. The evolution of the tRNA system, which separated codons from direct interaction with amino acids, then allowed reassignment of codons and hence adaptive evolution. Traditionally, these forces have been assumed to be antagonistic (Model 1), but they may actually have been complementary (Model 2) [6]; for example, current codon assignments may assign biosynthetically similar amino acids to similar codons meeting both stereochemical and adaptive criteria.

specific interactions between amino acids and RNA. As outlined in Section 4, arginine, and perhaps isoleucine, interact specifically with their codons in the context of RNA aptamers. The next two chapters will argue that the genetic code is at or near a global optimum for error minimization, especially when translation error and/or possible historical constraints are taken into account. More sequences of aptamers to amino acids, including amino acids not in the genetic code, are needed to determine the extent to which chemical factors determined the codon assignments and the choice of amino acids.

We envisage a series of definite, though perhaps overlapping, stages in the evolution of the code (Fig. 5) [6]. In the earliest stages, in the RNA world, RNA sequence tags would have paired specifically with amino acids by direct stereochemical interactions. Early peptides produced by this "direct templating" [11] need not have had catalytic function. Short, positively-charged arginine repeats could have neutralized the phosphate backbones of RNA molecules, allowing them to pass through membranes [80] or helping them to refold into active structures [81]. The next stage, coevolution between the

early codes and the set of amino acids, would be most important at the onset of the RNA-protein world, as amino acid and peptide cofactors became
more prevalent. At this point, the code might have expanded on the basis of
metabolic relatedness of amino acids, with precursor amino acids ceding some
of their codons to their biosynthetic products [14]. To preserve continuity
with the original templated proteins, this expansion would have to maintain
the rules originally established by stereochemical interactions. In the final
stage, after the evolution of the mRNA–tRNA–aminoacyl–tRNA synthetase
system, there would no longer be any direct interaction between amino acids
and codons. Consequently, codon swapping in different lineages would have
allowed the code to become optimized through codon reassignment.

Code optimization was not necessarily limited to this final stage, however.
Error minimization could have been synergistic with both stereochemical considerations and biosynthetically driven code expansion, eventually resulting
in the present code (Fig. 5). The fact that the structure of the code is at or
near a global optimum in this respect [21–24,82] emphasizes the crucial gap
in understanding code evolution: we are only now beginning to elucidate the
pattern of direct chemical interactions between RNA and amino acids. Only
in the light of this knowledge can we resolve the time of action and relative
importance of the three main facets of code evolution: history, chemistry, and
selection.[1]

References

1. Szathmáry, E. & Maynard Smith, J. (1995). The major evolutionary transitions. *Nature* 374:227–232.
2. Miller, S.L. (1953). Production of amino acids under possible primitive earth conditions. *Science* 117:528–529.
3. Miller, S.L. (1987). Which organic compounds could have occurred on the pre-biotic earth? *Cold Spring Harbor Symposia on Quantitative Biology* LII:17–27.
4. Gánti, T. (1975). Organisation of chemical reactions into dividing and metabolizing units: the chemotons. *Biosystems* 7:189–195.
5. Crick, F.H.C. (1968). The origin of the genetic code. *J. Mol. Biol.* 38:367–379.
6. Knight, R.D., Freeland, S.J. & Landweber, L.F. (1999). Selection, history, and chemistry: the three faces of the genetic code. *TiBS*, 24:241-247.
7. Dunnill, P. (1966). Triplet nucleotide-amino acid pairing: A stereochemical basis for the division between protein and nonprotein amino acids. *Nature* 210:1267–1268.
8. Pelc, S.R. & Welton, M.G.E. (1966). Stereochemical relationship between coding triplets and amino-acids. *Nature* 209:868–872.
9. Gilbert, W. (1986). The RNA world. *Nature* 319:618.
10. Yarus, M. (1991). An RNA-amino acid complex and the origin of the genetic code. *New Biologist* 3:183–189.
11. Yarus, M. (1998). Amino acids as RNA ligands: A direct-RNA-template theory for the code's origin. *J. Mol. Evol.* 47:109–117.

---

[1] For updates to this chapter, please see [87] and [88] and references therein.

12. Woese, C.R. (1967). *The Genetic Code: The Molecular Basis for Genetic Expression.* New York: Harper & Row.

13. Wong, J.T.-F. (1975). A co-evolution theory of the genetic code. *Proc. Natl. Acad. Sci. USA* 72:1909–1912.

14. Dillon, L.S. (1975). The origins of the genetic code. *The Botanical Review* 39:301–345.

15. Miseta, A. (1989). The role of protein associated amino acid precursor molecules in the organization of genetic codons. *Physiol. Chem. Phys. Med. NMR* 21:237–242.

16. Taylor, F.J.R. & Coates, D. (1989). The code within the codons. *Biosystems* 22:177–187.

17. Di Giulio, M. (1989). Some aspects of the organization and evolution of the genetic code. *J. Mol. Evol.* 29:191–201.

18. Di Giulio, M. (1998). The historical factor: the biosynthetic relationships between amino acids and their physiochemical properties in the origin of the genetic code. *J. Mol. Evol.* 46:615–621.

19. Sonneborn, T.M. (1965). Degeneracy of the genetic code: extent, nature, and genetic implications. In *Evolving Genes and Proteins*, V. Bryson and H.J. Vogel, eds. New York: Academic Press. pp. 377–297.

20. Zuckerkandl, E. & Pauling, L. (1965). Evolutionary divergence and convergence in proteins. In *Evolving Genes and Proteins*, V. Bryson and H.J. Vogel, eds. New York: Academic Press.

21. Ardell, D.H. (1998). On error minimization in a sequential origin of the standard genetic code. *J. Mol. Evol.* 47:1–13.

22. Haig, D. & Hurst, L.D. (1991). A quantitative measure of error minimization in the genetic code. *J. Mol. Evol.* 33:412–417.

23. Freeland, S.J. & Hurst, L.D. (1998). The genetic code is one in a million. *J. Mol. Evol.* 47:238–48.

24. Freeland, S.J. & Hurst, L.D. (1998). Load minimization of the code: history does not explain the pattern. *Proc. Roy. Soc. Lond. B* 265:1–9.

25. Ring, D., Wolman, Y., Friedmann, N. & Miller, S.L. (1972). Prebiotic synthesis of hydrophobic and protein amino acids. *Proc. Natl. Acad. Sci. USA* 69:765–768.

26. Wolman, Y., Haverland, W.J. & Miller, S.L. (1972). Nonprotein amino acids from spark discharges and their comparison with the Murchison meteorite amino acids. *Proc. Natl. Acad. Sci. USA* 69:809–811.

27. Weber, A.L. & Miller, S.L. (1981). Reasons for the occurrence of the twenty coded protein amino acids. *J. Mol. Evol.* 17:273–284.

28. Kvenvolden, K., Lawless, J.G., et al. (1970). Evidence for extraterrestrial amino-acids and hydrocarbons in the Murchison meteorite. *Nature* 228:923–926.

29. Kvenvolden, K.A., Lawless, J.G. & Ponnamperuma, C. (1971). Nonprotein amino acids in the Murchison meteorite. *Proc. Natl. Acad. Sci. USA* 68:486–490.

30. Crothers, D.M. (1982). Nucleic acid aggregation geometry and the possible evolutionary origin of ribosomes and the genetic code. *J. Mol. Biol.* 162:379–391.

31. Trifonov, E. & Bettecken, T. (1997). Sequence fossils, triplet expansion, and reconstruction of earliest codons. *GENE* 205:1–6.

32. Lehmann, U. (1985). Chromatographic separation as selection process for prebiotic evolution and the origin of the genetic code. *Biosystems* 17:193–208.
33. Nagyvary, J. & Fendler, J.H. (1974). Origin of the genetic code: a physical-chemical model of primitive codon assignments. *Orig. Life* 5:357–362.
34. Woese, C.R., Dugre, D.H., Dugre, S.A., Kondo, M. & Saxinger, W.C. (1966). On the fundamental nature and evolution of the genetic code. *Cold Spring Harb. Symp. Quant. Biol.* 31:723–736.
35. Woese, C.R., Dugre, D.H., Saxinger, W.C. & Dugre, S.A. (1966). The molecular basis for the genetic code. *Proc. Natl. Acad. Sci. USA* 55:966–974.
36. Weber, A.L. & Lacey, J.C., Jr. (1978). Genetic code correlations: amino acids and their anticodon nucleotides. *J. Mol. Evol.* 11:199–210.
37. Jungck, J.R. (1978). The genetic code as a periodic table. *J. Mol. Evol.* 11:211–224.
38. Lacey, J.C., Jr. & Pruitt, K.M. (1969). Origin of the genetic code. *Nature* 223:799–804.
39. Saxinger, C. & Ponnamperuma, C. (1971). Experimental investigation on the origin of the genetic code. *J. Mol. Evol.* 1:63–73.
40. Raszka, M. & Mandel, M. (1972). Is there a physical chemical basis for the present genetic code? *J. Mol. Evol.* 2:38–43.
41. Saxinger, C. & Ponnamperuma, C. (1974). Interactions between amino acids and nucleotides in the prebiotic milieu. *Orig. Life* 5:189–200.
42. Lacey, J.C., Jr., Weber, A.L. & White, W.E., Jr. (1975). A model for the coevolution of the genetic code and the process of protein synthesis: review and assessment. *Orig. Life* 6:273–283.
43. Reuben, J. & Polk, F.E. (1980). Nucleotide-amino acid interactions and their relation to the genetic code. *J. Mol. Evol.* 15:103–112.
44. Podder, S.K. & Basu, H.S. (1984). Specificity of protein-nucleic acid interaction and the biochemical evolution. *Orig. Life* 14:477–484.
45. Porschke, D. (1985). Differential effect of amino acid residues on the stability of double helices formed from polyribonucleotides and its possible relation to the evolution of the genetic code. *J. Mol. Evol.* 21:192–198.
46. Lacey, J.C., Jr., Wickramasinghe, N.S.M.D., Cook, G.W. & Anderson, G. (1993). Couplings of character and of chirality in the origin of the genetic system. *J. Mol. Evol.* 37:233–239.
47. Lacey, J.C., Jr. & Mullins, D.W., Jr. (1983). Experimental studies related to the origin of the genetic code and the process of protein synthesis—a review. *Orig. Life* 13:3–42.
48. Lacey, J.C., Jr. (1992). Experimental studies on the origin of the genetic code and the process of protein synthesis: a review update. *Orig. Life Evol. Biosph.* 22:243–275.
49. Szathmary, E. (1993). Coding coenzyme handles: a hypothesis for the origin of the genetic code. *Proc. Natl. Acad. Sci. USA* 90:9916–9920.
50. Maizels, N. & Weiner, A.M. (1987). Peptide-specific ribosomes, genomic tags, and the origin of the genetic code. *Cold Spring Harbor Symp. Quant. Biol.* LII:743–749.
51. Maizels, N. & Weiner, A.M. (1993). The genomic tag hypothesis: modern viruses as molecular fossils of ancient strategies for genomic replication. In *The RNA World*, R.F. Gesteland and J.F. Atkins, eds. New York: Cold Spring Harbor Laboratory Press. pp. 577–602.

52. Ralph, R.K. (1968). A suggestion on the origin of the genetic code. *Biochem. Biophys. Res. Comm.* 33:213–218.
53. Hopfield, J.J. (1978). Origin of the genetic code: a testable hypothesis based on tRNA structure, sequence, and kinetic proofreading. *Proc. Natl. Acad. Sci. USA* 75:4334–4338.
54. Root-Bernstein, R.S. (1982). Amino acid pairing. *J. Theor. Bio.* 94:885–894.
55. Root-Bernstein, R.S. (1982). On the origin of the genetic code. *J. Theor. Bio.* 94:895–904.
56. Shimizu, M. (1982). Molecular basis for the genetic code. *J. Mol. Evol.* 18:297–303.
57. Hendry, L.B. & Whitham, F.H. (1979). Stereochemical recognition in nucleic acid-amino acid interactions and its implications in biological coding: a model approach. *Perspect. Biol. Med.* 22:333–345.
58. Alberti, S. (1997). The origin of the genetic code and protein synthesis. *J. Mol. Evol.* 45:352–358.
59. Keeling, P.J. & Doolittle, W.F. (1997). Widespread and ancient distribution of a noncanonical genetic code in diplomonads. *Mol. Biol. Evol.* 14(9):895–901.
60. Lagerkvist, U. (1978). "Two out of three": an alternative method for codon reading. *Proc. Natl. Acad. Sci. USA* 75:1759–1762.
61. Lagerkvist, U. (1980). Codon misreading: a restriction operative in the evolution of the genetic code. *American Scientist* 68:192–198.
62. Knight, R.D. & Landweber, L.F. (1998). Rhyme or reason: RNA-arginine interactions and the genetic code. *Chem. Biol.* 5(9):R215–20.
63. Ellington, A.D. & Szostak, J.W. (1990). In vitro selection of RNA molecules that bind specific ligands. *Nature* 346:818–822.
64. Robertson, D.L. & Joyce, G.F. (1990). Selection in vitro of an RNA enzyme that specifically cleaves single-stranded DNA. *Nature* 344:467–468.
65. Tuerk, C. & Gold, L. (1990). Systematic evolution of ligands by exponential enrichment: RNA ligands to bacteriophage T4 DNA polymerase. *Science* 249:505–510.
66. Landweber, L.F., Simon, P.J. & Wagner, T.A. (1998). Ribozyme engineering and early evolution. *BioScience* 48:94–103.
67. Famulok, M. & Szostak, J.W. (1992). Stereospecific recognition of tryptophan agarose by in vitro selected RNA. *J. Am. Chem. Soc.* 114:3990–3991.
68. Majerfeld, I. & Yarus, M. (1994). An RNA pocket for an aliphatic hydrophobe. *Nature Struct. Biol.* 1:287–292.
69. Zinnen, S. & Yarus, M. (1995). An RNA pocket for the planar aromatic side chains of phenylalanine and tryptophane. *Nucleic Acids Symp. Ser.* 33:148–151.
70. Famulok, M. (1994). Molecular recognition of amino acids by RNA-aptamers: an L-citrulline binding RNA motif and its evolution into an L-arginine binder. *J. Am. Chem. Soc.* 116:1698–1706.
71. Majerfeld, I. & Yarus, M. (1998). Isoleucine: RNA sites with essential coding sequences. *RNA* 4:471–478.
72. Burgstaller, P., Kochoyan, M. & Famulok, M. (1995). Structural probing and damage selection of citrulline- and arginine-specific RNA aptamers identify base positions required for binding. *Nucleic Acids Res.* 23:4769–4776.
73. Connell, G.J., Illangsekare, M. & Yarus, M. (1993). Three small ribooligonucleotides with specific arginine sites. *Biochemistry* 32:5497–5502.
74. Connell, G.J. & Yarus, M. (1994). RNAs with dual specificity and dual RNAs with similar specificity. *Science* 264:1137–1141.

75. Tao, J. & Frankel, A.D. (1996). Arginine-binding RNAs resembling TAR identified by in vitro selection. *Biochemistry* 35:2229–2238.
76. Yang, Y., Kochoyan, M., Burgstaller, P., Westhof, E. & Famulok, F. (1996). Structural basis of ligand discrimination by two related RNA aptamers resolved by NMR spectroscopy. *Science* 272:1343–1346.
77. Tao, J. & Frankel, A. (1992). Specific binding of arginine to TAR RNA. *Proc. Natl. Acad. Sci.* 89:2723–2726.
78. Yarus, M. (1989). Specificity of arginine binding by the tetrahymena intron. *Biochemistry* 28:980–988.
79. Maizels, N. & Weiner, A.M. (1994). Phylogeny from function: Evidence from the molecular fossil record that tRNA originated in replication, not translation. *Proc. Natl. Acad. Sci. USA* 91:6729–6734.
80. Jay, D.G. & Gilbert, W. (1987). Basic protein enhances the incorporation of DNA into lipid vesicles: model for the formation of primordial cells. *Proc. Natl. Acad. Sci. USA* 84:1978–1980.
81. Herschlag, D., Khosla, M., Tsuchihashi, Z. & Karpel, R.L. (1994). An RNA chaperone activity of non-specific RNA binding proteins in hammerhead ribozyme catalysis. *Embo J.* 13:2913–24.
82. Alff-Steinberger, C. (1969). The genetic code and error transmission. *Proc. Natl. Acad. Sci. USA* 64:584–591.
83. Osawa, S. (1995). *Evolution of the Genetic Code*. Oxford: Oxford University Press.
84. Tourancheau, A.B., Tsao, N., Klobutcher, L.A., Pearlman, R.E. & Adoutte, A. (1995). Genetic code deviations in the ciliates: evidence for multiple and independent events. *EMBO J.* 14:3262–3267.
85. Hayashi-Ishimaru, Y., Ehara, M., Inagaki, Y. & Ohama, T. (1997). A deviant mitochondrial genetic code in prymnesiophytes (yellow-algae): UGA codon for tryptophan. *Curr. Genet.* 32:296–299.
86. Hayashi-Ishimaru, Y., Ohama, T., Kawatsu, Y., Nakamura, K. & Osawa, S. (1996). UAG is a sense codon in several chlorophyceas mitochondria. *Curr. Genet.* 30:29–33.
87. Knight, R.D. & Landweber, L.F. (2000). Guilt by Association: The Arginine Case Revisited. *RNA.* 6:499–510.
88. Knight, R.D., Freeland, S.J. & Landweber, L.F. (2001). Rewiring the Keyboard: Evolvability of the Genetic Code. *Nature Reviews Genetics.* 2:49–58.

Imposing Specificity by Localization: Mechanism and Evolvability

Mark Ptashne and Alexander Gann

Abstract. Cells detect extracellular signals by allostery and then give those signals meaning by "regulated localization". We suggest that this formulation applies to many biological processes and is particularly well illustrated by the mechanisms of gene regulation. Analysis of these mechanisms reveals that regulated localization requires simple molecular interactions that are readily used combinatorially. This system of regulation is highly "evolvable", and its use accounts, at least in part, for the nature of the complexities observed in biological systems.

1   Introduction

Two broad classes of enzymes are distinguished by their modes of regulation. Members of the first class, exemplified by the enzymes of intermediary metabolism, recognize one or a few specific substrates, and are regulated by substrate concentration and by allosteric effects exerted by other small molecules. In contrast, members of the second class can recognize a large array of related substrates, the concentrations of which do not vary. This class includes, for example, RNA polymerase, protein sorting and degrading enzymes, and the kinases and phosphatases of signal transduction pathways.

We shall discuss a common and widely used strategy by which enzymes in this second class are regulated; how, for example, one extracellular signal leads to one pattern of gene expression or protein phosphorylation, whereas another directs the same enzymatic machinery to produce a different pattern. As a great deal of recent work has revealed, this strategy entails the regulated localization of the enzyme with the appropriate substrate. Thus, in response to one signal, an enzyme is directed to one substrate on which it then acts spontaneously: in response to a different signal, it is directed to, and works on, a different substrate. The term localization is used here in the sense of "apposition", and does not necessarily imply sequestration to particular sites or compartments within the cell.

We shall argue that gene regulation presents a particularly well-characterized example of the localization strategy. In this case, localization is often effected by "locator" proteins–transcriptional activators–that bring the enzyme, RNA polymerase, to specified genes (or more precisely, to specific promoter sequences found there). Specificity can be, and typically is, imposed by simple binding interactions between a locator, the transcriptional machinery, and the DNA. We suggest that much of the complexity of gene regulatory systems has been acquired by the accretion of evolutionary "add-ons" to this basic mechanism, a scenario that accounts, at least in part, for the nature of that complexity. The system is highly "evolvable" [1,2]: new patterns of gene expression are readily generated, often using new combinations of existing activators. After discussing these matters as they apply to gene regulation, we consider a few examples from signal transduction that illustrate common strategies for imposing specificity in these disparate systems (for related discussions see [3–6]).

2   Cooperative Binding of Proteins to DNA

Much of gene regulation depends upon the cooperative binding of proteins to DNA. Cooperative binding is used to direct proteins to specific sites on DNA, that is, to properly locate them. Figure 1 shows a simple example. As is typical of a DNA-binding protein, the depicted protein recognizes related sequences with different affinities. At its cellular concentration, the protein spontaneously binds to certain sites ("strong" sites) but leaves others ("weak" sites) unfilled. The protein can be directed to, or located at, a weak site by interacting with a second protein binding simultaneously at a nearby DNA site. The second protein has located the first at a specific weak site by increasing the local concentration of the first protein in the vicinity of that site.

Effective use of cooperative binding requires that the concentration of interacting proteins be controlled. This requirement arises because, usually, rather weak interactions between pairs of cooperatively binding proteins– interaction energies on the order of one or a few kilocalories–dictate the reaction. Simply raising the concentration of a protein–as little as ten-fold–often suffices, therefore, to promote spontaneous binding to weak sites. Consequently, if cooperative binding is the way of regulating localization, the interacting proteins must be maintained below levels at which their interactions become unnecessary for binding.

In the simplest scenario, neither partner of a pair of cooperatively binding proteins needs to undergo a modification or a conformational change; rather, the interaction between the proteins, as well as that between the proteins and DNA, need only provide binding energy. Therefore, these kinds of interactions–which can of course be highly specific–need only be adhesive (glue- or velcro-like).

3   Activators as Locators

Escherichia coli RNA polymerase illustrates several of these general features of DNA-binding proteins (Fig. 2). At the concentration of polymerase found in the bacterium, certain promoters constitute "strong" sites and are therefore recognized spontaneously at high frequency, whereas others are "weak", and are recognized only infrequently. Genes with either category of promoter,

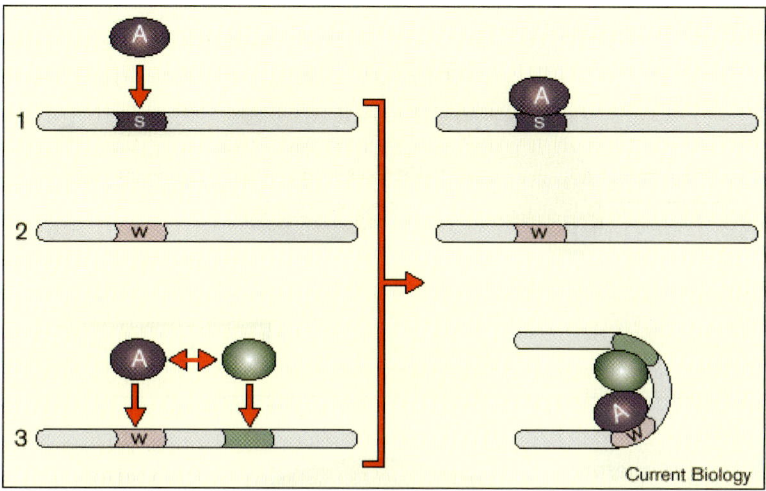

Fig. 1. Localization by cooperative binding to DNA. Protein A binds to the strong (s) site on DNA molecule 1, but not to the weak (w) site on molecule 2. Protein A does, however, bind to the weak site on molecule 3 by virtue of an interaction with, and hence cooperative binding with, another protein that binds to a site nearby. At the appropriate concentrations and affinities, the "helping" protein could be another molecule of protein A binding to a second binding site. Another way to have protein A fill the weak sites on molecules 2 and 3 would be to raise the concentration of the protein, in which case no cooperativity would be needed. Note that the two sites on molecule 3 are separated by an unspecified number of base pairs, and the DNA has formed a loop to accommodate the binding of the two proteins.

however, can be regulated using the principles described above so as to produce equally high (or low) levels of transcription, and to do so only when appropriate. For example, RNA polymerase can be directed to a specific weak promoter, and the gene thereby activated, by binding cooperatively to DNA with another protein, called an activator.

The typical activator bears two essential surfaces: one that recognizes a specific site on DNA, and another, the "activating region", that interacts with RNA polymerase. In this scenario, the specificity of the reaction, i.e., which promoter is chosen, is dictated by specific binding of the activator to a site near one (or another) promoter. From our present perspective, as mentioned in the introduction, activators would appropriately be called "locators". We have reviewed elsewhere [7] the various strands of evidence demonstrating that many (but not all) genes, in both prokaryotes and eukaryotes, are designed so that they can be regulated by localization (see Box 1).

The simple scheme for gene activation illustrated in Fig. 2 readily lends itself to modulation by further cooperative binding. For example, in many instances the activator itself interacts, and binds cooperatively to DNA, with other proteins. Those additional proteins may or may not be activators them-

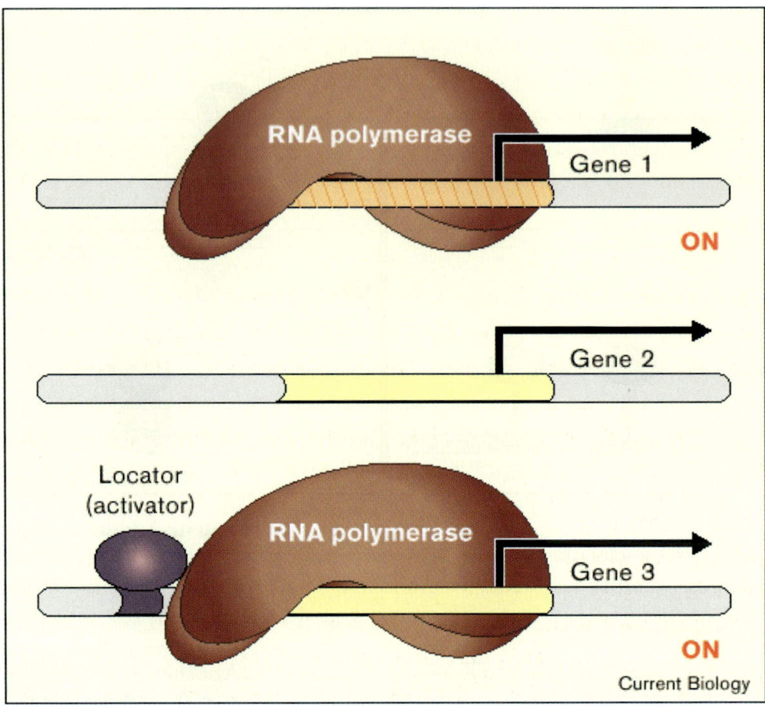

Fig. 2. Gene activation as an example of cooperative binding to DNA. The promoter sequence of gene 1 binds polymerase sufficiently tightly that the gene is "on" in the absence of any activator (and of any repressor that would otherwise prevent polymerase binding). Genes 2 and 3 have weak promoters, and polymerase binds only if helped to do so by an activator (locator), as illustrated for gene 3.

selves, but in either case the result is to make the effect of any given activator dependent upon cooperative binding with other proteins. As we shall see, these kinds of auxiliary interactions can be used to make activation of a given gene dependent upon more than one physiological signal, and to make sensitive switches. There is a further source of cooperativity implicit in this scheme, one that makes it easy to see how activators that do not interact with each other can nevertheless work synergistically. Any DNA-bound activators that can simultaneously touch the transcriptional machinery would work synergistically, because each would contribute binding energy to the recruitment reaction. The observations of unrelated activators working synergistically when placed near a gene are consistent with this expectation, and they suggest facile evolutionary pathways for modifying the regulation of genes.

Other proteins, called repressors, prevent access to the promoter and turn off transcription. Many genes are controlled by a combination of repressors and activators. This strategy plausibly follows from the notion that activators merely increase the local concentration of polymerase at a promoter, and so in their absence there will inevitably be a basal level of transcription at a rate that will vary depending on the strength of the promoter. Thus where genes are controlled by activators of the sort we are describing, they are often maintained in the off state by repressors in the absence of those activators. In eukaryotes, nucleosomes would be expected to contribute to this effect (see Box 2).

4   Allostery–The Rest of the Story

Extracellular signals that regulate genes are not generally detected by the simple binding interactions of the sort we have been describing; rather, each such signal is often accompanied by an allosteric change in a target protein. For example, the Lac repressor of *E. coli* undergoes a structural transition, upon binding to a metabolic derivative of lactose, that prevents it from binding DNA [8]. This and many other examples suggest a generalization: allosteric-like interactions are typically used to reveal the presence of an extracellular signal, but the specific interpretation of that signal is then dictated by the localization mechanisms we have discussed. Moreover, the meaning of any given signal can be changed or expanded without changing the allosteric response itself. For example, Lac repressor detects the presence of lactose, but that condition can be used to repress any gene depending upon the disposition of the repressor binding sites on DNA.

5   Examples of Gene Regulation in Bacteria

We shall consider the action of two well-studied bacterial transcriptional activators: catabolite activator protein (CAP) and lambda repressor (despite its name). The ability of each to bind DNA, and hence to work, is determined by extracellular signals that induce changes in the proteins: CAP functions only in the absence of glucose [9,10], and lambda repressor is inactivated when DNA is damaged by agents such as ultraviolet light [11]. The specificity of action of each protein, i.e., which gene it regulates, is determined by its DNA-binding address. CAP ordinarily binds to sites near, and activates, genes encoding enzymes required for metabolism of various sugars, and lambda repressor ordinarily activates its own gene [11]. If a CAP site is introduced upstream of the promoter of the lambda repressor gene, in the absence of glucose, CAP will activate that gene [12]. The meaning of the physiological signal–in this case the absence of glucose–can thus be "reinterpreted" simply by introducing the relevant DNA site in front of a gene.

The activities of CAP and lambda repressor illustrate two additional features expected of locators that work as outlined above. First, if both CAP and lambda repressor are positioned adjacent to a promoter so that each can make its natural contact with polymerase, the two activators work synergistically, as expected if the proteins simultaneously contact polymerase and, adding those energies of interaction, work together to recruit polymerase [12]. Second, each protein can, and at certain promoters does (as one of the names suggests) work as a repressor; all that is required is that the protein be positioned so that, rather than making a fruitful contact with polymerase, it blocks polymerase binding [10,11].

5.1   Sugar Metabolism Genes in *E. coli*: Multiple Signals and Combinatorial Control

This case shows one way that regulators that do no more than help or hinder polymerase localization can make expression of a gene dependent on two signals, and it also illustrates how regulators can be used in different combinations. *E. coli* bears separate sets of genes, each of which encodes enzymes that direct metabolism of one or another of a wide array of sugars. The biological problem is to ensure that any given set of such genes is expressed if, and only if, two conditions hold: first, that the relevant sugar, such as lactose or galactose, is present in the medium, and second, that glucose, a better carbon source, is absent.

Figure 3 shows how this is achieved for the *lac* genes. These genes are activated by CAP, which as we have noted is only active in the absence of glucose, but only if lactose, which inactivates the Lac repressor, is also present. Regulators that work as described in Fig. 3 readily lend themselves to being used in different combinations. For example, a CAP site is also located upstream of the *gal* genes, where CAP works with the Gal repressor to control transcription. Thus CAP activates the *gal* genes in the absence of glucose, provided that galactose is simultaneously present to inactivate the Gal repressor. CAP works in combination with many other regulators at some 100 genes in *E. coli* [10].

It is not difficult to imagine how systems such as this evolved by "tinkering" [13] with a rudimentary system that worked but was inefficient. Thus, for example, in the absence of binding sites for the regulatory proteins, the weak *lac* promoter would be read at a constant and low level. The bacterium would be able to use lactose, but it would make the enzymes even when there was no lactose substrate and also when the superior carbon source glucose was present. The first improvement would be addition of a CAP-binding site, positioned so that CAP would contact polymerase at the promoter and hence bind cooperatively to DNA with it. This would not be difficult because, evidently, many of a wide range of activator-polymerase contacts suffice for activation (see Box 1). The system would now provide high levels of the enzyme in the absence of glucose, and lower levels in its presence, without regard to the presence of lactose. A further refinement would be addition of a binding site for Lac repressor, which would ensure that transcription is off in the absence of lactose.

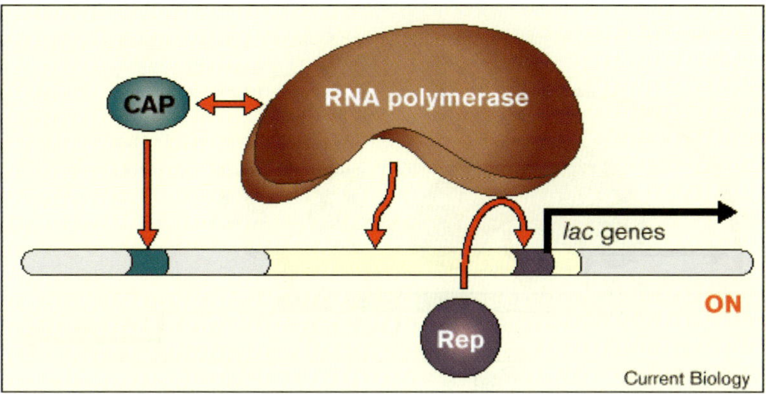

Fig. 3. The *lac* promoter in *E. coli*. In the absence of any controlling factors, and at concentrations typically found in the cell, polymerase transcribes the genes at a low level. Transcription is increased some 50-fold by CAP, which binds just upstream of the polymerase and, by simultaneously contacting polymerase with its "activating region", binds cooperatively with it. The Lac repressor (Rep) has the opposite effect: it binds to a site in the promoter that overlaps sequences that otherwise would be contacted by RNA polymerase and thereby prevents transcription. CAP and Lac repressor respond to separate physiological signals allosterically; CAP binds DNA only when complexed with cyclic AMP, which is depleted by growth in glucose; and when complexed with a metabolite of lactose, Lac repressor cannot bind DNA.

5.2 Phage Lambda: Using Simple Binding Interactions to Make a Sensitive Switch

The following example shows how simple binding interactions can create a switch that responds in an all-or-none fashion to an extracellular signal. The biological problem is that the genes of the bacterial virus lambda within a host *E. coli* cell must be maintained in a silent state, known as lysogeny, until an inducing signal is detected, whereupon they must be efficiently activated, leading to lytic growth [11]. This regulatory problem has been solved by constructing a biphasic switch involving two adjacent promoters that are controlled according to the rule that when one is on, the other is off. Here we find two forms of cooperativity in addition to that involving an activator and RNA polymerase, and these additional features are crucial to the efficiency of the switch. The details of how the switch works are explained below and illustrated in Fig. 4; it is not difficult to imagine how this switch might have evolved by a series of "add-ons", and a possible scenario for this is detailed in Box 3.

The key regulator is the lambda repressor, a protein that simultaneously activates transcription of its own gene as it turns off other genes. As shown in Fig. 4, two DNA-bound repressor dimers are positioned so that they cover and turn off the strong rightwards promoter, P_R, which controls the lytic

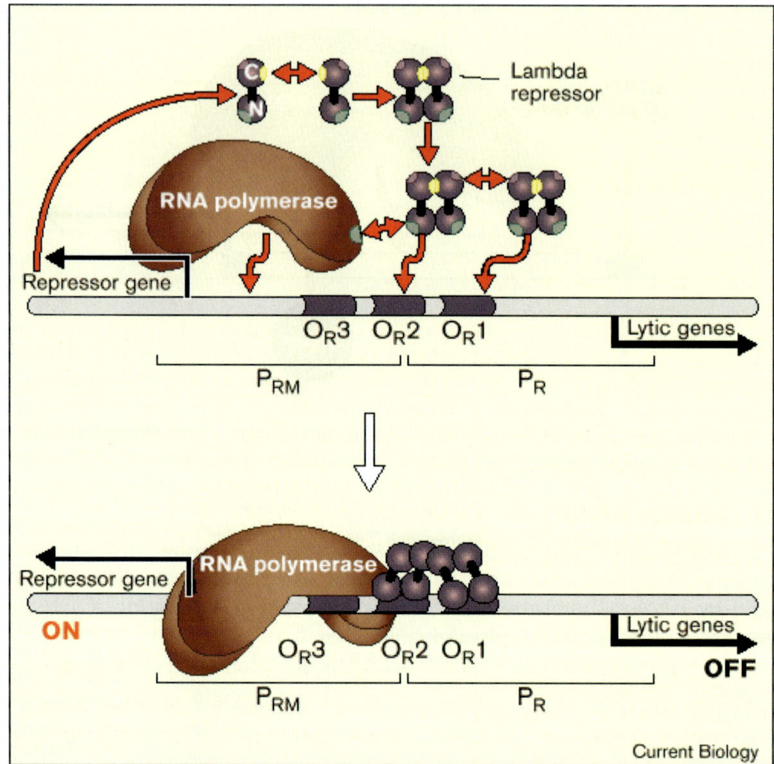

Fig. 4. The phage lambda switch. Repressor monomers, comprising two domains separated by a linker, are in equilibrium with dimers, the DNA-binding species. Two repressor dimers bind cooperatively to the adjacent operator sites O_R1 and O_R2. Repressor at these two sites represses the lytic promoter P_R, a strong promoter that works at a high level spontaneously unless repressed; simultaneously, repressor activates the weak promoter of the repressor gene itself, P_{RM} (by virtue of a contact between repressor at O_R2 and polymerase at P_{RM}). At higher concentrations repressor also binds to O_R3 and turns off P_{RM}, and thereby negatively regulates repressor synthesis. The three surfaces on repressor involved in the three examples of cooperativity–repressor dimerization, interaction between dimers, and interaction with polymerase to activate P_{RM}–are shaded. As described in the text, repressor is cleaved in response to ultraviolet radiation, and as a consequence transcription from P_R is turned on as that from P_{RM} is turned off.

genes; simultaneously, one of these repressors contacts RNA polymerase and activates transcription of the weak leftwards promoter, P_{RM}. This activation ensures that, once repressor synthesis has been initiated (an event that requires a separate promoter and activator), the repressor maintains its own synthesis. The phage genome is thereby stably maintained in a near-silent state, the only active gene being that of repressor itself. The system stably perpetuates itself until the cell encounters the signal that triggers the switch mechanism. Then, as repressor is inactivated, the rate of further repressor synthesis also drops. The first gene transcribed upon induction, *cro*, encodes a repressor that turns off P_{RM}, thus further ensuring that induction of lytic growth is an "all-or-none" effect.

The two additional forms of cooperativity in the lambda switch alluded to above mediate cooperative binding of the repressor to DNA. Thus, in the cell, repressor monomers are in concentration-dependent equilibrium with dimers, the DNA-binding species, and two repressor dimers bind cooperatively to the adjacent operator sites, as shown in Fig. 4. These repressor–repressor interactions ensure that the operator sites are filled as a highly sigmoidal function of the repressor concentration, providing both a buffer against minor fluctuations in repressor concentration and a dramatic change in state when some significant but readily obtainable proportion of repressor (approximately 90%) is inactivated.

A remarkable feature of the switch is that it depends upon a series of weak protein–protein interactions. Thus, under physiological conditions, P_{RM} is only activated by a factor of about ten, and cooperative binding to the two adjacent sites also has just a ten-fold effect. Each of these interactions therefore requires only a kilocalorie or two of binding energy, an amount easily provided by a simple protein–protein interaction. The requirement for each of the three protein-protein interactions, analyzed separately, can be dispensed with simply by increasing the concentration of one of the components. For example, increasing the concentration of polymerase *in vitro* is sufficient to elicit activated levels of transcription from P_{RM} in the absence of repressor [14]. Also, although binding of repressor to the site adjacent to polymerase ($O_R 2$) ordinarily depends upon interaction with another repressor dimer binding to the auxiliary site ($O_R 1$), merely increasing the repressor concentration some ten-fold obviates the need for this interaction–repressor then binds spontaneously to $O_R 2$ and performs both of the required functions (activation and repression). Thus, although repressor at $O_R 1$ also helps repress P_R, its uniquely required function is to impose cooperativity on the system.

As illustrated in Fig. 4, the three protein–protein interactions seen in the switch–repressor dimerization, cooperative binding of repressor dimers, and interaction with polymerase–involve separate patches on the surface of repressor. Nevertheless, it is likely that, as expected for a series of simple binding interactions, they are interchangeable: for example, the protein–protein interaction between repressor and polymerase responsible for activation can be replaced by the one that normally mediates cooperative binding. This is an example of an "activator bypass" experiment of the kind described in Box 1. In this case, polymerase is modified so as to bear a pair of lambda repressor carboxyl domains; interaction of these carboxyl domains with those of a lambda repressor bound to DNA nearby suffices for gene activation. The interaction that ordinarily mediates cooperative binding of lambda repressors can thus equally well mediate transcriptional activation [15].

The importance of the two repressor–repressor interactions that promote cooperative DNA binding–repressor dimerization and interaction of repressor dimers–is demonstrated by the fact that induction works simply by eliminating these functions. We noted above that ultraviolet irradiation inactivates repressor; this inactivation is mediated by a protein, RecA, which recognizes DNA and undergoes an allosteric transition that activates its protease function. Repressor is cleaved at a specific site in the peptide sequence that links the two domains, amino and carboxyl, of the protein. The amino domain is capable of carrying out the essential functions of the intact repressor–DNA binding, and hence repression of one set of genes, and contact with polymerase, and hence activation of the repressor gene–but at the concentration found in cells, it fails to do so in the absence of the cooperative effects mediated by the carboxyl domain. The sole function of the carboxyl domain is to promote dimer formation and interaction between dimers (using separate surfaces); separating the amino from the carboxyl domain, which eliminates both of these forms of cooperativity, is sufficient to trigger induction.

And herein lies the problem incurred by relying upon relatively weak binding interactions to impose specificity by localization: the components must be maintained over a relatively narrow range of concentration. This is accomplished here by the imposition of a third repressor binding site, O_R3, that overlaps P_{RM}; repressor bound to O_R3 blocks polymerase binding to P_{RM} and thus negatively regulates its own synthesis. Repressor binds (cooperatively) to O_R1 and O_R2 with an affinity some ten-fold higher than that with which it binds to O_R3, and so O_R3 becomes relevant only at higher repressor concentrations. This simple governing mechanism ensures that repressor never reaches a concentration at which it can bind to O_R2 without dimerizing and interacting cooperatively with another repressor dimer binding to O_R1. As might be expected from this line of analysis, genes encoding many transcriptional regulators–and indeed those encoding the subunits of RNA polymerase–are regulated so as to ensure that the concentrations of their products are maintained below specified levels (see for example [10,16]).

6   Gene Regulation in Eukaryotes

We noted above an experiment in which a bacterial gene was brought under control of a heterologous activator (CAP) merely by introducing the binding site for that regulator near the gene. Similar experiments have been performed with many activators and genes in many eukaryotes; the experiment is actually easier to perform in eukaryotes, as a successful outcome is much less dependent upon precise positioning of the activator relative to the gene. Two factors evidently contribute to this greater flexibility of the eukaryotic system: a typical eukaryotic activator apparently binds more tightly to its targets in the transcriptional machinery than does a typical bacterial activator, and hence will work from further upstream; and a typical eukaryotic activating region evidently can contact several, perhaps many, sites on the transcriptional machinery. The latter property may be particularly important in allowing an activator to work at a wide array of promoters. For example, it may be that, depending on the position of the activator on DNA in relation to the transcriptional start site in any given case, certain contacts are used in place of others (L. Gaudreau, J. Nevado, M. Keaveney, Z. Zaman, G. Bryant, M. Adam, K. Struhl, and M.P., unpublished observations).

Eukaryotes have widely exploited combinatorial strategies to create gene regulatory networks. Many eukaryotic genes, especially in higher organisms, respond in a switch-like fashion to multiple signals. That is, the gene is "on" if, and only if, several physiological signals are detected simultaneously. The following example shows how the mechanisms we have been discussing are used to create such a switch for the human interferon-β gene (Fig. 5). Here we find that three separate activators–NFκB, ATF/Jun, and IRF3/7–bind DNA cooperatively to form a structure called an enhanceosome. Because of the co-operativity, formation of this complex requires that each of the activators receives its appropriate physiological signal, rendering it capable of binding to DNA. Virus infection, which produces all three signals, thus triggers formation of the enhanceosome and activates the interferon-β gene [17,18]. Once the enhanceosome has formed, the activating regions carried on its various constituents simultaneously contact the transcriptional machinery and thereby work synergistically to activate transcription.

Optimal functioning of the β-interferon enhanceosome requires rather precise spacings between the binding sites for the components listed above, as well as for certain auxiliary proteins. Those spacings ensure that the various components can simultaneously touch one another, DNA, and the transcriptional machinery. The precise positioning of the enhanceosome with respect to the promoter is not critical, however, and the enhanceosome functions when positioned at any of many sites within hundreds of base pairs of the gene.

7 Localization in Signal Transduction

We noted in the introduction that many biological systems use the principle of imposing specificity by localization. Here we give a few examples from signal transduction, each chosen to illustrate one or another aspect encountered in our discussion of gene regulation.

7.1 STATs and Smads

We suggest an analogy between the workings of a receptor, in this case we shall consider here a cytokine receptor, and transcriptional activators like

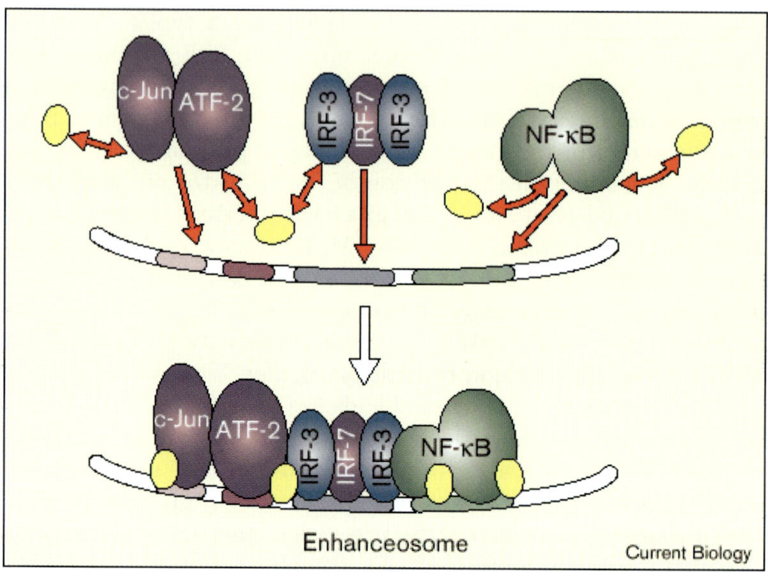

Fig. 5. The human interferon-β enhanceosome. Three transcriptional activators – NF-κB, ATF/Jun, and IRF-3/7–are activated in response to virus infection. The mechanism of activation is different for each transcription factor. Thus, for example, NF-κB is released from a bound inhibitor and allowed to enter the nucleus, and the DNA-binding function of ATF/Jun is activated by phosphorylation. These transcriptional activators - interacting with each other and with auxiliary proteins such as HMG-Y (yellow) - bind cooperatively to DNA to form the enhanceosome.

CAP. The latter, as we have seen, detect external signals and interpret them by working as locators, bringing together an enzyme, RNA polymerase, with one or another of its potential substrates, the promoters of target genes. In the eytokine system, the receptor responds to its signal by bringing together an enzyme–a kinase–with one or another of its potential substrates, the so-called STAT proteins. As with the activators, the specificity of the response, i.e., which STAT is phosphorylated, is determined by simple binding interactions and therefore is readily changed (see [19] and references therein).

The cytokine system works, in brief, as illustrated in Fig. 6, which shows two different cytokines interacting with their respective receptors and activating two different STATs. The first step, receptor recognition, brings together two receptor chains and triggers phosphorylation of receptor tyrosine residues. (In this case, detection of the signal may itself be regarded as a relocation process [4,20].) This phosphorylation creates a specific STAT-binding site. The bound (relocated) STAT is apposed to a kinase (a JAK) which phosphorylates it, thereby activating the STAT. The activated STAT, now a dimer, moves to the nucleus and activates specific genes.

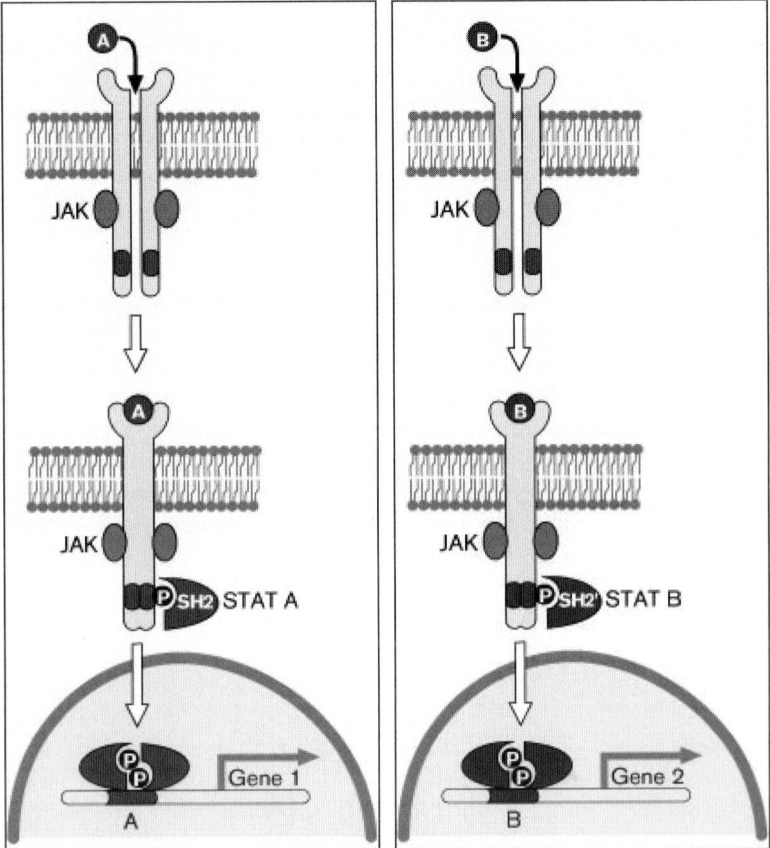

Fig. 6. STAT activation. Cytokine A activates gene 1 by inducing phosphorylation of STAT A, whereas cytokine B activates gene 2 by inducing phosphorylation of STAT B. If the STAT A binding site on the cytokine A receptor is replaced by a site that binds STAT B, cytokine A activates gene 2 instead of gene 1.

The identity of the STAT activated by a given cytokine is determined by which STAT binds the receptor. That specificity is readily altered: interchanging STAT-binding sites between receptors, or receptor-binding sites–so-called Src homology 2 (SH2) domains–between STATs, suffices to swap the specificity of the responses. Thus, the specificity of the response to a given cytokine is determined, not by the inherent specificity of the kinase, but by the identity of the STAT that is brought into the vicinity of the specific receptor and its associated kinase.

We encounter a similar theme with signaling by the growth and differentiation factors of the tumour growth factor β (TGF-β) family, TGF-β itself and the bone morphogenetic proteins (BMPs) [21]. In these cases, the receptor phosphorylates, on serines or threonines, one of a subset of so-called

Smad proteins, the receptor-regulated or R-Smads. Phosphorylated R-Smad binds the related protein Smad4 to form a complex which moves into the nucleus, where it regulates gene expression by interacting with specific DNA-binding proteins. The genes regulated by activation of a given receptor are determined by the particular Smad that it phosphorylates.

Here again, specificity is determined solely by localization. The TGF-β receptor binds and phosphorylates Smad2 but not Smad1, whereas the BMP receptor binds and phosphorylates Smad1 but not Smad2. *In vitro*, however, the kinase associated with either receptor can phosphorylate both Smads, and *in vivo*, swapping the Smad-docking sites between receptors, or the receptor-binding domains between Smads, switches specificity, just as we saw in the STAT system. Changing just four residues in the receptor, or as few as two residues in the Smad, is sufficient to effect such a switch in specificity [22].

Because specificity is imposed by localization in these signaling pathways, they are particularly "evolvable". That is, it is easy to see how the meaning a cell ascribes to a given cytokine or TGF-β family member can be changed or expanded by attaching binding sites for the appropriate STAT or Smad, respectively, to its receptor. New responses can thus be generated without the need to evolve new enzymatic activities or specificities, a requirement that would presumably be more taxing.

7.2   Ras

An important aspect of the "localization" idea is that once the enzyme (RNA polymerase, for example) is brought to the substrate (in this instance, a specific promoter), the enzymatic activity (transcription) proceeds spontaneously. Experiments in which RNA polymerase was artificially brought to the gene ("activator bypass" experiments; see Box 1), with subsequent activity, have been crucial in formulating our ideas. The same experimental approach has revealed the sufficiency of localization in another signal transduction pathway, that involving the small GTPase Ras.

Many receptor tyrosine kinases, such as the epidermal growth factor receptor, exert their effects through the Ras pathway, a series of interactions between components widely conserved in eukaryotic evolution. Once again, phosphorylation of sites on the receptor in response to the extracellular signal creates a binding site for, and thus recruits, another protein, in this case Grb2. The "adaptor" protein Grb2 in turn binds and recruits to the membrane Sos, which then interacts with and activates, by promoting exchange of GTP for bound GDP, membrane-bound Ras. Ras in turn recruits and activates Raf, a kinase that initiates the so-called mitogen-activated protein (MAP) kinase cascade that results finally in activation of various proteins, including a number of transcription factors.

In an experiment analogous to an activator bypass experiment, Sos was artificially tethered to the membrane (by myristoylation), and Ras was found to be activated as a result [23]. Thus an important, and perhaps the sole, role of the upstream components in this pathway is to recruit Sos to the membrane in response to the appropriate signal, where it can work on Ras. Once again, simple binding interactions, in this case involving SH2 and SH3 domains, are involved. As would be predicted from this result, over-production of a fragment of Sos, without specific recruitment to the membrane, also activates the Ras pathway, albeit weakly [24].

7.3 MAP Kinase Pathways in Yeast

One consequence of using localization to impose specificity is that the same enzyme can be used in many different pathways–in the case of RNA polymerase, to transcribe, in a regulated fashion, many different genes. This requires that the enzyme work in combination with many different regulators. In this section we see an example where the specificity of a kinase depends upon its location, which in turn is determined by interactions with different partners.

In the yeast *Saccharomyces cerevisiae*, two separate MAP kinase pathways, one activated by mating pheromones and the other by changes in osmolarity, use a common kinase, Ste11. In one case, mating pheromones activate Ste11, which then phosphorylates Ste7. In contrast, changes in osmolarity trigger Sho1 to activate Ste11, which in this case then phosphorylates Pbs2. Despite the shared component, there is ordinarily no crosstalk between the pathways, because they are isolated from each other by sequestration on separate scaffolds: Ste11 binds with Ste7 and other components of that pathway to the scaffold protein Ste5, whereas Ste11 binds with Sho1 and other components to Pbs2, itself a component of that pathway and the scaffold. Activation of Ste11 can thus have at least two "meanings", depending upon which other components it is co-localized with [25]. (For an interesting mutant in which crosstalk does occur, see [26].)

8 An Alternative World

Why is the strategy of imposing specificity by localization found so widely in nature? Consider, for example, control of transcription. One could imagine a system in which specificity is determined purely by allosteric control. In such a system, there would be a separate RNA polymerase for each promoter, transcription being triggered only upon integration of the required signals that would together induce an allosteric transition in the appropriate polymerase. Such a system might appear more simple, in some regards at least, than that which is observed. For example, there would be no need for locators nor the elaborate use of cooperativity of the type we have described.

The first difficulty in constructing such a purely allosteric world would be to design polymerases that would each integrate the effects of multiple signals. For example, at the *lac* promoter the polymerase would have to be active if, and only if, lactose were present and glucose absent. The problem would be magnified in higher eukaryotes where, as we have seen, the presence or absence of multiple signals is often integrated in the decision as to whether a given gene is transcribed. Even if these design problems were solvable (see [27]), it seems likely that it would be more difficult to use the principle of combinatorial control in designing new polymerases that responded to new combinations of signals. That is, whereas locators, can readily be used combinatorially, as we have seen, it is difficult to imagine that allosteric modules (if they existed) could be used so flexibly.

Box 1. Gene Activation: a Changed Perspective

Our license to describe gene activation as a process of RNA polymerase localization, or recruitment [7], depends upon a series of developments that over the past few years have "uncomplicated" our view of the process. We reviewed these matters recently [7], and here outline a few of these developments (references are given only for papers that appeared since publication of [7]).

As of a few years ago, at least three apparent problems confounded attempts to formulate a unified model of gene activation. First, in eukaryotes, transcription initiation seemed to require the multi-step assembly of a complicated machine, and activators were imagined to affect various steps in this process; this would be in striking contrast to bacteria, where we find a pre-assembled RNA polymerase molecule. Second, the requirement for specific additional proteins for activators to work *in vitro* raised the possibility that such proteins acted as "signal transducers", converting the machinery into an activated form by an allosteric transition. This would again contrast strikingly with the situation in bacteria. Third, the bacterial regulatory proteins lambda repressor and *E. coli* catabolite activator protein (CAP) appeared to affect kinetically distinguishable steps in transcription initiation, raising the possibility that even these bacterial activators can work by different mechanisms that require different kinds of activator–polymerase interactions.

These conceptual difficulties appear much less formidable now because of a number of developments. First, it was found that in yeast, although the transcriptional machinery may be even larger than previously imagined, many of the proteins occur in the cell in large complexes, possibly just one large complex. Complications remain of course: there may be different forms of these complexes, more than one recruiting event may be required, and the specific requirements may differ at different genes (see [28,29]), but the emerging picture much more closely resembles the situation in bacteria than did the previous scenario.

Two different kinds of experiment argue against the notion that activators allosterically modify polymerase. First, the so-called activator bypass experiments show that bacterial and eukaryotic genes can both be activated in the absence of any typical (classical) activator. For example, a DNA-binding domain fused to a component of the transcriptional machinery can activate transcription very efficiently at promoters bearing the appropriate DNA-binding site, in either bacteria or yeast, as can an arbitrary contact between a DNA-tethered peptide and the machinery. In addition, simply increasing the concentration of the bacterial or yeast transcriptional machinery *in vitro* suffices to mimic the effects of activators. Second, a variety of experiments, both *in vivo* and *in vitro*, show eukaryotic genes can be activated by typical activators in the absence of certain proteins that had previously been described as specifically required for activation ([30–32]; see also [33]).

Finally, despite the apparent differences between the ordinary actions of CAP and lambda repressor mentioned above, it is not necessary to postulate that they contact polymerase in importantly different ways. For example, activator bypass experiments show that arbitrary interactions between a DNA-tethered peptide and polymerase activate transcription at the two promoters ordinarily activated by lambda repressor and CAP, respectively. The experiments also show that CAP can efficiently activate transcription when artificially positioned at the promoter normally activated by lambda. For these and other reasons, it now seems likely that, despite the kinetic differences noted above, CAP and lambda repressor both activate transcription by simple adhesive or glue-like interactions with polymerase. Because of this, and because the essential effect of these activators is to stabilize the polymerase at the promoter, we refer in the text to the process interchangeably as recruitment/cooperative binding. (For a fuller discussion see [7].)

In bacteria, a typical activator, such as lambda repressor or CAP, touches one or two specific sites on polymerase. But many sites are potential targets for different activators and it is not difficult to create, by mutation, new interactions that mediate activation. In eukaryotes, the typical activating region evidently contacts any of several, perhaps many, sites on the transcriptional machinery. When tethered to DNA, many vaguely related peptides (reminiscent in this regard of sorting signals on proteins) can function as activating regions. We know of one case in bacteria in which recruitment does not suffice for activation. In that case the activator (NTRC) presumably contacts a unique site on polymerase (which bears a special sigma subunit) and, in an energy-dependent process, induces a conformational change that triggers transcription (see [7]).

Box 2. Chromatin and Gene Regulation

To some extent there have been two cultures studying the problem of eukaryotic gene regulation: one focused on the properties of nucleosomal DNA (chromatin), and the other, inspired by the bacterial paradigms, studied the actions of specific regulatory proteins without regard for chromatin structure. The main theme of this article brings these two approaches into congruence. Thus, the idea that activation merely involves locating the transcription machinery at the gene implies that any factors that inhibit or facilitate that relocation process can have an effect on gene expression. In principle, for example, simply removing histones would suffice to activate a gene whose promoter had a high affinity for the eukaryotic transcription machinery, similar to the high affinity of *E. coli* RNA polymerase for strong bacterial promoters.

Two recent experiments have examined the effect of depletion, in yeast, of histone H4 on the expression of a variety of genes. In one case, amongst a wide array of genes assayed, 70% were unaffected; of the remainder, some showed increased and some decreased transcription (R. Young, personal communication). In the other case, there was no detectable increase in transcription of the genes *CUP1*, *GAL1*, or *ADH1;* a modest (few-fold) increase in transcription from several heat-shock genes was observed, but the level of transcription reached was far below that elicited by the physiological activator of these genes (M. Green, personal communication). These studies indicate that for many, perhaps all, genes, histone removal does not suffice to achieve full activation, but they do not argue against the idea that histone modification or removal might help activators work.

Recent experiments show that histones are subject to a number of modifications - acetylation, phosphorylation, ubiquitination - any of which might affect transcription by modulating the accessibility of DNA to regulatory proteins and the transcription machinery [34]. Thus, for example, Gcn5, a protein that facilitates activation of a set of genes in yeast, encodes an acetylase that can act on histones, and its removal decreases activation of a few genes. It is suggested that acetylation "loosens" the histones and thereby facilitates access. Similarly, a deacetylase, targeted to a gene in yeast, can have an inhibitory effect. A variety of other protein complexes have been described in yeast and higher eukaryotes that may modify the configuration of nucleosomes [35,36]. It would seem, *a priori*, that there are three ways that these complexes might work: they might be incorporated as part of the transcription machinery brought to the promoter by the activators; they might be brought separately to the DNA by activators; and/or they might work constitutively as a background function in cells. Which, if any, of these applies in any given case is a subject of current investigation.

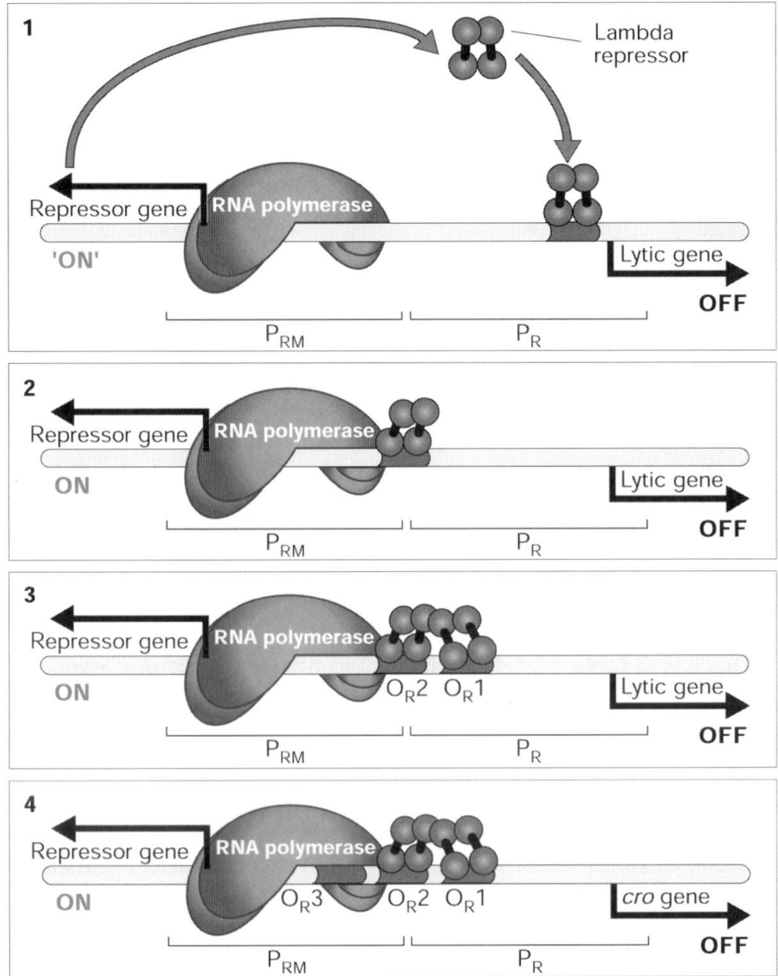

Fig. 7.

Box 3. Hypothetical Stages in the Evolution of the Lambda Switch

Stage 1. The primitive lambda genome bears two promoters, one for the lytic genes (P_R) and one for the repressor gene (P_{RM}). A single lambda repressor binding site overlaps P_R, and repressor bound at this site turns off the lytic genes. But the bound repressor has no stimulatory effect on P_{RM}, and so repressor synthesis is unregulated. If sufficient repressor were made, the lysogen would be stable, but induction would be inefficient: the repressor gene would continue to be transcribed at the same rate before and after induction, and the newly made repressor would impede lytic growth.

Stage 2. The single repressor-binding site has been moved close to P_{RM}, so that repressor bound there would contact polymerase at P_{RM} and thereby stimulate that promoter at the same time as it represses P_R. In the process we might imagine P_{RM} to have been weakened, so that high levels of repressor synthesis depend upon that stimulation. As discussed in the text for the *lac* case, positioning an activator, in this case lambda repressor, so that it can activate transcription is not a difficult task. This improvement facilitates induction because as repressor is destroyed, its rate of synthesis drops. Despite this improvement, the switch mechanism would remain inefficient. Among other problems (see below) the curve describing the binding of a single protein to a single site on DNA as a function of repressor concentration bears no steep inflection. The switch therefore would lack the all-or-none quality that ensures stable lysogeny in the absence of a signal but efficient induction upon receipt of the signal.

Stage 3. At this stage, an additional repressor-binding site has been introduced, so that the system now resembles that of Fig. 4, except that O_R3 is missing. A new protein–protein interaction surface has also been introduced, which mediates cooperative binding of repressor dimers to the adjacent sites. This additional cooperativity increases the efficiency of the switch mechanism, as described in the text, but only if the repressor concentration does not fluctuate to a higher level at which binding occurs without cooperativity.

Stage 4. The third repressor-binding site (O_R3) has been introduced, allowing repressor to negatively regulate its own synthesis. Thus, the repressor concentration never exceeds a critical level, which helps ensure an efficient switching mechanism. The final refinement is the introduction of *cro*, the first gene transcribed from P_R upon induction. Cro protein binds tightly to O_R3 and abolishes repressor synthesis as the lytic cycle begins. (At a ten-fold higher concentration, Cro also binds to O_R1 and O_R2 and down-regulates its own synthesis later in the lytic cycle.)

None of the complex elements of the switch we have described is "accidental". We surmise this from the fact that lambda is but one of a group of bacterial viruses, each of which bears the key features we have just described, although the molecular details differ in each case. For example, in phage P22, repressor at O_R2 touches polymerase at P_{RM}, but because the positioning of these elements is different from that found in lambda, a rather different surface of the repressor contacts polymerase. We do not know whether this represents convergent or divergent evolution, but it would seem that despite differences in molecular detail, the main features of the switch have been either re-invented or retained in the face of evolutionary pressure.

Acknowledgements. Many people discussed these matters with us and/or commented on the manuscript. We thank in particular Pete Broad, Jim Darnell, Dale Dorsett, Richard Ebright, David Evans, Robert Fletterick, John Gerhardt, Grace Gill, Michael Green, Steve Harrison, Ann Hochschild, Sandy

Johnson, Marc Kirschner, Tom Maniatis, Noreen Murray, Tony Pawson, Michael Rosen, Jim Rothman, Stuart Schreiber, David Thaler, Don Wiley, and Rick Young. We also thank Renate Helmiss for the illustrations, which were reprinted from *Current Biology* (vol. 8, pp. 812–822, 1998).

References

1. Kirschner M, Gerhart J: Evolvability. *Proc Natl Acad Sci USA* 1998, *95*, 8420–8427.
2. Gerhart J, Kirschner M: *Cells, Embryos and Evolution*. Oxford: Blackwell Scientific, 1997.
3. Pawson T, Scott JD: Signaling through scaffold, anchoring, and adaptor proteins. *Science* 1997, *278*, 2075–2080.
4. Austin AJ, Crabtree GR, Schreiber SL: Proximity versus allostery: the role of regulated of protein dimerization in biology. *Chem Biol* 1994, *1*, 131–136.
5. Irvine R: Insolitol phospholipids: translocation, translocation, translocation. *Curr Biol* 1998, *8*, R557–R559.
6. Patton EE, Willems AR, Tyers M: Combinatorial control in ubiquitin dependent proteolysis: don't Skp the F-box hypothesis. *Trends Genet* 1998, *14*, 236–243.
7. Ptashne M, Gann A: Transcriptional activation by recruitment. *Nature* 1997, *386*, 569–577.
8. Muller-Hill B: *The lac Operon: a Short History of a Genetic Paradigm*. Walter De Gruyter, 1996.
9. Savery N, Rhodius, Busby S: Protein-protein interactions during transcription activation: the case of *Escherichia coli* cyclic AMP receptor protein. *Phil Trans R Soc Lond [Biol]* 1996, *351*, 543–50.
10. Busby S, Kolb A: The CAP modulon. In *Regulation of Gene Expression in Escherichia coli*. Edited by Lin ECC. Georgetown, Texas: RG Landes, 1996, 255–279.
11. Ptashne M: *A Genetic Switch, Phage Lambda and Higher Organisms, 2nd edn; revised printing 1998*. Cambridge, Massachusetts: Cell and Blackwell Scientific, 1992.
12. Joung JK, Koepp DM, Hochschild A: Synergistic activation of transcription by bacteriophage λcl protein and *E. coli* cAMP receptor protein. *Science* 1994, *265*, 1863–1866.
13. Jacob F: Evolution by tinkering. *Science* 1977, *196*, 1161–1167.
14. Meyer BJ, Ptashne M: Gene regulation at the right operator (OR) of bacteriophage lambda III: lambda repressor directly activates gene transcription. *J Mol Biol* 1980, *139*, 195–205.
15. Hochschild A, Dove SL: Protein-protein contacts that activate and repress prokaryotic transcription. *Cell* 1998, *92*, 597–600.
16. Dykxhoorn DM, St. Pierre R, Van Ham O, Linn T: An efficient protocol for linker scanning mutagenesis: analysis of the translational regulation of an *Escherichia coli* RNA polymerase subunit gene. *Nucleic Acids Res* 1997, *25*, 4209–4218.
17. Wathelet MG, Lin CH, Parekh B, Ronco LV, Howley PM, Maniatis T: Virus infection induces the assembly of coordinately activated transcription factors on the IFN-β enhancer *in vivo*. *Mol Cell* 1998, *1*, 507–518.

18. Carey M: The enhanceosome and transcriptional synergy. *Cell* 1998, *92*, 5–8.

19. Darnell JE Jr: Stats and gene regulation. *Science* 1997, *277*, 1630–1635.

20. Weiss A, Schlessinger J: Switching signals on or off by receptor dimerization. *Cell* 1998, *94*, 277–280.

21. Massague J, Hata A, Liu F: TGF-β signaling through the Smad pathway. *Trends Cell Biol* 1997, *7*, 187–192.

22. Chen YG, Hala H, Lo RS, Wotton D, Shi Y, Pavelitch N, Massague J: Determinants of specificity in TGF-β signal transduction. *Genes Dev* 1998, *12*, 2144–2152.

23. Aronheim A, Engelberg D, Li N, Al-Alawi N, Schlessinger J: Membrane targeting of the nucleotide factor sos is sufficient for activating the ras signaling pathway. *Cell* 1994, *78*, 949–961.

24. Wang W, Fisher EMC, Jia Q, Dunn JM, Porfiri E, Downward J, Egan SE: The GRB2 binding domain of SOS1 is not required for down-stream signal transduction. *Nature Genet* 1995, *10*, 294–300.

25. Elion EA: Routing MAP kinase cascades. *Science* 1998, *281*, 1625–1626.

26. O'Rourke SM, Herskowtiz I: The Hog1 MAPK prevents cross talk between the HOG and pheromone response in *Saccharomyces cerevisiae*. *Genes Dev* 1998, *12*, 2874–2886.

27. Liu X, Guy HI, Evans DR: Identification of the regulatory domain of the mammalian multifunctional protein CAD by the construction of an *Escherichia coli* hamster carbamyl-phosphate synthetase. *J Biol Chem* 1994, *269*, 27747–27755.

28. McNeil JB, Agah H, Bentley: Activated transcription independent of the RNA polymerase II holoenzyme in budding yeast. *Genes Dev* 1998, *12*, 2510–2521.

29. Chang M, Jaehning JA: A multiplicity of mediators: alternative forms of transcription complexes communicate with transcriptional regulators. *Nucleic Acids Res* 1997, *25*, 4861–4865.

30. Shen W, Green MR: Yeast TAF 145 functions as a core promoter selectivity factor, not a general coactivator. *Cell* 1997, *90*, 615–624.

31. Oelgeschlager T, Tao Y, Kang YK, Roeder RG: Transcription activation via enhanced preinitiation complex assembly in a human cell-free system lacking TAFs. *Mol Cell* 1998, *1*, 925–931.

32. Gaudreau L, Adam M, Ptashne M: Activation of transcription *in vitro* by recruitment of the yeast RNA polymerase II holoenzyme. *Mol Cell* 1998, *1*, 913–916.

33. Keaveney M, Struhl K: Activator-mediated recruitment of the RNA polymerase II machinery is the predominant mechanism for transcriptional activation in yeast. *Mol Cell* 1998, *1*, 917–924.

34. Kuo MH, Allis CD: Roles of histone acetyltransferases and deacetylases in gene regulation. *BioEssays* 1998, *20*, 615–626.

35. Schnitzler G, Sif S, Kingston RE: Human SWI/SNF interconverts a nucleosome between its base state and a stable remodeled state. *Cell* 1998, *94*, 17–28.

36. Lorch Y, Cairns BR, Zhang M, Kornberg RD: Activated RSC-nucleo-some complex and persistently altered form of the nucleosome. *Cell* 1998, *94*, 29–34.

Towards a Predictive Biology:
The Example of Bacteriophage T7

Drew Endy

Abstract. I examine a relatively simple and well-characterized virus, bacterio-phage T7, as a platform for advancing the development of a predictive system-level biology. This examination results in a non-fitted mechanistic simulation capable of predicting the virus' growth cycle resolved at the level of unique intracellular species. From this effort I hope to approach the following questions. How good are the predictions from such a simulation? Can we evaluate our level of understanding for a biological system by comparing such quantitative predictions to observations? What new questions regarding evolved biological systems become addressable using such a simulation? Finally, if the behavior of an evolved biological system can be predicted, can the same abilities be applied to design novel biological systems?

1 Introduction

At present, biology's dominant mode of inquiry remains the characterization of biological systems at ever-finer levels of detail. This mode of inquiry has led to the discovery of new frontiers involving molecular and submolecular phenomena, and new technologies for their characterization. Our success in exploring these frontiers has produced a large body of information describing the components of biological systems. At the same time there exists a smaller effort devoted to synthesizing information from each component into higher level representations of these systems. When these two approaches are successfully combined they should result in a detailed molecular level description of a biological system and provide a tool for furthering our understanding.

Because the success of such a synthetic approach is predicated on a sufficient body of data from which synthesis can occur, its application has traditionally *followed* the development of the database itself. Today however, the critical mass of data exists to develop a synthetic framework (if not a detailed representation) for many biological systems. Because new experimental techniques (e.g., mRNA and protein monitoring chips, electrospray mass spectrometry, genome-scale yeast-two hybrid, genome-scale protein structure models) are beginning to produce large amounts of data, it is possible that biology may become understanding-limited at the system-level (as opposed to the detail level). It therefore seems prudent that the two approaches should be pursued concurrently. The work presented here briefly summarizes the development of a simulation for the bacteriophage T7 growth cycle and evaluates our level of understanding for this particular biological system.

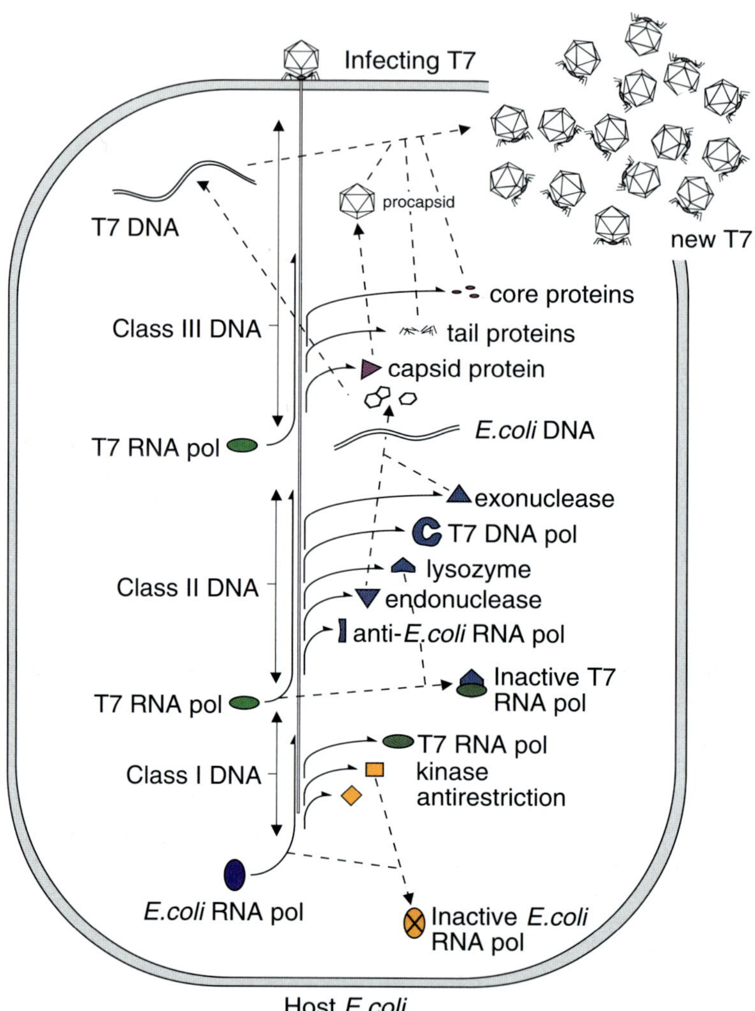

Fig. 1. The T7 growth cycle. T7 DNA enters the cell in a linear fashion leading to the sequential expression of T7 genes. The solid lines with half arrows indicate transcription and translation, the dashed lines denote reaction, and the solid lines with full arrows mark the three classes of T7 DNA. Expression of class I, II, and III DNA is shown.

2   Bacteriophage T7

T7 is a lytic phage that infects *E. coli* and produces approximately 100 progeny per infected cell within thirty minutes at 30C. It was first character-

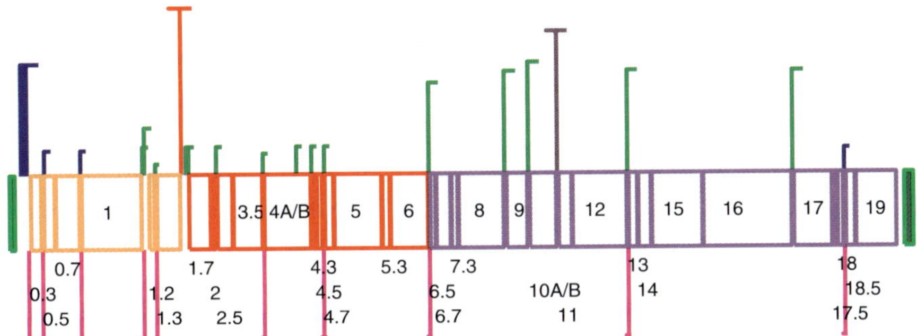

Fig. 2. The wild type T7 genome. The boxes represent coding regions (gene numbers are given as space permits). The vertical lines above the genes represent promoters (half cross bars) and terminators (full cross bars). Line height is proportional to promoter or terminator strength. The vertical lines below the genes represent RNaseIII sites.

ized by Demerec and Fano (1944); described in detail by Studier and Dunn (1983); Dunn and Studier (1983). The T7 genome is divided into three classes based on the function and timing of gene expression. Class I genes, which enter the host and are transcribed first, moderate the transition in metabolism from host to phage. Class II genes are expressed next and are primarily responsible for T7 DNA replication. Lastly, class III genes enter the cell and are expressed, leading to the production of proteins required for the phage particle, its maturation, and packaging of phage DNA (Figures 1,2).

3   T7 simulation

I have simulated the T7 growth cycle using a coupled system of equations that are integrated numerically. Deterministic kinetics are used to represent all reactions (accurate simulation of certain genetic circuits requires the use of stochastic kinetics, see McAdams and Arkin, 1997, 1998 and Arkin *et al.*, 1998 for examples). All mechanisms and parameters utilized in the simulation are taken from the biological literature (and direct communication with T7 biologists). The simulation starts with the adsorbed phage particle and mechanistically represents translocation of the infecting DNA, transcription from this DNA in accordance with all known genetic regulatory elements, translation of each protein, DNA replication, procapsid formation, DNA packaging, and particle assembly. From this, the simulation predicts the *in vivo* concentrations of all molecular species produced by the phage as well as several host species. Detailed descriptions of the simulation are available in the literature (Endy *et al.*, 1997 and Endy, 1998).

4   Results

After adsorption, the T7 growth cycle is initiated by translocation of its linear DNA into the host. Although other phages such as lambda inject their DNA within one minute of binding, the entry of T7 DNA takes about eight minutes (Garcia & Molineux, 1995a,b). Except for the first 850bp, T7 DNA entry is mediated by either the *E. coli* or T7 RNA polymerase. This unique mode of entry influences the sequential expression of T7 genes. Class I genes are transcribed by the *E. coli* RNA polymerase, which recognizes three promoters positioned near the leading end of T7 DNA. Class I mRNAs direct the synthesis of T7 RNA polymerase, which then transcribes the class II and class III genes. A class I gene product (gp0.7) and a class II gene product (gp2) effect the inhibition of the *E. coli* RNA polymerase as it is replaced by the T7 RNA polymerase (Figure 3). The observed transition from host-to phage-based transcription is characteristic of the simulation's prediction. Note the experimental data do *not* confirm the spike in phage-based transcription at 4 to 5 minutes post-infection.

Transcription via the T7 RNA polymerase is directed by 15 promoters distributed across the T7 genome (the two replication promoters are not believed to compete for RNA polymerase at a significant level). In vitro data (Ikeda, 1992) suggests the 5 class III promoters are stronger than the 10 class II promoters. Further, a class II protein, T7 lysozyme, binds T7 RNA polymerase and effects a reduction in T7 RNA polymerase-mediated transcription (perhaps ensuring competition for the remaining active polymerase among the class II and class III promoters) about the time T7 DNA replication beings (Summers, 1970; Zhang & Studier, 1995). This change in promoter strengths, when coupled to T7 DNA entry is thought to provide a mechanism for "switching" transcription from the class II to the class III genes (Figure 4). This mechanism is quantitatively supported by the simulation.

Following the central dogma, synthesis of 59 protein species follows via translation from the viral mRNA (Figure 5). While transcription regulation is relatively well understood, T7 translation regulation remains poorly characterized. This can be observed by noting that the time at which the synthesis of each protein is predicted to *begin* correlates well with the observed time. Further, the relative levels of protein synthesis predicted by the simulation tend to agree with the observed levels (there are some notable exceptions, e.g., gp0.3 and gp5.5). This agreement is largely due to the accurate representation of DNA entry and transcription in the simulation. Conversely, the predicted time courses for protein synthesis, defined largely by regulation of translation, do not agree as well with observations. Most notable is the

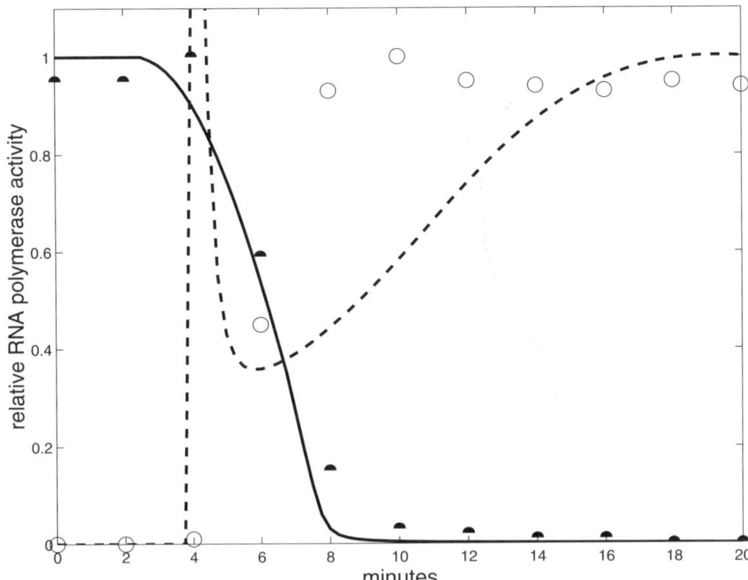

Fig. 3. Shift in RNA polymerase activity, from host (RNAP) to phage (T7 RNAP). Experimental (Hesselbach and Nakada, 1977) and simulated mRNA synthesis capacity. Filled circles (experimental) and solid line (simulated) are RNAP. Empty circles (experimental) and dashed line (simulated) are T7 RNAP. Each series is scaled relative to its maximum value.

unsupported prediction that synthesis of several class I and class II proteins (e.g., gp0.3, gp1, gp1.3, gp1.7, gp2.5, gp5) continues late into the growth cycle.

As a final product, the simulation predicts the formation of intracellular phage particles. Comparing this prediction to experimental data (Figure 6) reveals that the predicted burst of virus occurs earlier and plateaus above the observed curve. As the production of progeny phage is the result of a series of reactions (DNA entry, transcription, translation, replication, particle assembly, DNA packaging) it should be expected that discrepancies between prediction and observation for processes early in the growth cycle will be reflected in the burst curve. For example, the continued synthesis of metabolic proteins (e.g., the T7 DNA polymerase, gp5) will indirectly increase the simulated rate of virus production. However, the discrepancy in progeny yield indicates another poorly understood mechanism. In this case, the phage appear to replicate enough genomes to support a burst of 180 progeny however only 100 are produced. Whether this difference is because the excess replicated DNA can not physically be packaged, the host cell is lysed prior to completion of packaging, or because of another mechanism remains an open question.

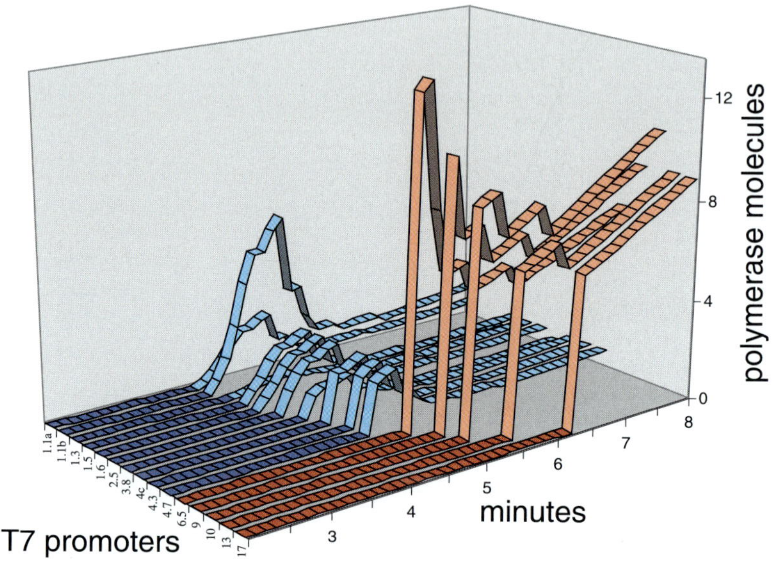

Fig. 4. Predicted allocation of active T7 RNAP to the 10 class II and 5 class III promoters during DNA entry. The x-axis is minutes post infection. The height of each ribbon indicates the number of T7 RNAP molecules initiating transcription from a particular promoter. Note the predicted switch in T7 RNAP allocation from the class II to class III promoters between 4 and 6 minutes.

5   Directions

By integrating what is known about bacteriophage T7 into a simulation; using the simulation to predict what results; comparing these predictions to observation, we summarize and highlight the strengths and failings in our current level of understanding for T7. For example, transcription regulation, as it effects the wild type growth cycle, appears to be well characterized while translation regulation, DNA packaging, and cell lysis still pose major questions. By tightly coupling the predictive power of the simulation to hypothesis formulation and experimental testing it should be possible to accelerate the rate at which these questions are answered.

Given such a simulation what new questions become approachable? I submit the chief value of such a simulation will be in taking the answers to the questions *"what* is present in this organism?" and *"what* is happening?" and applying them to systematically and quantitatively answer *"why* are these

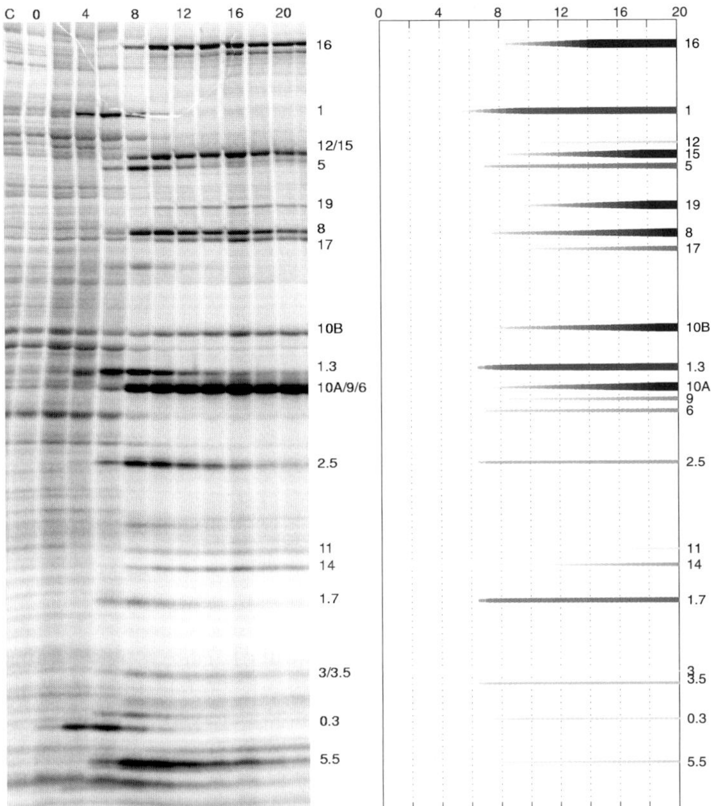

Fig. 5. Observed (left) and predicted (right) protein synthesis rates for wild type T7. The proteins bands on the experimental gel are named where known. The 'C' lane is uninfected host protein synthesis. The simulated gel does not show the synthesis of host proteins. To observe protein synthesis, cultures of BL21 were grown aerobically at 30C in B2 glucose medium (Studier, 1975) to a cell density of 2E8 per ml and infected at a multiplicity of 10 using a purified phage stock. At various times 100 l samples were labeled with 30 C/ml [35S]methionine for 90 sec and then chased for 45 sec using 1 ml of LB. Labeled proteins were separated by electrophoresis through a linear gradient of 7.5 to 17.5 percent polyacrylamide in the presence of 0.1 percent sodium dodecyl sulfate and visualized by phosphorimaging with ImageQuant (Molecular Dynamics) software. The simulated output for the synthesis rate of each T7 gene product was numerically integrated over the course of the phage growth cycle in 30 second intervals. Only gene products clearly identifiable by experiment are portrayed on the simulated gels. Protein synthesis rates were weighted based on the number of methionine and cysteine residues in each protein (Dunn & Studier, 1983). A maximum signal threshold was chosen for the simulated gel and band intensity (minimum as white, maximum as black) and thickness were both scaled linearly from zero to the maximum. Simulated band positions were based on the positions defined by experiment.

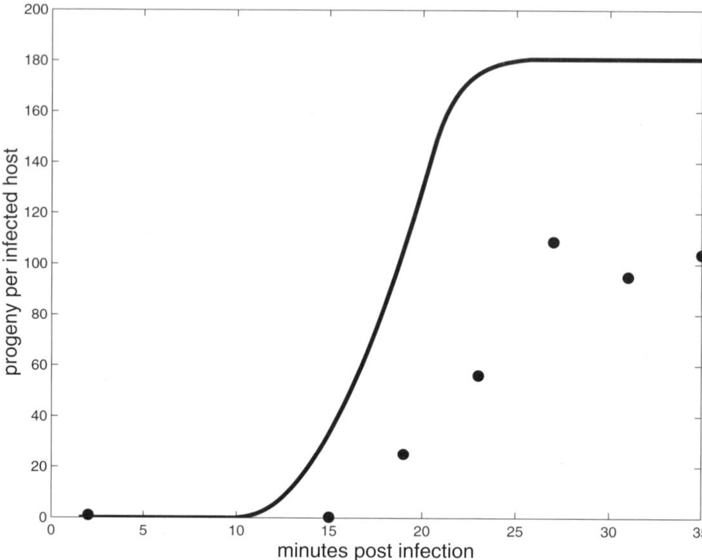

Fig. 6. Predicted (solid line) and observed (circles) intracellular phage production. Cultures of BL21 were grown aerobically at 30C in LB medium to a cell density of 2E8 per ml and infected at a multiplicity of 0.01 using a purified phage stock. Five min after infection, the culture was diluted 100-fold into fresh medium to prevent further phage adsorption; greater than 85 percent of the infecting particles formed an infective center. Unadsorbed and progeny phage were determined after CHCl3 treatment of samples; the initial titer of phage added was measured without exposure of the sample to CHCl3. Plaques were counted after 5 hr incubation at 37C and phage bursts are expressed as progeny per infective center.

components present?" and "*how* are they effecting their ends?" What happens when the distribution of T7 RNA polymerase strengths is changed from that of the wild type? Can distributions be selected *in silico* for faster than wild type growth? Can different distributions be selected for to optimize alternative objective functions (e.g., distributions of protein expression over time)? As another example, why has evolution provided us the particular order of genetic elements seen in the "wild type" virus? How will changes in genetic element order effect the viral growth cycle (Endy *et al.*, in preparation)? With more work we should be able to accurately predict the outcome of such changes. From such experiences it will become possible to understand and design complex biological systems.

6   Acknowledgements

The T7 simulation was originally developed at the Thayer School of Engineering, Dartmouth College, Hanover, NH by the author and Drs. John

Yin (now at the University of Wisconsin-Madison Department of Chemical Engineering) and Ian Molineux (University of Texas-Austin Department of Microbiology and Institute of Cellular and Molecular Biology). Significant revisions to the simulation have been effected by Lingchong You (University of Wisconsin-Madison Department of Chemical Engineering). The author is currently supported by a grant from the Office of Naval Research and the Molecular Sciences Institute.

References

1. Arkin A, Ross J, McAdams HH (1998). Stochastic kinetic analysis of developmental pathway bifurcation in phage lambda-infected Escherichia coli cells. *Genetics* 149(4):1633-48.
2. Demeric M, Fano U (1944). Bacteriophage-resistant Mutants in Escherichia coli. *Genetics* 30:119-136.
3. Dunn JJ, Studier FW (1983). Complete Nucleotide Sequence of Bacteriophage T7 DNA and the Locations of T7 Genetic Elements. *J. Mol. Biol.* 166:477-535.
4. Endy D, Kong D, Yin J (1997). Intracellular Kinetics of a Growing Virus: A Genetically Structured Simulation for Bacteriophage T7. *Biotech. Bioeng.* 55:375-389.
5. Endy D (1998). Dissertation: Development and Application of a Genetically-Structured Simulation for Bacteriophage T7. Dartmouth College, Hanover, NH 03755.
6. Endy D, You L, Yin J, Molineux, IJ (2000). Computation, prediction, and experimental tests of fitness for bacteriophage T7 mutants with permuted genomes. *Proc Natl Acad Sci U S A.* 97(10):5375-80.
7. Garc RL, Molineux IJ (1995). Rate of translocation of bacteriophage T7 DNA across the membranes of Escherichia coli . *J. Bact.* 177:4066-4076.
8. Garc RL, Molineux, IJ (1995). Incomplete entry of bacteriophage T7 DNA into F plasmid-containing Escherichia coli strains. *J. Bact.* 177:4077-4083.
9. Hesselbach BA, Nakada D. (1977). I protein: bacteriophage T7-coded inhibitor of Escherichia coli RNA polymerase. *J. Virology* 24: 746-760.
10. Ikeda RA (1992). The efficiency of promoter clearance distinguishes T7 class II and class III promoters. *J. Bio. Chem.* 267:11322-11328.
11. McAdams HH, Arkin A (1997). Stochastic mechanisms in gene expression. *Proc. Natl. Acad. Sci. USA* 94(3):814-819.
12. McAdams HH, Arkin A (1998). Simulation of prokaryotic genetic circuits. *Annu. Rev. Biophys. Biomol. Struct.* 27:199-224.
13. Studier FW (1975). Gene 0.3 of bacteriophage T7 acts to overcome the DNA restriction system of the host. *J. Mol. Biol.* 94:283-95.
14. Studier FW, Dunn JJ (1983). Organization and Expression of Bacteriophage T7 DNA, *CSH Quant. Biol.* 47:999-1007.
15. Summers WC (1970). The process of infection with coliphage T7 IV. Stability of RNA in bacteriophage-infected cells. *J. Mol. Bio.* 51:671-678.
16. Zhang X., Studier FW (1995). Isolation of transcriptionally active mutants of T7 RNA polymerase that do not support phage growth. *J. Mol. Biol.* 250:156-68.

Using Artificial Reagents to Dissect Cellular Genetic Networks

Roger Brent

Abstract. I describe work from the laboratory that promises to improve our ability to analyze the networks of genes that govern biological phenomena. By deepening our understanding of the genetic networks that govern gene expression and signal transduction, these experiments should speed the day when we can quantitatively predict the behaviors of these systems and understand how the particular ways that cells use to perform computations arose. In the process, the technologies we use to explore these systems may provide useful starting points to help build new ones.

1 Introduction

We have been working to define the function of individual genes and of genetic networks in eukaryotes. This line of work began when we realized that the accumulation of interaction data from interaction mating two-hybrid experiments (Finley and Brent, 1994) should allow us to develop computational tools to search these data for patterns of protein interactions of functional significance (Lok *et al.*, 1998). In fact, we have now developed such tools, which allow a user to search through our database of protein interactions ("Interaction 1.0") for patterns of binary interactions that may have functional significance, for example, those that might signify protein complexes or regulated protein kinases (Lok *et al.*, unpublished).

The next computational challenge is to extend these algorithms so that they conjoin connection data with gene sequence and other kinds of genomic data. However, it is now clear that the functional inferences we will be able to draw in the near term from such systematically generated biological data will often be disappointing, in that these inferences will not be of sufficient insight or predictive quality to interest the majority of contemporary biologists. A long-term approach to this problem is to bring into being technologies for the systematic generation of new types of biological information.

In the interim, we have been working to develop techniques that embody some of the reach and power of classical manipulative transmission genetics, but which extend these to important but genetically intractable systems such as human cells (Colas *et al.*, 1996, Cohen *et al.*, 1998). Here, I will describe some of the most important results of these experiments.

2 Results

These studies rely on our ability to select from combinatorial protein libraries molecules, which we refer to as peptide aptamers. These molecules are comprised of an amino acid loop of variable sequence, encoded by a combinatorial library, protruding from the surface of a platform protein. These molecules recognize biological molecules with high (in some cases, low picomolar; Fabbrizio *et al.*, 1999) affinity and specificity (Colas *et al.*, 1996). The design of peptide aptamers was inspired by that of antibodies, but the engineered molecules have only a single hypervariable region and are designed to function inside cells.

Here, I will review published studies in which we use these reagents to dissect the importance of specific protein–protein interactions and to establish the function of polymorphisms by specifically affecting the function of specific allelic variants. I will allude to unpublished work in which we use aptamers to functionally derivatize targeted proteins, to change the subcellular localization of targeted proteins, and to identify the genes responsible for definable phenotypes.

2.1 Dissecting specific protein–protein interactions

The first apparent application of peptide aptamers was to use them as agents that bound to different surfaces of a target protein to disrupt the interactions between that surface of the targeted protein and its partner(s). In fact, many of the first peptide aptamers we selected against human Cdk2 did disrupt the ability of that protein kinase to phosphorylate a widely used model substrate, histone H1. These early experiments suggested that the inhibition of kinase activity was competitive, perhaps due to the binding of the aptamer in the vicinity of the kinase's active site, and blocking its interaction with the H1 substrate (Figure 1). Subsequent experiments (Cohen *et al.*, 1998) demonstrated that our initial guess that inhibition by anti-Cdk2 aptamers was competitive proved to be correct. They also revealed an important and unanticipated result: that most of the aptamers that inhibited Cdk2-dependent phosphorylation of histone H1 had no effect on its ability to phosphorylate another substrate, the human Retinoblastoma (Rb) protein. The simplest interpretation of this result is that these anti-Cdk2 aptamers bind to sites near the vicinity of the kinase active site needed for recognition of H1, but not of Rb.

Cohen then introduced these aptamers into human cells, and observed that, as predicted, anti-Cdk2 aptamers retarded the passage of cells through the cell cycle. Cell cycle inhibition was not complete, however. Taken together, Cohen's results suggested that Cdk2 has at least two kinds of substrates, operationally distinguished by the anti-Cdk2 aptamers, and that introduction of the aptamers into mammalian cells blocks phosphorylation of at least one substrate useful for cell cycle transit, but that Cdk2 has at least one other substrate not inhibited by these aptamers whose phosphorylation is sufficient to cause the cell to travel through the cell cycle. No Cdk inhibitory protein with this substrate specificity exists in nature. These results thus show that

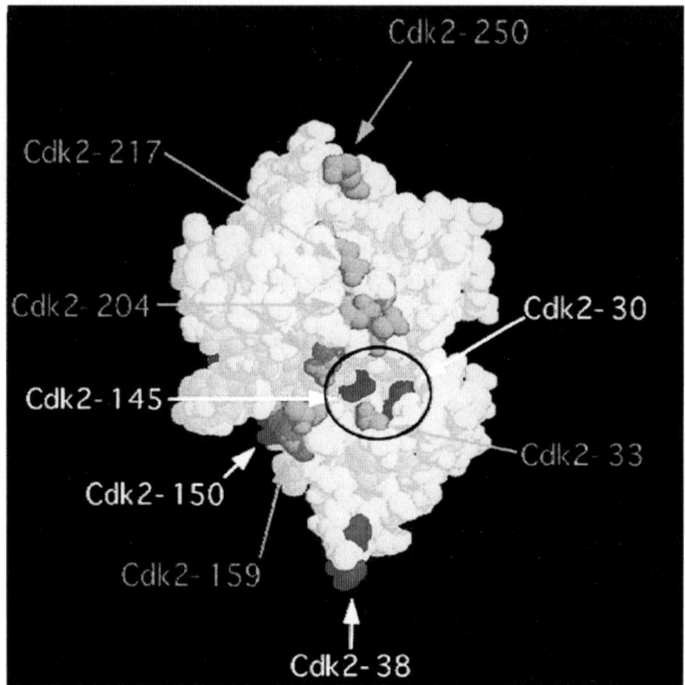

Fig. 1. A Rasmol representation of Cdk2 with mutations that affect binding of peptide aptamer pep8 marked in dark gray and labeled in white, and those that do not affect pep8 binding marked in light gray (from Cohen *et al.*, 1998). The circle denotes the vicinity of the active site; only one mutation that does not affect pep8 binding, Cdk2-K33A, lies within the circle.

if evolution has not provided a protein with the properties required for us to explore a particular biological question, it is now possible to select one.

3   Establishing the function of polymorphisms in coding regions

We developed a "two-bait" interaction trap two-hybrid system (Xu *et al.*, 1997). This system allows one to select proteins that interact with one protein but not with another. One application of this system, in fact our first, was to use it to select peptide aptamers that interacted with one allelic form of the Ras oncoprotein but not the other. Subsequent experiments (Xu *et al.*, submitted), increased the number of these allele-specific (or at least allele-preferring) anti-Ras aptamers. Xu has shown recently that some of these aptamers block Ras function in human cells. These experiments are important because the frequency of human polymorphisms is now variously estimated to be between 1 in 1000 and 1 in 300. Many of these polymorphisms are in coding regions and have functional consequences. It is likely that peptide aptamers will be useful as reagents to probe the function (by inactivation) of specific allelic variants in the diploid background of human and other higher eukaryotic cells.

We have very recently used peptide aptamers and aptamer derivatives in three new ways, all of which should be useful for the analysis of protein function. First, we have used aptamer derivatives that contain ubiquitin ligase domains to ubiquitinate target molecules in living cells. Such decoration with ubiquitin moieties is likely to inactivate the target protein, and other proteins with which it may be physically associated (Colas *et al.*, 2000). Second, we have used aptamers fused to a nuclear localization sequence to drive the target protein into the nucleus. Studies of the effects of regulated mislocalization of cellular proteins will in some cases offer clues to their function (Colas *et al.*, 2000). Finally, and most recently, we have demonstrated that we can use peptide aptamers as dominant genetic agents to cause a phenotype, and then use the aptamers to identify the proteins that contribute to that phenotype. This ability to fish out genes that make up genetic networks will be of great value in the study of now poorly understood phenotypes in human cells and other systems that lack manipulative genetics.

4   Discussion

This paper has reviewed the use of peptide aptamers – engineered proteins, selected from combinatorial libraries – to shed light on the function of proteins in genetic networks. I can frame this work in two ways that may be relevant to the purposes of this volume.

First, note that in the vernacular of the workshop, the peptide aptamer work, by helping us understand protein interactions, is helping us understand the protein interactions that perform cellular computations, particularly those involved in the computations cells perform during signal transduction. We are working on ways to trace the flow of information through these cellular signalling pathways at higher throughput, and eventually, to simulate them.

Second, note that the peptide aptamer work represents the use of combinatorial-synthetic and selective (i.e., "evolutionary") techniques to achieve quite specific molecular recognition and even new biological function. Many interesting biological systems are likely to be approachable only by dominant genetic techniques. The ability to select molecules with new functions moves us closer to the day when, if nature does not provide a given biochemical function useful to analyze a genetic circuit, it will be possible to evolve one (Cohen *et al.*, 1998).

Third, note that, although I have not discussed it here, both a deeper understanding of information flow in genetic networks, and the pieces of technology we must build to understand this flow, ought to provide useful ideas and tools for the construction of biological computational devices. In fact, the work by Xu *et al.* (1997) represents a possible path to such devices. Figure 2 shows that DNA bound Ras in two-bait cells functions as a protein switch that turns on or turns off transcription based on logical protein inputs.

Logical operations on protein inputs

Raf = 1

Sos = 0

GAP = 1

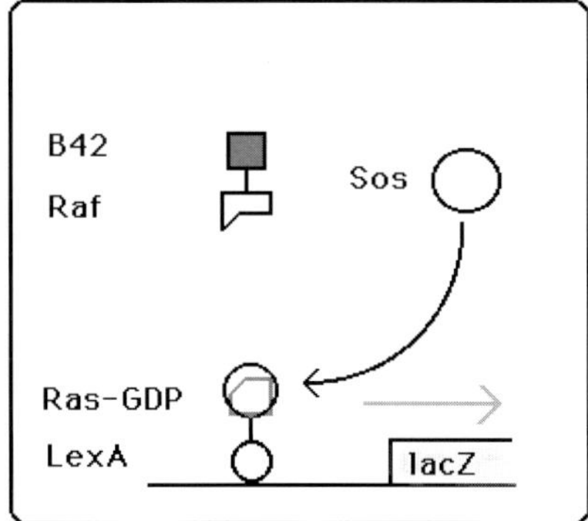

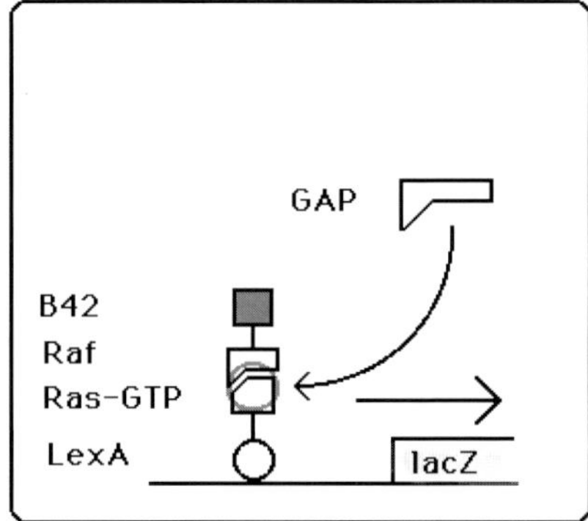

Fig. 2. A protein-based logical switch. Here, changes in transcription result from induced changes in the conformation of a DNA bound LexA–Ras fusion protein. Transcription, which depends in this case on a B42–Raf fusion, is Off if the input to the cell is 0 (Sos) but On if the input is 1 (Gap). Here, at least the switch is fast, even if the input and output are slow. We have proposed (Xu *et al.*, 1998) that elaboration of this protein-based technology, especially if the input and output can be made to depend on processes other than gene transcription, might provide one route to the construction of cell-based logical devices.

In conclusion, we hope that this work will keep us usefully occupied while we move toward a future in which the data in our possession will enable the analytical (computational) prediction of the behavior of living systems. We expect that the development of such analytical frameworks will be one of the major creative enterprises of the early 21st century.

Acknowledgements. The laboratory's work on peptide aptamers is supported by a grant from the National Institute of General Medical Sciences. Work on the computational aspects of protein interactions is supported by a grant from the National Human Genome Research Institute. Simulation work is supported by a grant from the Defense Advanced Research Projects Agency and the Office of Naval Research.

References

1. Cohen, B., Colas, P., and Brent, R. 1998. A novel cell cycle inhibitor isolated from a combinatorial library. *Proc. Natl. Acad. Sci. USA, 95*, 14272-14277.
2. Colas, P., Cohen, B., Jessen, T., Grishina, I., McCoy, J., and Brent, R. 1996. Genetic selection of peptide aptamers that recognize and inhibit Cyclin-dependent kinase 2. *Nature, 380*, 548-550.
3. Colas, P., Cohen, B., Ko Ferrigno, P., Silver, P.A., and Brent, R. 2000. Targeted modification and transportation of intracellular proteins. *Prof. Natl. Acad. Sci. USA, 97*, 13720–13725.
4. Fabbrizio, E., Le Cam, L., Polanowska, J., Lamb, N., Brent, R., and Sardet, C. 1999. Inhibition of mammalian cell proliferation by genetically selected peptide aptamers that functionally antagonize E2F activity. *Oncogene, 18*, 4357–4363.
5. Finley, R. L., and Brent, R. 1994. Binary and ternary interactions between Drosophila cell cycle regulators. *Proc. Natl. Acad. Sci. USA 91*, 12980-12984.
6. Lok, W. L., Cohen, R., and Brent, R. 1998. Interaction 1.0, downloadable from `www.molsci.org`.
7. Xu, C. W., Mendelsohn, A., and Brent, R. 1997. Cells that register logical relations among proteins. *Proc. Natl. Acad. Sci. USA, 94*, 12473-12478.
8. Xu, C. W., Lao, Z., and Brent, R. Inactivation of Ras function by allele specific peptide aptamers. *submitted.*

Computational Aspects of Gene (Un)Scrambling in Ciliates

Andrzej Ehrenfeucht, David M. Prescott, and Grzegorz Rozenberg

Abstract. Ciliates, a very ancient group of organisms, have evolved extraordinary ways of organizing, manipulating, and replicating the DNA in their micronuclear genomes. The way that ciliates transform genes from their micronuclear (storage) form into their macronuclear (expression) form constitutes a very interesting case of "DNA computing *in vivo*". In this paper we investigate in detail one aspect of this transformation, viz., *gene (un)scrambling*. In particular, we use the formal framework of *pointer reduction systems* to investigate the computational aspects of gene (un)scrambling.

1 Introduction

DNA computing is an interdisciplinary research area concerned with computational processes where data are represented by DNA molecules and the processing of data is achieved through microbiological operations. The area has been initiated by the (now famous) paper by Adleman [1]. The research in this area has led to novel paradigms for computation, e.g., [9], and to very interesting experimental research testing basic biological principles of novel methods of computing, see, e.g., [4]. The fast growing research in this area includes both *in vitro* and *in vivo* computing. A recent series of papers by Landweber and Kari, e.g., [5], and [6], has brought to the attention of DNA computing community the beauty of the computational processes taking place (*in vivo*) in ciliates during the transformation of genes from their micronuclear to their macronuclear form. In this paper we continue the investigation of these computational processes, and in particular we investigate the computational aspects of gene (un)scrambling.

The paper is organised as follows.

In the first part (Sects. 1–4) we survey the main aspects of the organization and manipulation of the DNA in the micronuclear and macronuclear genomes. In Sects. 5 through 11 of the second part we introduce pointer reduction systems as a formal framework for the investigation of gene unscrambling, and discuss within this framework the computational aspects of unscrambling using the operation of loop excision on direct repeats, and the (new) operation of hairpin excision/reinsertion on inverted repeats. In Sect. 12, closing the second part, we postulate some possible origins of scrambling. Then in the last section of this paper, Sect. 13, we discuss research directions that continue the work presented in this paper.

We made a special effort to present our formal framework in an informal fashion, so that it is understandable for a motivated biologist interested in gene processing in ciliates.

2   Micronuclear and Macronuclear DNA in Ciliates

Ciliates are a very ancient group of organisms. A recent estimate places their origin at $\approx 2 \times 10^9$ years ago [14]. During their evolution the ciliates have diversified into many groups, containing tens of thousands of genetically different organisms, but retaining two uniting features: cilia and nuclear dualism. During evolution, the genomes of the hypotrich group of ciliates have undergone profound modifications in DNA organization. These modifications, in turn, require extraordinary genome processing, consisting of cutting, elimination, splicing, and reorganization of DNA sequence, when a germline nucleus develops into a somatic nucleus after cell mating.

A ciliate contains two kinds of nuclei (nuclear dualism) in the same cell: a micronucleus, which serves as a germline nucleus and is used in cell mating, and a macronucleus, which serves as a somatic nucleus and provides the RNA transcripts to operate the cell. During mating, two cells adhere to one another and form a connecting cytoplasmic channel. The micronuclei in the two cells undergo meiosis, and the two cells exchange haploid micronuclei through the cytoplasmic channel. An exchanged haploid micronucleus fuses with a resident haploid micronucleus to form a new diploid micronucleus in each cell. The channel closes, the cells separate, and the new diploid micronucleus in each cell divides by mitosis. One of the daughter micronuclei develops into a new macronucleus during the next 60 hours, and concomitantly, the old macronucleus and unused haploid micronuclei are destroyed.

A major DNA processing event in the development of a micronucleus to a macronucleus is the excision of all genes from the chromosomes. The DNA molecules in the chromosomes of the micronucleus are very long, consisting of hundreds of kilobase pairs (kb). Genes occur individually or in groups dispersed along a DNA molecule, separated by long stretches of spacer DNA (Fig. 1).

Micronuclear Chromosome

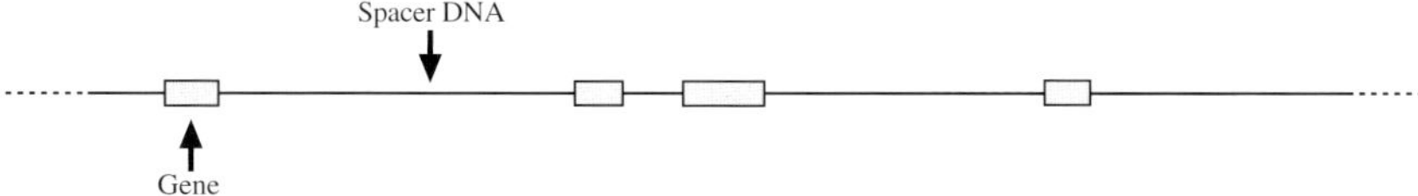

Fig. 1. Arrangement of genes in a micronuclear chromosome. Single genes and groups of genes are separated by very long spacers of AT-rich, noncoding DNA.

Macronuclear development begins with multiple replications of these long molecules to form polytene chromosomes with 64 or 128 identical DNA copies. The polytene chromosomes are then degraded, releasing multiple copies of $\approx24,000$ different gene-size DNA molecules. During chromosomal degradation all the spacer DNA, which accounts for $\approx95\%$ of the germline DNA sequence complexity, is destroyed. How the organism recognizes and excises genes from the polytene chromosomes and differentially destroys spacer DNA is not known. Examination of the DNA region around the junctions of several micronuclear genes with their flanking spacer DNA has revealed no signal sequences in the *Oxytricha/Stylonychia* group, e.g., consensus sequences, that might direct the excision process.

When gene-size molecules are released, telomeric sequences containing $5'G_4T_4/C_4A_43'$ are synthesized on their ends. These molecules undergo further replication to reach a copy number of $\approx1,000$ on average in the mature macronucleus. As a result of these events the genome in the somatic macronucleus consists entirely of highly amplified, gene-size molecules with the generalized structure shown in Fig. 2.

Fig. 2. A generalized model of a macronuclear DNA molecule, consisting of a $5'$ nontranslated leader, a coding sequence, and a $3'$ nontranslated trailer. The molecule is capped at both ends by telomere sequence and $3'$ single-strand tails.

A gene-size molecule consists of a $5'$ nontranslated leader segment that functions in transcriptional control, a gene-coding region (open reading frame, or ORF, for protein encoding genes), and a $3'$ nontranslated trailer that presumably contains a transcriptional stop signal. Telomeric sequences of $3'G_4T_45'/5'C_4A_43'$ are present at each end, with a 16-base, $3'$ single stranded tail of $(3'G_4T_45')_2$. The telomeric sequences are required to replicate the ends of molecules and to protect molecular ends from nuclease digestion.

3   Internal eliminated segments in micronuclear genes

Comparisons of macronuclear DNA molecules with their micronuclear precursors reveal that micronuclear precursors are interrupted by short, noncoding, AT-rich (75% to 100% AT) segments. These are called internal eliminated segments, or IESs, because they are excised and destroyed in the formation of macronuclear gene–size molecules. An example is the gene encoding β telomere binding protein (βTP). The micronuclear version of this gene in *Oxytricha nova* contains three IESs, one toward the end of the ORF and two in the $3'$ trailer (Fig. 3).

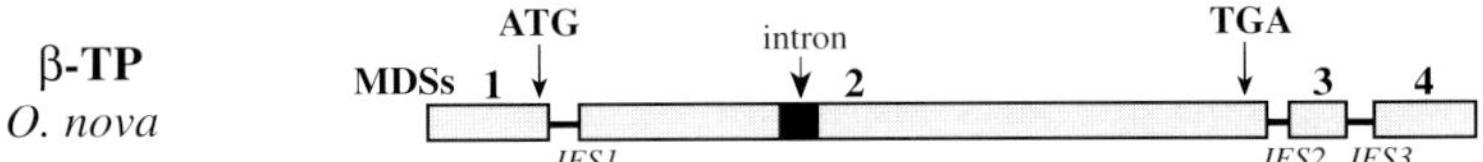

Fig. 3. The micronuclear βTP gene of *O. nova*. Three IESs create four MDSs (blocks). *ATG* = the start codon. *TGA* = the stop codon.

These IESs are respectively 32, 24, and 39 base pairs (bp) long and are composed of completely different sequences. The four regions in the βTP gene separated by IESs are called macronuclear-destined segments, or MDSs. The IESs are flanked by short repeat sequences in the adjoining MDSs (Fig. 4).

MDS 1   MDS 2

—-ATCAAAAT$\underline{GTCC}$Actcacatgcaaaataatatattatattgtta
$\underline{GTCC}$AAAGGCGCAT—-

Fig. 4. IES 1 in the micronuclear βTP gene of *O. nova*. The 32-bp IES (81% AT) is flanked by a repeat of *GTCCA* in MDSs 1 and 2. The *ATG* start codon is in bold letters. From [7].

IES 1 is flanked by *GTCCA*, IES 2 by *TAAAGT*, and IES 3 by *AGTC*. One copy in each pair of repeats is excised along with the IES during macronuclear development. The other copy is found at the splice site of the adjoining MDSs in the mature macronuclear gene.

During evolution of hypotrich species, IESs change in number, length, position, and sequence. This is illustrated by a comparison of the βTP gene in three species: *O. nova*, *O. trifallax*, and *Stylonychia mytilus*. *O. nova* contains three IESs, *O. trifallax* contains six, and *S. mytilus* contains two (Fig. 5).

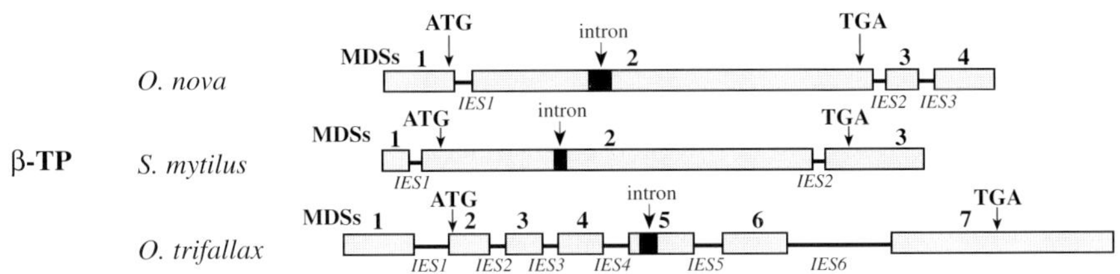

Fig. 5. Diagrams of the micronuclear genes encoding βTP, illustrating the variation in number and position of internal eliminated segments (IESs) in a gene from species to species. From [11].

None of the IESs corresponds in length, position, or sequence from species to species. This means that IESs can be added to or lost from the βTP gene during evolution of these species from a common ancestor. The repeat sequences that flank IESs are presumed to have an essential role in IES excision and/or MDS splicing, but they are too short to define by themselves the ends of an IES or IES excision. For example, the 292-bp IES 6 in the βTP gene of *O. trifallax* is flanked by the repeat sequence *AGT*. This trinucleotide also occurs five times within the IES, at bp 38, 57, 129, 187, and 213, again 44 bp upstream of the IES, and three times in the coding sequence, at 5, 81, and 91 bp following the IES. Thus, additional information is needed to identify the appropriate *AGT* repeats in IES excision and/or MDS splicing events. The source and nature of the additional information is currently unknown, but experiments in the very distantly related ciliate, *Paramecium*, clearly implicate the old macronucleus as a source of a template for guiding IES excision [2].

4   Scrambled Micronuclear Genes

Nine micronuclear genes and their macronuclear counterparts have been sequenced in *Oxytricha* and *Stylonychia* species. All are interrupted by IESs, and in three of the nine genes, the MDSs are in scrambled disorder.

The actin I gene of *O. nova* contains eight IESs and nine MDSs in the disorder 3–4–6–5–7–9–2–1–8 (Fig. 6).

actin I

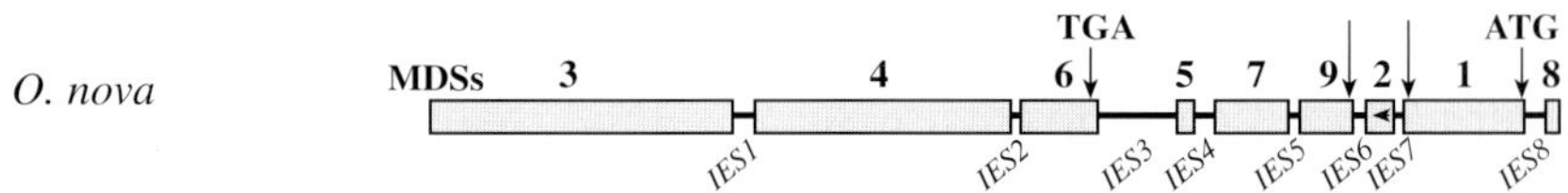

Fig. 6. The scrambled micronuclear version of the actin I gene in *O. nova*. MDSs are blocks, and IESs are lines connecting MDSs. *ATG* = the start codon. *TGA* = the stop codon. 3′TAS and 5′TAS are the 5′ and 3′ telomere addition sites. From [11].

MDSs 3 and 4 are not scrambled relative to each other, and the intervening IES is flanked by repeats with the sequence *AATC*. The other seven MDSs are scrambled. In the case of scrambled MDSs, the IESs are flanked by longer repeats, ranging from 9 to 13 bp (Table 1).

MDS 2 is inverted with respect to all other MDSs, and the repeat pair in MDS 2 and 1 and the repeat pair in MDS 2 and 3 arc inverted sequences, i.e., 5′*CTT ACT AC AC AT* 3′ and 5′*ATG TGT AGT AAG* 3′ for MDSs 1 and 2. We have proposed [13] that the IESs in the micronuclear actin I gene are

Left end of MDS	MDS	Right End of MDS
5′ telomere addition site	1	*CTTACTACACAT*
CTTGACGACTCC	2	*ATGTGTAGTAAG*
GGAGTCGTCAAG	3	*AATC*
AATC	4	*CTCCCAAGTCCAT*
CTCCCAAGTCCAT	5	*GCCAGCCCC*
GCCAGCCCC	6	*CAAAACTCTA*
CAAAACTCTA	7	*CTTTGGGTTGA*
CTTTGGGTTGA	8	*AGGTTGAATGA*
AGGTTGAATGA	9	3′ telomere addition site

Table 1. Direct Repeats at MDS/IES Junctions in the Actin I Gene

excised and MDSs unscrambled and spliced by folding a DNA molecule so as to align all the pairs of repeats (Fig. 7).

Folding includes inverting MDS 2 so that the inverted repeats now align as direct repeats. Recombination between the repeats in a pair would remove the IES between them, remove one copy of the repeat, and splice the flanking MDSs. This puts the nine MDSs in the correct order for transcription of the gene in the macronucleus. In fact, the start codon, *ATG*, is reconstituted by joining *AT* at the 3′ end of MDS 1 to G at the 5′ end (after inversion) of MDS 2.

The gene encoding α telomere binding protein (αTP) is interrupted by 13 IESs, creating the nonrandom, scrambled pattern of MDSs: 1–3–5–7–9–11–2–4–6–8–10–12–13–14 (Fig. 8).

MDSs 12–13–14 are not scrambled relative to one another; the IESs that separate them are flanked by pairs of short repeats with the sequences *CCCAA* and *ACT*. The 11 remaining IESs are scrambled in a nonrandom, odd/even pattern. None of the MDSs is inverted. The sequences of the repeats are given in Table 2.

Folding of the molecule to align pairs of repeats, followed by recombination between the repeats in each pair, would remove all IESs and splice MDSs in the orthodox order (Fig. 9).

The third scrambled gene encodes the large catalytic subunit of DNA polymerase α (DNA pol α). In *O. nova* it consists of 44 IESs that divide the gene into 45 nonrandomly scrambled MDSs in the largely odd/even pattern of the type in the αTP gene: ——27–26–24–22–20–18–16–14–12–10–8–6–4–1–2–3–5–7– 9–11–13–15–17–19–21–23–25–27–28–30–32–34–36–38–40–42– 44–45 (Fig. 10).

MDSs 29–31–33–35–37–39–41–43 are missing from the main body of the gene, and their location has not been identified. We suspect that the eight missing MDSs have been displaced by an inversion between flanking DNA and the IES between MDSs 27 and 29. The repeats flanking IESs between nonscrambled MDSs 1–2–3, 26–27, and 44–45 are short, i.e., respectively, 2,

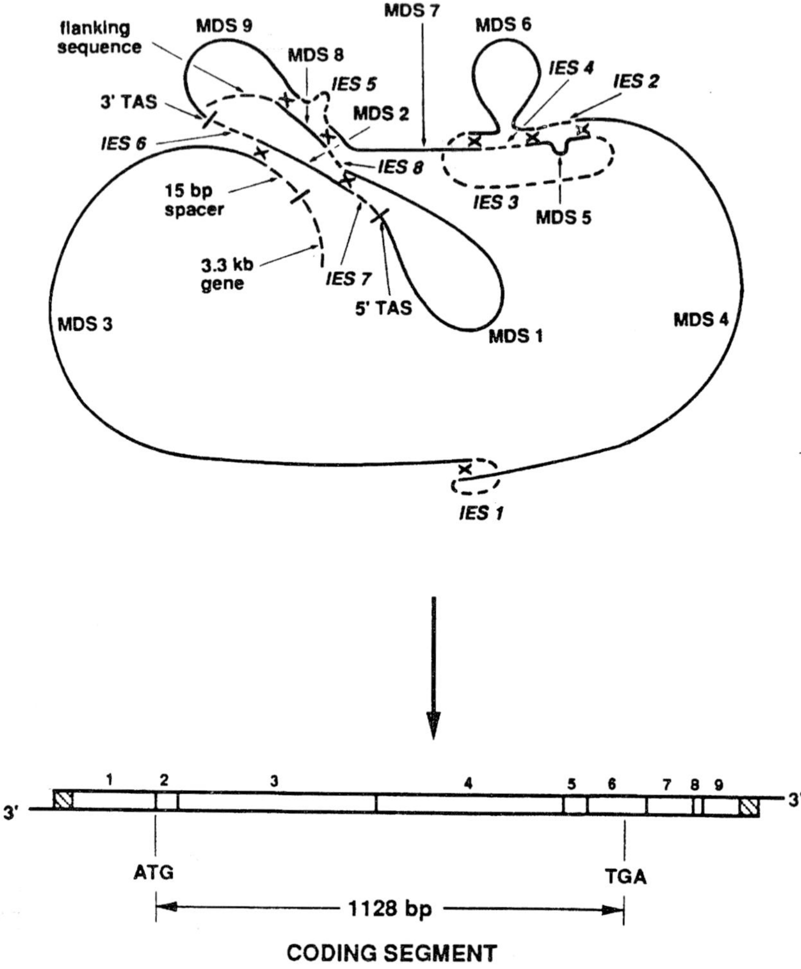

Fig. 7. In the recombination model for unscrambling of the actin I gene in *O. nova*, the micronuclear DNA is folded to align pairs of repeats at IES-MDS junctions (top). Excision of the actin I gene and addition of telomeres produces the exact sequence determined for the version in the macronucleus (bottom). From [13].

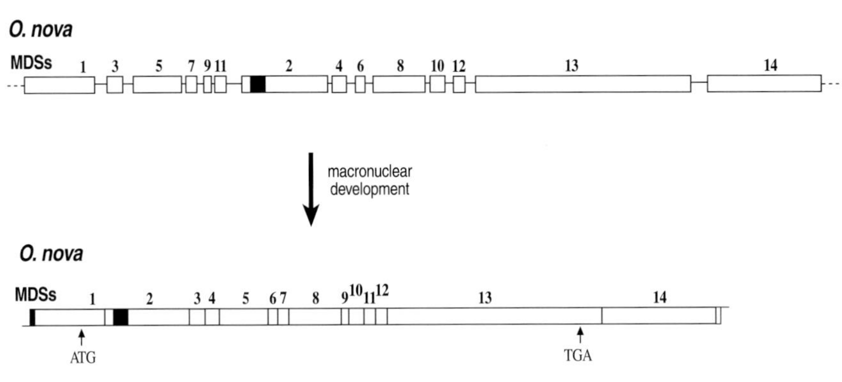

Fig. 8. The scrambled micronuclear αTP gene of *O. nova*. The 14 MDSs are in a nonrandom, odd/even pattern. During macronuclear development the MDSs are spliced in the orthodox order. MDSs are blocks. IESs are lines connecting MDSs. The black box in MDS 2 is an intron. Revised from [8].

Left end of MDS	MDS	Right End of MDS
5′ telomere addition site	1	*GAAGGCGCTGC*
GAAGGCGCTGC	2	*GCCACCCTC*
GCCACACACA	4	*AGAGCTACCCTC*
AGAGCTACCCTC	5	*TCAAGCAAG*
TCAAGCAAG	6	*TTGAGAAGAACGA*
TTGAGAAGAATGA	7	*AGAACCTGA*
AGAACCTGA	8	*AAGGAC*
AAGGAC	9	*AAGTGTTCT*
AAGTGTTCT	10	*AGAACT*
AGAACT	11	*GAATCAGATCAGCCACTTA*
GAATGAGATCAGCCACTTA	12	*CCCAA*
CCCAA	13	*ACT*
ACT	14	3′ telomere addition site.

Table 2. Direct Repeats at MDS/IES Junctions in the Micronuclar αTP Gene. Underlined nucleotides indicate base-pair mismatches between repeats.

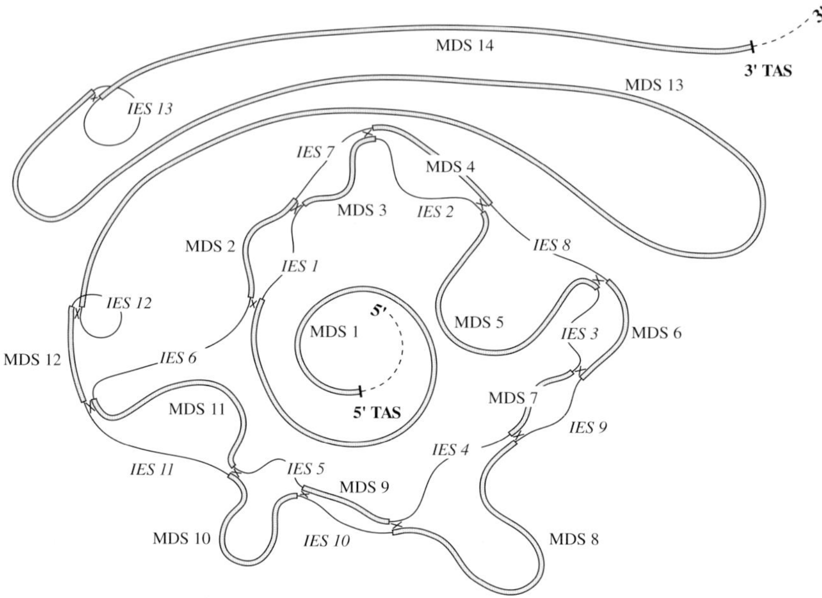

Fig. 9. Folded model of micronuclear DNA containing the gene encoding αTP. The nonrandom scrambling pattern gives rise to concentric circles when pairs of repeats are aligned. Recombination between the two members in a pair of repeats removes IESs 1 to 11 as a single circle, removes IESs 12 and 13 as separate circles, and splices the 14 MDSs in the order present in the functional macronuclear gene. From [8].

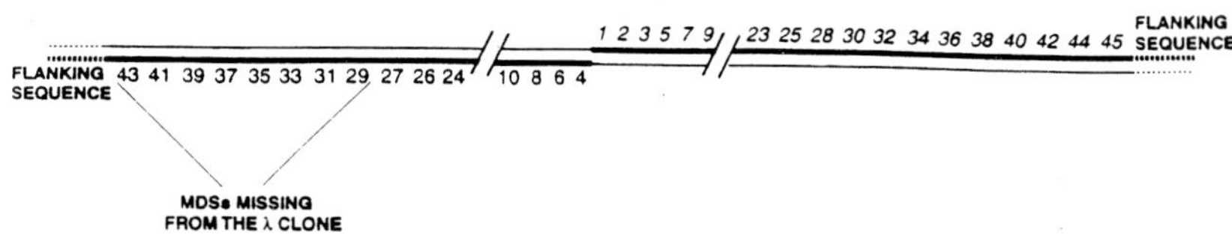

Fig. 10. MDSs in the scrambled micronuclear DNA pol α gene of *O. nova*. The odd numbered MDSs 29–43 are absent from the clone and their arrangement is hypothetical. The thin strand is transcribed. From [3].

4, 4, and 4. Repeats flanking IESs between scrambled MDSs range in length from 6 to 19 bp (Table 3).

Left end of MDS	MDS	Right End of MDS
5′ telomere addition site	1	TA
TA	2	$GAGT$
$GAGT$	3	$GAATGGCAATA$
$GAATGGCAATA$	4	$AAGAATGC$
$AAGAATGC$	5	$TAAGAAT$
$TAAGAAT$	6	$ACTCTTTCTTT$
$ACTCTTTCTTT$	7	$TGAAGAAAATA\underline{C}CG$
$TGAAGAAAATA\underline{A}CG$	8	$TGCTCT$
$TGCTCT$	9	$AGAGCTATT$
$AGAGCTATT$	10	$AAGAATA\underline{GG}CAA$
$AAGAATA\underline{T}GCAA$	11	$TTTGAG\underline{C}TG$
$TTTGAGTTG$	12	$TTAAGATCA$
$TTAAGATCA$	13	$GAGCTAA$
$GAGCTAA$	14	$TGGTTCAA$
$TGGTGCAA$	15	$ATGATTTGAA$
$ATGATTTGAA$	16	$AAAACTT$
$AAAACTT$	17	$GA\underline{G}ATAGCAAT$
$GA\underline{A}ATAGCAAT$	18	$ATGGAAAG$
$ATGGAAAG$	19	$AGAGACA$
$AGAGACA$	20	$TGGCTCATAAT$
$TGGCTCATAAT$	21	$AACAAGAA$
$AACAAGAA$	22	$AATTATG$
$ATTATG$	23	$ATTCCAA$
$ATTCCAA$	24	$TAGCTAAG$
$TAGCTAAG$	25	$TTGACTTTG$
$TTGACTTTG$	26	$AGAC$
$AGAC$	27	$CTAACAGTATGTATGG$
$CTAACAGTATGTATGG$	28	Not determined
Not determined	29	$TCTT$
$TCTT$	30	3′ telomere addition site

Table 3. Direct Repeats at MDS/IES Junctions in the Micronuclear DNA pol α Gene. MDSs 29, 31, 33, 35, 37, 39, 41 and 43 have not yet been located in the micronuclear genome and the relevant repeat sequences have not been determined, although MDSs 28, 30, 32, 34, 36, 38, 40 and 42 are present in the λ micronuclear clone and have been sequenced. Mismatches are underlined. From [3].

Thus, the rule holds that scrambled MDSs contain longer (avg. length = 11 bp) repeats than do nonscrambled MDSs (avg. length = 4 bp). The longer repeats are theoretically of adequate length to guide unscrambling unambiguously. The folding/recombination model for unscrambling is shown in Fig. 11.

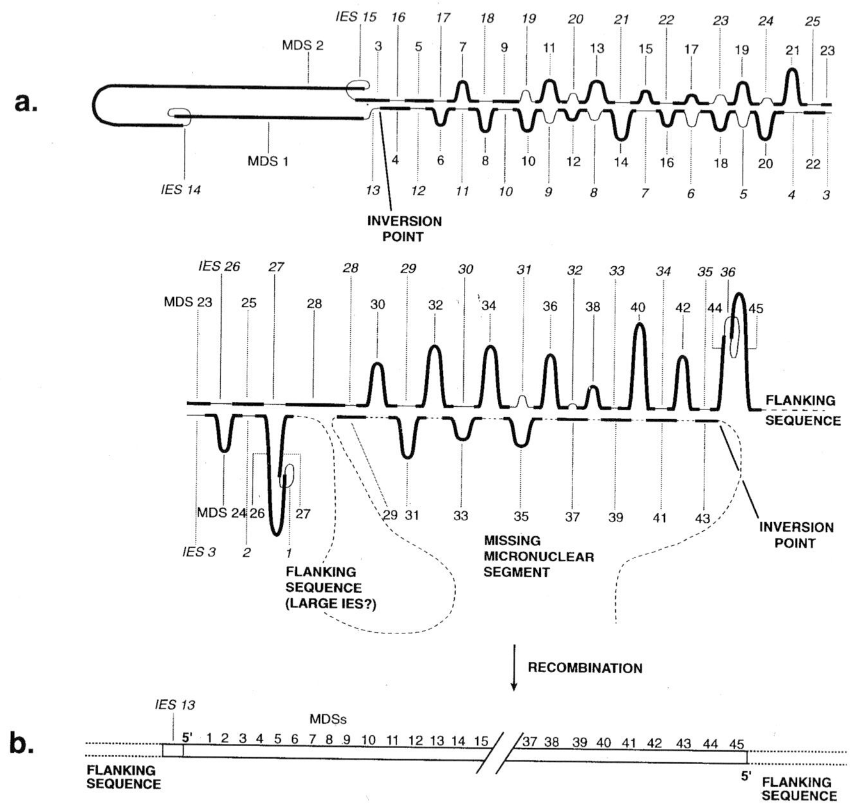

Fig. 11. Recombination model for unscrambling of the micronuclear DNA pol α gene in *O. nova*. (a) Folding of the micronuclear DNA to bring the two members of each repeat pair into parallel alignment. Repeats occur where MDSs overlap at their ends. The MDSs and IESs are drawn to approximately proportional lengths. The missing, odd-numbered MDSs 29–43 are hypothesized to be contained in a single segment of DNA elsewhere in the genome. These could possibly be separated from MDS 27 in the main body of the gene by a very long (> 2000 bp) IES. The lengths of the missing MDSs are calculated from the macronuclear gene sequence, knowing the sequences of even-numbered MDSs 28–44 in the micronuclear clone. The unknown IESs between the missing MDSs are indicated as short dotted lines. An inversion point is present between MDS 1 and MDS 4. The second inversion point is presumably present in the flanking DNA that extends from MDS 27 or 43. (b) Splicing of MDSs in the orthodox order by recombination removes all IESs from the micronuclear gene without excision of the gene from the chromosome. From [3].

The folding/recombination model is based on intramolecular folding and recombination. But, because unscrambling occurs in polytene chromosomes, it is formally possible that unscrambling is an intermolecular process among the 64 or 128 identical chromosomal DNA molecules. Intermolecular unscrambling could still be guided by the repeat pairs in the ends of MDSs. Extrapolating from the nine micronuclear genes studied so far, we estimate that at least 150,000 recombinations between repeat sequences are needed per haploid genome to remove all IESs and splice all MDSs in the orthodox order, both in intramolecular and intermolecular models of unscrambling.

5   Concerted and Sporadic Insertion of IESs

The odd/even patterns of MDSs in the αTP and DNA pol α genes can be accounted for by the simultaneous insertion of multiple IESs, as illustrated in Fig. 12 for the αTP gene in *O. nova*.

In addition to concerted insertion of multiple IESs, individual IESs have also been inserted sporadically in the αTP gene, possibly before or after creation of the odd/even scrambled pattern of MDSs. In addition to concerted insertion of IESs in the DNA pol α gene, sporadic insertions also occurred, creating nonscrambled MDSs. Subsequent to the concerted insertion of IESs, the gene underwent the inversion between MDSs 1 and 4 and probably a second inversion between MDSs 27 and 29.

The significance of IESs and MDS scrambling is not known, but these phenomena may reflect a malleability of DNA that facilitates and accelerates gene evolution, possibly by shuffling of MDSs into new combinations.

6   Basic Operations

In Sects. 2 through 5 we have surveyed some of the basic concepts and issues concerning the organization and manipulation of the DNA in the micronuclear and macronuclear genomes of ciliates. We move now to a formal study of this processing of the DNA. We will be especially interested in the computational aspects of gene (un)scrambling which we will study in the formal framework of pointer reduction systems introduced in this paper. *We made a special effort to present this formal framework in an informal fashion, so that it is understandable for a motivated biologist interested in gene processing in ciliates.*

From the computational point of view, the operations of insertion and excision are crucial for the forming of IESs and their removal.

The operation of *insertion* is illustrated in Fig. 13.

Thus the operation of insertion in this form creates a direct repeat (α, α).

Here and throughout the paper $\bar{\alpha}$ denotes the *inversion* of a string α. The inversion of α is obtained by the composition of two operations: *reverse* and

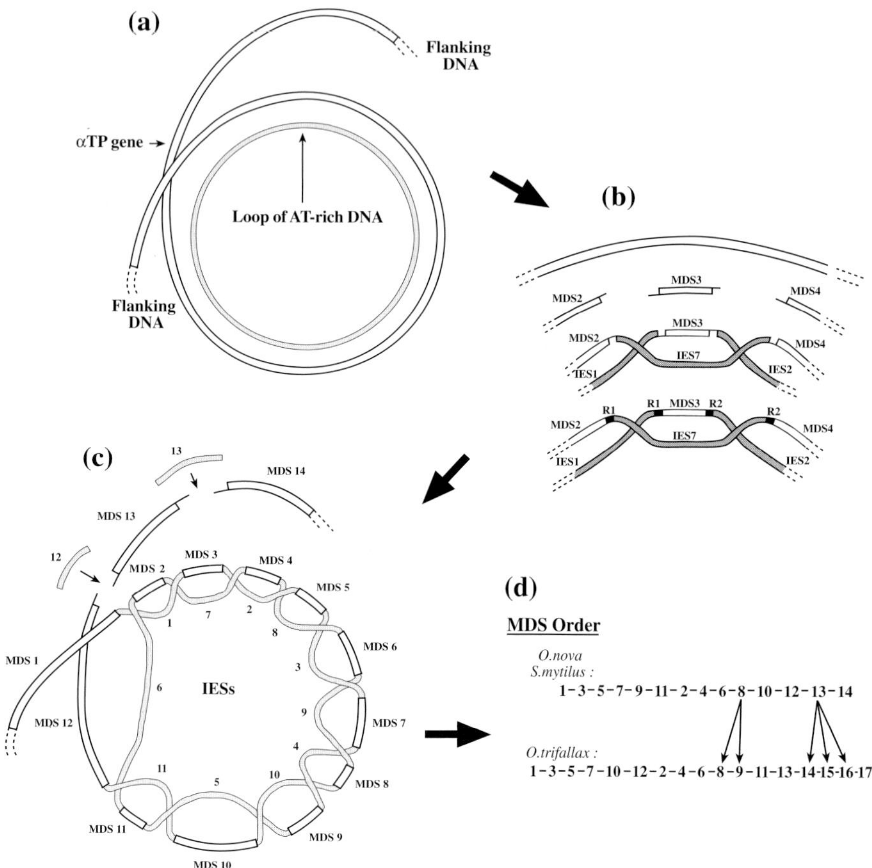

Fig. 12. A hypothesis to account for the origin of the nonrandomly scrambled structure of the micronuclear αTP genes in *O. nova, S. mytilus,* and *O. trifallax.* (a) A loop of AT-rich DNA is aligned with an αTP gene that contains no IESs. (b) Staggered cuts are made in the αTP gene, followed by recombination of sections of AT-rich DNA with the single-stranded overhangs resulting from staggered cuts. Single-strand gaps are filled in to create repeat pairs (R1, R2, etc.) (c) Insertion of multiple IESs at staggered cuts in the αTP gene (as shown in b) creates 11 IESs separating MDSs 1 through 12. MDSs 13 and 14 are created by separate (sporadic) insertions. (d) The regular odd/even pattern of MDSs in *O. nova* and *S. mytilus* evolve into the *O. trifallax* pattern by insertion of three additional IESs, dividing MDS 8 into MDSs 8 and 9, and dividing MDS 13 into MDSs 14, 15, and 16. MDS 14 in *O. nova/S. mytilus* becomes MDS 17 in *O. trifallax.* From [10].

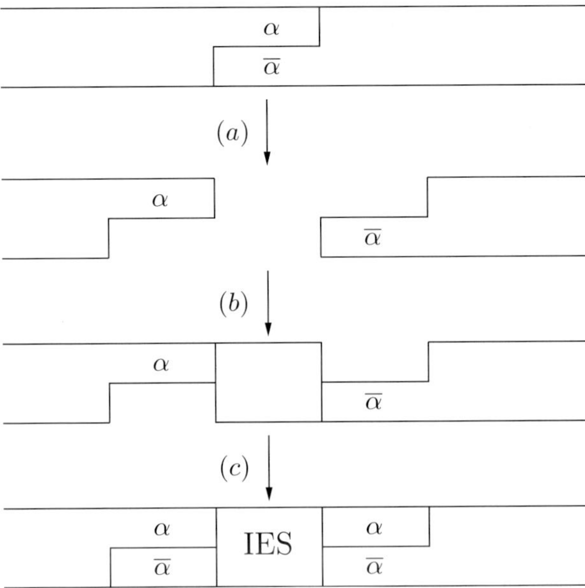

Fig. 13. An insertion of an IES into a micronuclear gene. (a) A staggered cut is introduced. (b) A segment of AT-rich DNA (the nascent IES) is ligated to the single stranded overhangs. (c) Fill-in of the gaps creates a pair of direct repeats (α, α) in the MDSs that flank the new IES.

complement (applied in either order). The reverse of α (called also the *mirror image* of α) results from reading α backwards: thus for $\alpha = CGT$ its reverse is TGC. The (Watson–Crick) complement of a string results by replacing each letter (each nucleotide) by its Watson–Crick complement: thus for the string TGC its complement is ACG; consequently the inversion of $\alpha = CGT$ is $\bar{\alpha} = ACG$.

The operation of *excision* is illustrated in Fig. 14.

Thus the operation of excision in this form creates two molecules, a linear and a circular one, while it takes as its argument a loop-folded molecule aligned on a direct repeat. Hence we refer to this operation of excision as the *(loop, direct repeat)-excision*, or *ld-excision* for short.

We will consider now the operation of excision (and then reinsertion) on different arguments: hairpin-folded molecules aligned on an inverted repeat. This operation is referred to as the *(hairpin, inverted repeat)-excision/reinsertion*, or *hi-excision/reinsertion* for short. An inverted repeat is illustrated in Fig. 15.

Then the molecule from Fig. 15, hairpin-folded and aligned on the indicated inverted repeat is illustrated in Fig. 16.

We have defined above the inversion $\bar{\alpha}$ for a string (single stranded molecule) α. This definition extends naturally to "double strings" (double stranded

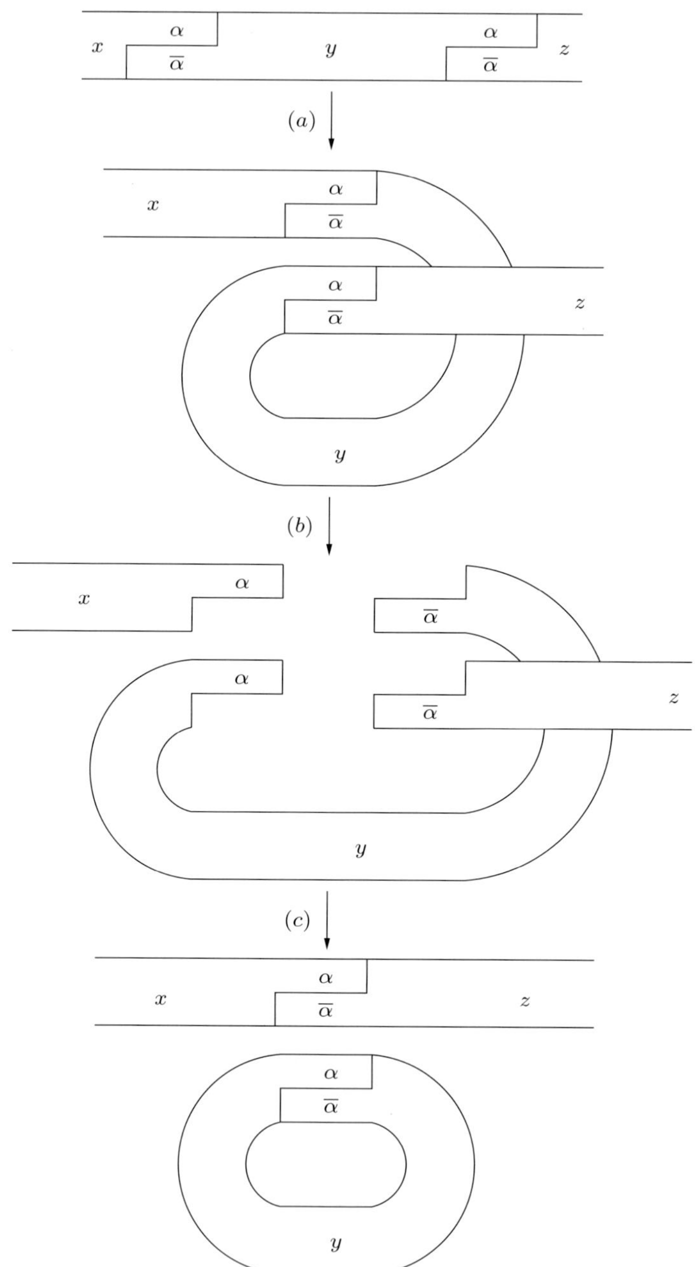

Fig. 14. An excision on a loop of an IES from a micronuclear gene. (a) A loop is formed. (b) Staggered cuts are introduced. (c) A homologous recombination takes place.

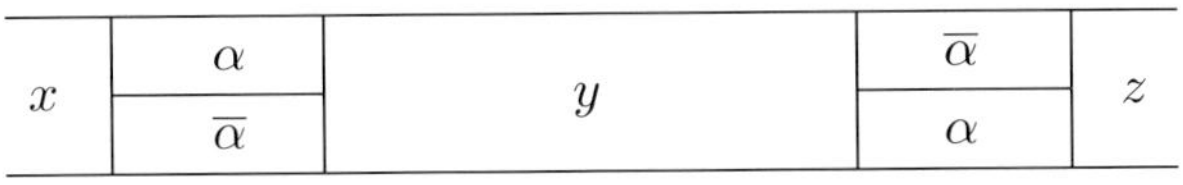

x	α	y	$\overline{\alpha}$	z
	$\overline{\alpha}$		α	

Fig. 15. An inverted repeat $(\alpha, \bar\alpha)$

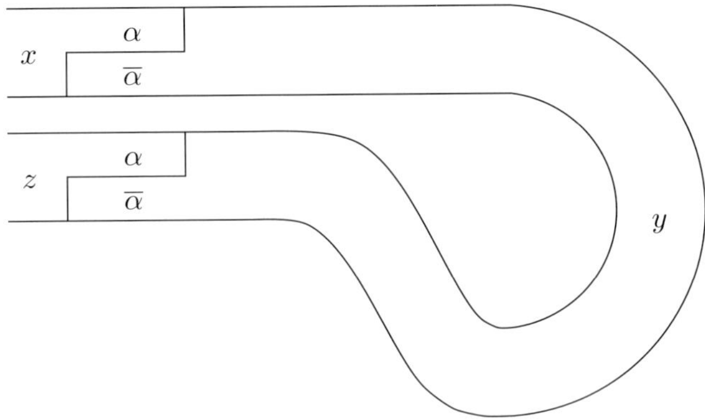

Fig. 16. Hairpin-folded molecule aligned on inverted repeat $(\alpha, \bar\alpha)$

molecules) - the inversion of a double string α is obtained by the composition of two operations: *reverse* and *exchange* (applied in any order). Again, the reverse of α results from reading α backwards: thus, for $\alpha = \frac{ACGAT}{TGCTA}$ the reverse of α is $\frac{TAGCA}{ATCGT}$. The exchange of a double string results by exchanging its two single strings for each other: thus for the double string $\frac{TAGCA}{ATCGT}$ its exchange is $\frac{ATCGT}{TAGCA}$; consequently the inversion of $\alpha = \frac{ACGAT}{TGCTA}$ is $\bar\alpha = \frac{ATCGT}{TAGCA}$. Note that for the double string corresponding to a perfect duplex, the exchange operation is the same as exchanging each of the single strings by its complement, in this sense the exchange operation generalizes the complement operation for single strings. As another example, consider the double string (double stranded molecule) with overhangs $\beta = \frac{ACGTCCGT}{TGCTGCAG}$.

Then, the inversion of β is $\bar\beta = \frac{GACGTCGT}{TGCCTGCA}$.

The operation of hi-excision/reinsertion is illustrated in Fig. 17. The inversion inverts here the entire loop, i.e., y becomes $\bar y$. The inverting effect of hi-excision/reinsertion is illustrated in "linear form" in Fig. 18. The key feature of the hi-excision/reinsertion is that it yields one molecule from one molecule, and so we deal here with *intramolecular processing* !

7   String Pointer Reduction Systems – Informal

The aim of our investigation is to formulate formal frameworks for the unscrambling process, where the unscrambling operations are the ld-excision and hi-excision/reinsertion. The formal framework that we discuss in this section is that of string pointer reduction systems. We begin with an example.

Assume that a macronuclear gene γ consists of seven MDSs: $M_1, M_2, ...,$ M_7, where the order $M_1 M_2 ... M_7$ is the transcription order (the orthodox order). This is illustrated in Fig. 19.

Each MDS M_i has the structure $(i, m_i, i+1)$, except that $M_1 = (b, m_1, 2)$ and $M_7 = (7, m_7, e)$ where b stands for "begin" and e stands for "end". We refer to i as the *incoming pointer* of M_i, and to $i+1$ as the *outgoing pointer* of M_i. Clearly, for $i < 7$, the outgoing pointer of M_i and the incoming pointer of M_{i+1} form a direct repeat pair – these pairs help to guide the unscrambling. For didactic reasons we have chosen here to present the guiding sequences of repeat pairs as integers – it is easier to follow the unscrambling process using this convention.

As a matter of fact, these seven MDSs are put together in γ by overlapping of direct repeats in the way indicated in Fig. 20.

Assume now that the scrambled order of the MDSs $M_1, ..., M_7$ in the micronuclear version of γ is $M_3 M_6 M_2 M_4 M_5 M_1 M_7$, where moreover M_2 and M_4 are inverted. This is illustrated in Fig. 21.

Note that the arrows above the MDSs indicate their polarity: the inverted arrows above M_2 and M_4 indicate that M_2 and M_4 are inverted (and thus have inverted polarities) in this scrambled order.

The key idea is that in order to follow the unscrambling process one does not have to follow what happens to the whole contents of MDSs (and the IESs in between). It suffices to follow what happens to pointers themselves. Successful unscrambling steps combine smaller MDSs into bigger MDSs, which really means removing pointers: a guiding sequence at an end of a smaller MDS, hence a pointer, gets shifted inside a new MDS, and so it no longer functions as a pointer!

We code now the sequence of MDSs from Fig. 21 as follows. Each noninverted M_i is coded as a pair of pointers $i\ i+1$, except that M_1 is translated as 2 and M_7 as 7. We simply omit symbols b and e from our translation – after all they do not correspond to guiding sequences: each guiding sequence has a "partner". Each inverted M_i is translated as a pair of pointers $\overline{i+1}$ $\overline{i}$, hence we reverse the order of pointers, and moreover we bar each of them (again we would omit $\bar{b}$ and $\bar{e}$ if they would be present, which is not the case in our example). Thus the sequence of MDSs from Fig. 21 is coded into the

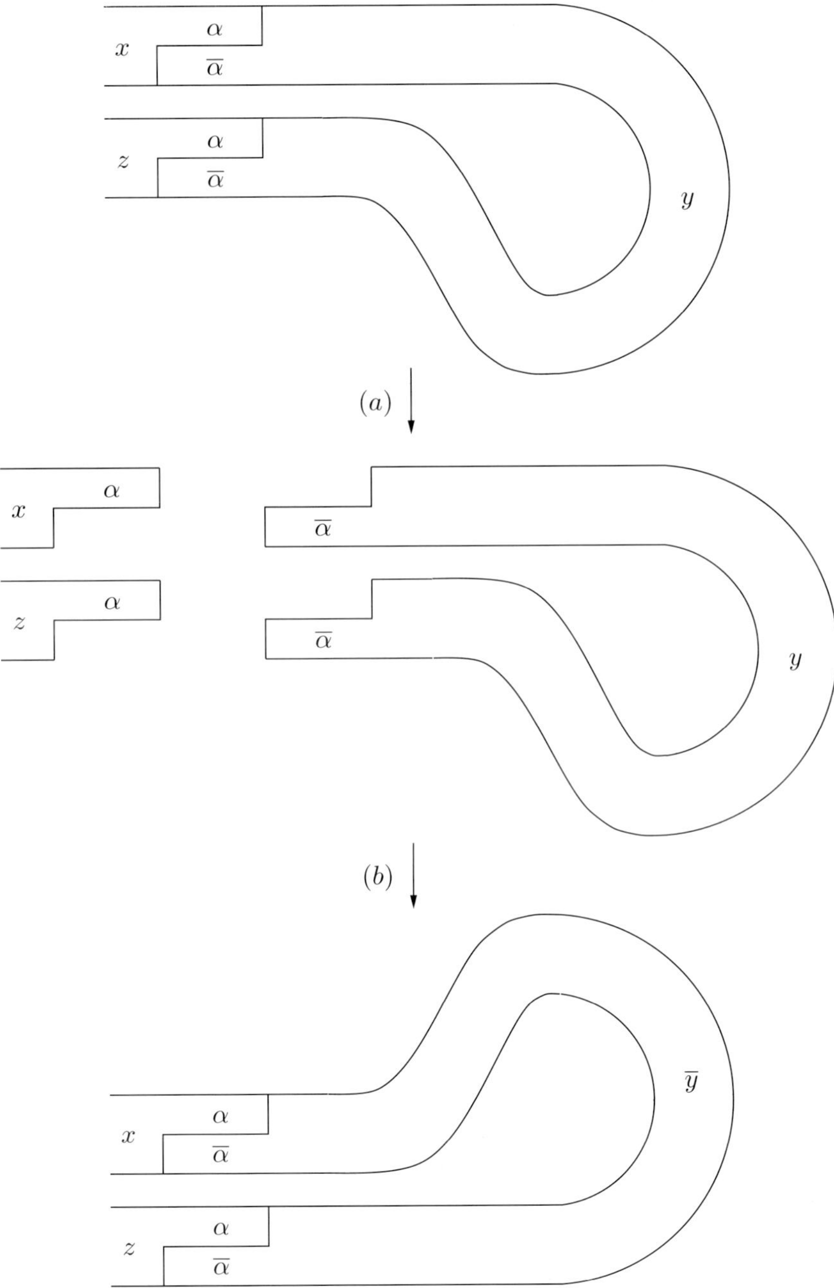

Fig. 17. A hi-excision/reinsertion in a micronuclear gene. (a) A staggered cut is introduced. (b) A cut-off fragment is inverted and then recombined with the rest (i.e., two other fragments).

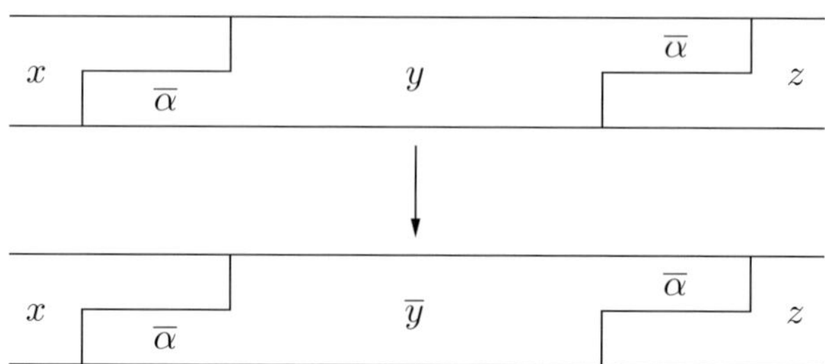

Fig. 18. The effect of hi-excision/reinsertion

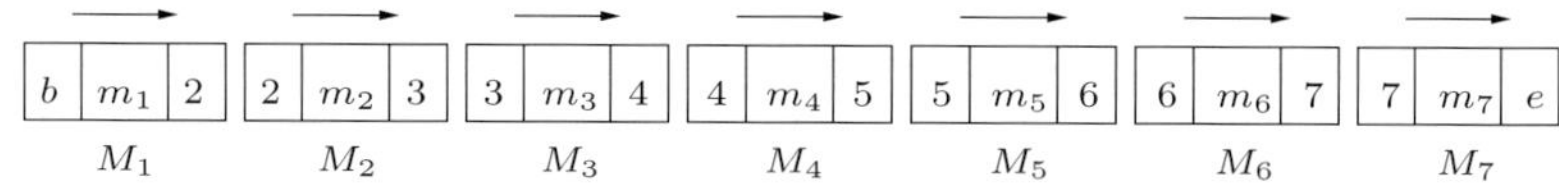

Fig. 19. The order in which MDSs will eventually occur in γ

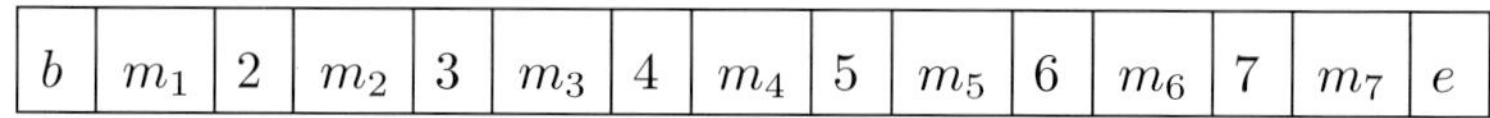

Fig. 20. $M_1...M_7$ assembled by overlapping pointers into the macronuclear version of γ

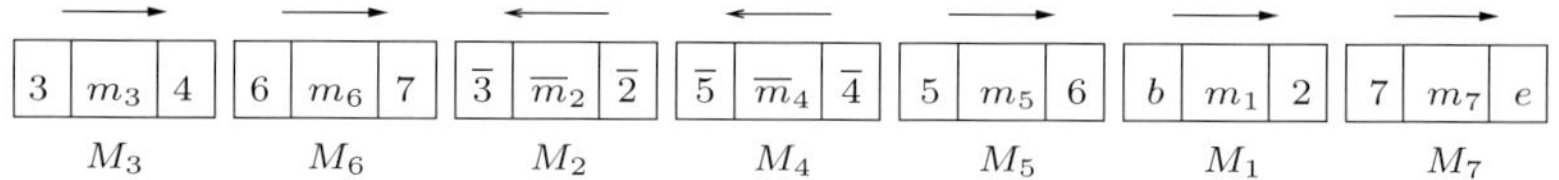

Fig. 21. The scrambled order in the micronuclear version of γ

string $w = 3\ 4\ 6\ 7\ \bar{3}\ \bar{2}\ \bar{5}\ \bar{4}\ 5\ 6\ 2\ 7$ of pointers and barred pointers. This string carries all the information that we will need in the first part of our study of the computational aspects of unscrambling, viz., the sequence of pointers and the indication for each of them whether or not it is inverted.

Note that if two MDSs in a micronuclear gene are next to each other in the good orthodox order, say M_i followed by M_{i+1}, then the ld-excision makes one MDS from these two. From the pointers point of view this means that the pointer $i+1$ gets removed – it will be present within the new MDS and hence it will not be a pointer anymore! Thus we get the following "pointer reduction rules": for any pointer i, the substring ii (hence two consecutive occurrences of i) as well as the substring $\bar{i}\,\bar{i}$ (hence two consecutive occurrences if $\bar{i}$) can be removed. In more formal way, these rules are written as: $ii \rightarrow \Lambda$ and $\bar{i}\,\bar{i} \rightarrow \Lambda$. Here Λ denotes the empty string, and so the rules say that the substrings ii and $\bar{i}\,\bar{i}$ can be erased. We refer to these rules as the *ld-rules*.

Each pointer is either an outgoing pointer or an incoming pointer. Thus if we have an inverted repeat (see Fig. 15) of a pointer i, say $i = 5$, then we can have two possible situations as illustrated in Fig. 22 (I_1 and I_2 are IESs).

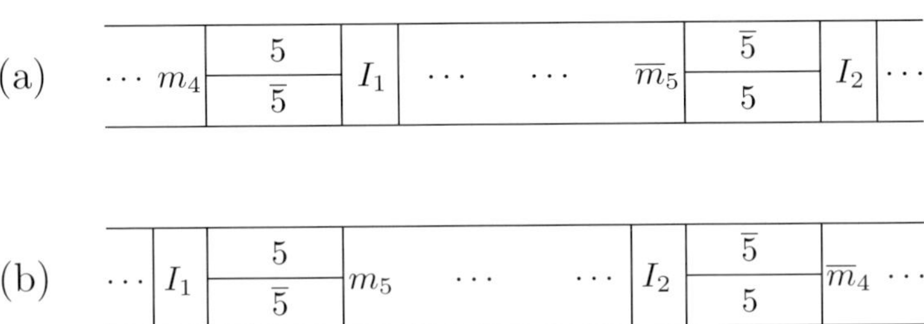

Fig. 22. Two possible situations for an inverted repeat. (a) The "left" pointer is the outgoing pointer (of M_4), and the "right" inverted pointer is the incoming pointer (of M_5). (b) The "right" inverted pointer is the outgoing pointer (of M_4) and the "left" pointer is the incoming pointer (of M_5).

Now if hi-excision/reinsertion aligned on this pointer, $i = 5$, is performed (as in Fig. 18), then two MDSs are fused together, and the pointer i disappears as illustrated in Fig. 23, for cases (a) and (b) from Fig. 22 (I_1 and I_2 are IESs).

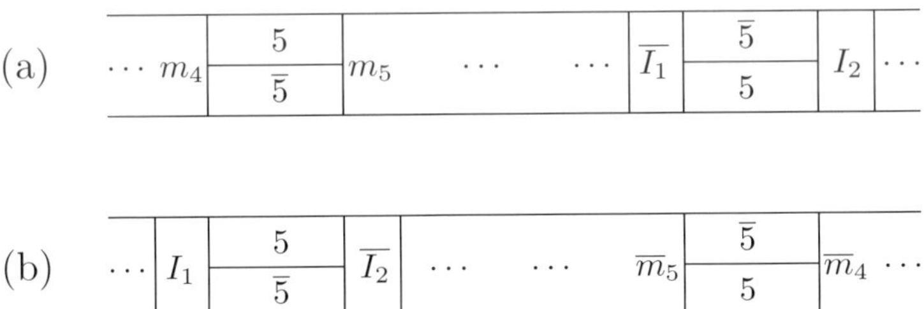

Fig. 23. The removal of a pointer by hi-excision/reinsertion

Note that in terms of pointers, the inversion of the string in between the pointer and its inversion leads to the following transformation: the order of all pointers is reversed, and moreover each pointer in this string becomes inverted, and each inverted pointer becomes a pointer. In terms of our bar notation this means that the string z of pointers in between the two pointers is reversed, and moreover the bar is removed from each barred pointer, while each pointer gets barred. Thus, if the string in between pointers is, e.g., $z = 7\ \bar{3}\ \bar{2}\ \bar{6}\ \bar{4}\ 6$, then after the application of hi-excision/reinsertion it becomes $\bar{6}\ 4\ 6\ 2\ 3\ \bar{7}$ – this is called the *reversed switch* of z and denoted by $rs(z)$.

This leads to the following pointer reduction rule: for any pointer i, the pair of pointers of the form $(i, \bar{i})$ or $(\bar{i}, i)$ can be removed, providing that the string in between is reverse switched. In a more formal way this is written as two rules $i \, y \, \bar{i} \to rs(y)$ and $\bar{i} \, y \, i \to rs(y)$. Hence i and $\bar{i}$ are removed and the string y in between is reverse switched. We refer to these rules as the *hi-rules*.

Now, using the above reduction rules our string $w = 3 \, 4 \, 6 \, 7 \, \bar{3} \, \bar{2} \, \bar{5} \, \bar{4} \, 5 \, 6 \, 2 \, 7$ can undergo the chain of transformations (reductions) illustrated in Fig. 24.

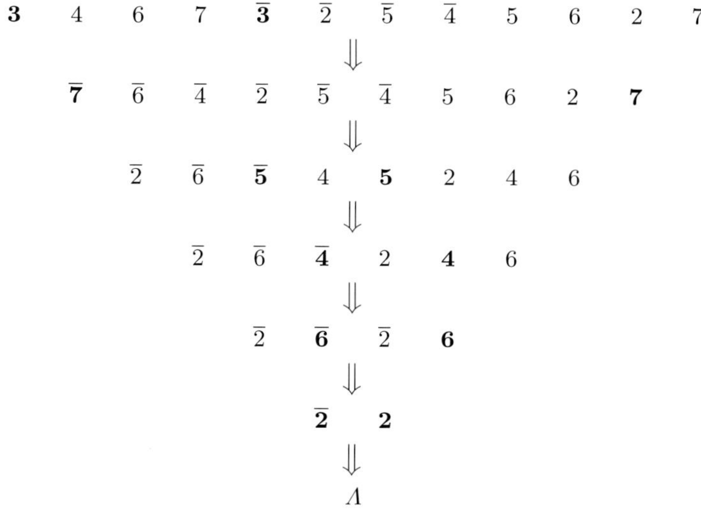

Fig. 24. A sequence of reductions for w

In each intermediate string, starting with w, we have given in bold face the pointer to be removed and its barred version - in this way the reader can clearly see the in between string that is going to be reverse switched.

The corresponding unscrambling chain for the sequence of MDSs from Fig. 21 is illustrated in Fig. 25.

This chain of transformations (a *reduction strategy*) is *successful* in the sense that it reduces w to the empty string Λ – it corresponds to a successful unscrambling of the micronuclear scrambled sequence of MDSs coded by w. The chain of applied transformations can be described by giving the sequence of pointers reduced at each step: in our example this sequence is $(3, 7, 5, 4, 6, 2)$. Note that only hi-rules were applied; as a matter of fact, in this case only hi-rules were applicable. Clearly, this is only one possible sequence of reduction rules applied to w. The application of another sequence $(4, 7, 5, 2)$ is illustrated in Fig. 26.

This strategy does not lead to success – we get stuck on the string 3636: no reduction rule can be applied here.

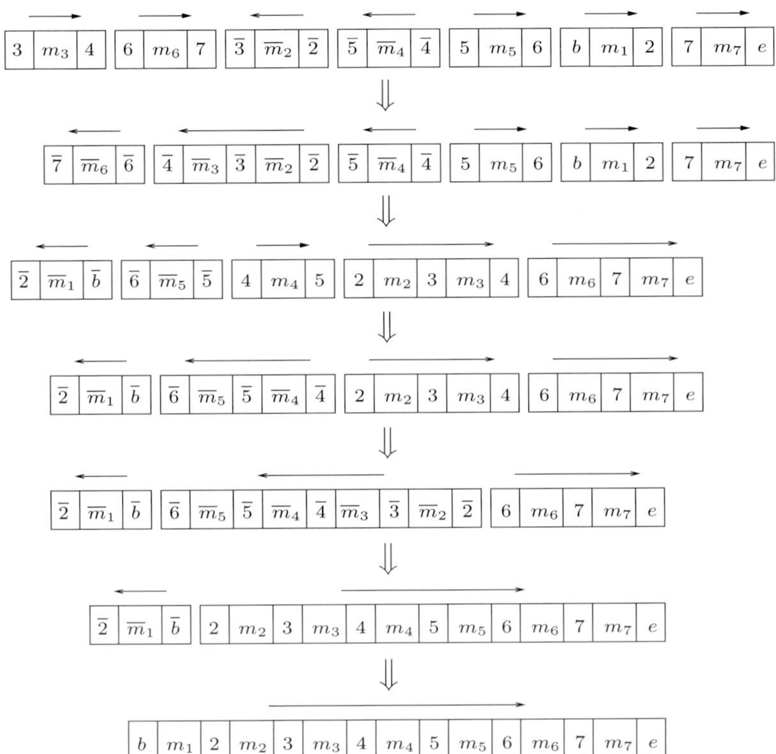

Fig. 25. The unscrambling chain for the sequence of MDSs

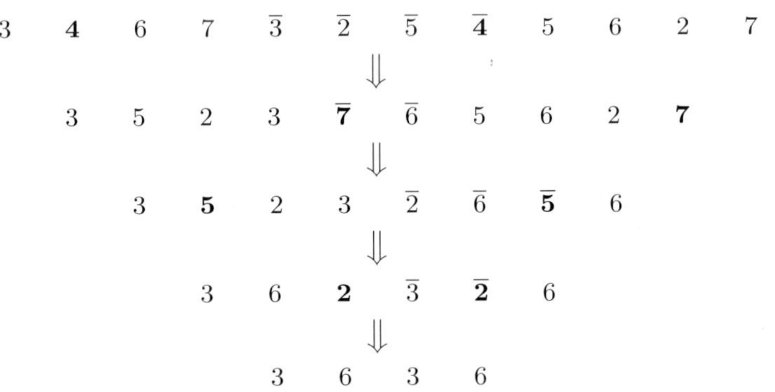

Fig. 26. Another reduction strategy for w

8   String Pointer Reduction Systems – More Formal

We move now to formalize the above - we will use here the standard terminology and notation of formal language theory. Hence, an *alphabet* is a finite set of *letters* (symbols), and a *word* (string) over an alphabet Σ is a sequence of letters over Σ ; the empty sequence is denoted by Λ, and it is referred to as the *empty word*. Σ^* is the set of all words over Σ. For a word w over Σ ($w \in \Sigma^*$), we use $alph(w)$ to denote the letters from Σ that actually occur in w. Thus, e.g., $w = abaab$ is a word over $\Sigma = \{a, b, c\}$, but $alph(w) = \{a, b\}$ – here w contains three occurrences of a and two occurrences of b, but no occurrence of c. For words x and y, we say that x is a *subword* of y, written x *sub* y if $y = uxz$ for some words u and z. Thus for $y = abacb$, both ab and ba are subwords of y, but bab is not a subword of y.

We write $x \in X$ for "x is a member of set X", $x \notin X$ for "x is not a member of X", and we use "$\subseteq$" to denote set inclusion ($X \subseteq Y$ if each member of X is also a member of Y), "$-$" to denote set difference ($X - Y = \{x \in X : x \notin Y\}$), and "$\cup$" to denote the union of sets ($X \cup Y = \{z : z \in X$ or $z \in Y\}$). Then $X_1 \cup X_2 \cup ... \cup X_m$, for $m \geq 2$, is the union of all the sets $X_1, ..., X_m$. We write "$\psi : X \to Y$" for "ψ is a function from X to Y". We use $\emptyset$ to denote the *empty set* (the set with no elements).

We will use the following, very specific alphabets. For a positive integer $n \geq 2$, $\Delta_n = \{2, 3, ..., n\}$, $\bar{\Delta}_n = \{\bar{2}, \bar{3}, ..., \bar{n}\}$ and $\Sigma_n = \Delta_n \cup \bar{\Delta}_n$. The elements from Δ are called *pointers*, and the elements from $\bar{\Delta}$ are called *barred pointers*. We will also use the following useful notation: for a letter $i \in \Delta$, $bar(i) = \bar{i}$, and for a letter $\bar{i} \in \bar{\Delta}, bar(\bar{i}) = i$.

We will be interested in legal words in Σ_n^*. A word $w \in \Sigma_n^*$ is *legal* if for every letter d in w, w has exactly two occurrences of letters from the set $\{d, bar(d)\}$: hence w may contain either two occurrences of d, or one occurrence of d and one occurrence of $bar(d)$. Thus, e.g., the words $2\,\bar{2}$ and $2\,4\,\bar{4}\,2$ are legal words over Σ_4 while $\bar{2}\,4\,\bar{4}\,3$ is not legal. We use the notation LW_n to denote the set of all legal words over Σ_n. For a legal word $x \in LW_n$ and a pointer i, we say that: (1) i is a *square in* x if x contains either ii or $\bar{i}\,\bar{i}$ as a subword, and (2) i is *good in* x if x contains both an occurrence if i and an occurrence of $\bar{i}$.

For a word $w \in \Sigma_n^*$, the *reverse switch* of w, denoted $rs(w)$, is the word obtained from w by reading w backwards and replacing each letter a in w by $bar(a)$. Thus, e.g., for $w = 2\,3\,\bar{2}\,\bar{4}\,4\,\bar{3}$, we get $rs(w) = 3\,\bar{4}\,4\,2\,\bar{3}\,\bar{2}$. More formally, the operation of reverse switch is the composition of two operations: reverse and switch. For a word $z = a_1...a_m$ where $m \geq 1$ and $a_1, ..., a_m$ are letters from Σ_n, the *reverse* of z, denoted $rev(z)$, is the word $a_m...a_1$, and the *switch* of z, denoted $sw(z)$, is the word $bar(a_1)...bar(a_m)$. Then for a word $w \in \Sigma_n^*$, the reverse switch of w is the word $rs(w) = sw(rev(w))$. Obviously, the order of composing the operations of switch and reverse is irrelevant, and so we also have $rs(w) = rev(sw(w))$.

Productions will be used to rewrite words by replacing a subword of a word by another subword. A production is written in the form $u \to v$, where u, v are words with $u \neq \Lambda$. The intuitive meaning of such a production is that one can rewrite a word x by replacing a subword u of x by v. More formally, we have the following definition.

For a production $\pi = u \to v$ and words x, y we say that x *directly derives* y *using* π, denoted by $x \Rightarrow_\pi y$, if $x = x_1 u x_2$ and $y = x_1 v x_2$ for some words x_1, x_2. Thus for the production $\pi = abc\bar{a} \to bc$, we have $bcabc\bar{a}cbc \Rightarrow_\pi bcbccbc$.

Here are the (*legal*) productions that we will use. For each $i \geq 2$:
τ_i is the production $ii \to \Lambda$,
$\bar{\tau}_i$ is the production $\bar{i}\,\bar{i} \to \Lambda$,
$Q_i = \{iz\bar{i} \to rs(z) : z \in \Sigma_n^*\}$, and
$\bar{Q}_i = \{\bar{i}zi \to rs(z) : z \in \Sigma_n^*\}$.

Applying production τ_i to a legal word x removes (erases) the subword ii, and applying production $\bar{\tau}_i$ removes the subword $\bar{i}\,\bar{i}$, hence both τ_i and $\bar{\tau}_i$ remove a square pointer in x, viz. i. Thus τ_i and $\bar{\tau}_i$ are called *square productions*, and they model applications of ld-excisions.

Applying production $iz\bar{i} \to rs(z)$ to a legal word x removes both the occurrence of i and the occurrence of $\bar{i}$ in x - in doing so this production replaces the in-between word z by its reversed switch $rs(z)$. The same holds for applying a production $\bar{i}zi \to rs(z)$ from $\bar{Q}_i$ to x. Hence these productions remove a good pointer, viz. i. Thus productions from Q_i and productions from $\bar{Q}_i$ are called *good productions* and they model applications of hi-excisions/reinsertions. Note that both Q_i and $\bar{Q}_i$ are infinite sets of productions.

For each $n \geq 2$:

$T_n = \{\tau_2, ..., \tau_n, \bar{\tau}_2, ..., \bar{\tau}_n\}$,
$R_n = Q_2 \cup Q_3 \cup ... \cup Q_n \cup \bar{Q}_2 \cup \bar{Q}_3 \cup ... \cup \bar{Q}_n$, and
$P_n = T_n \cup R_n$.

Hence T_n are all square productions for words in LW_n, and R_n are all good productions for words in LW_n. Then P_n are all productions (square and good) that we will use in rewriting words in LW_n.

Definition. For each $n \geq 2$, the *string n-pointer reduction system, n-sprs* for short, is the ordered pair $S_n = (\Sigma_n, P_n)$. We also say that S_n is a *string pointer reduction system, sprs* for short.

The n-sprs S_n defines the direct reduction relation $\Rightarrow_{S_n}$, and the reduction relation $\Rightarrow_{S_n}^*$ on the set LW_n of legal words over Σ_n as follows.

Let $x, y \in LW_n$. We say that x *directly reduces to* y *in* S_n, written $x \Rightarrow_{S_n} y$, if there exists a production $\pi \in P_n$ such that $x \Rightarrow_\pi y$ (note that since x is legal, so is y). Then we say that x *reduces to* y *in* S_n, written $x \Rightarrow_{S_n}^* y$, if either $x = y$ or there exists a sequence of words $x_0, x_1, ..., x_m$ for some $m \geq 1$, such that $x_0 = x$, $x_m = y$, and $x_j \Rightarrow_{S_n} x_{j+1}$ for all $0 \leq j \leq m - 1$.

Thus x reduces to y in S_n if either x equals y or x directly reduces to y or x reduces to y through a finite sequence of direct reductions. We say that x *successfully reduces in S_n* (x *is successfully reducible in S_n*) if x reduces to Λ in S_n (i.e., $x \Rightarrow^*_{S_n} \Lambda$). We do use here the word "reduces" rather than the word "derives" used before (for describing the effect of applying productions in general), because in each derivation step here we shorten the length of a word by reducing a pointer (i.e., by removing both of its occurrences, barred or nonbarred).

We move now back to the coding of sequences of MDSs by words consisting of pointers and barred pointers, as introduced in Section 6.

Let mi_G be the sequence of MDSs in the micronuclear version of γ and let *word* be our coding function. Thus $word(mi_\gamma)$ is a legal word. What interests us is the process of unscrambling of mi_γ into the macronuclear gene γ using ld-excision and hi-excision/reinsertion.

As a matter of fact, we are interested in the *general problem of unscrambling* of mi_γ. This means that mi_γ gets unscrambled to the orthodox order placed on either a circular molecule or a linear molecule. When the former takes place, we have *circular unscrambling*, and when the latter takes place, we have the *linear unscrambling*. Hence we allow unscrambling to any *cyclic conjugate* of the orthodox order presented in Fig. 19. A cyclic conjugate is formed by closing the pattern into a cycle and then reopening it at an arbitrary place again. Thus for the patterns $(m, b_1, 2)(2, m_2, 3)...(6, m_6, 7)(7, m_7, e)$ from Fig. 19, $(3, m_3, 4)(4, m_4, 5)...(7, m_7, e)(b, m_1, 2)(2, m_2, 3)$ is a cyclic conjugate, and so is $(6, m_6, 7)(7, m_7, e)(b, m_1, 2)(2, m_2, 3)...(5, m_5, 6)$.

As we have discussed already, the use of ld-excision to combine two MDSs into a new (bigger) MDS corresponds to the reduction of a pointer using a square production (from T_n). On the other hand, the use of a hi-excision/reinsertion corresponds to the reduction of a pointer using a good production (from R_n). Based on this observation one proves the following result.

Theorem. A sequence t of n MDSs, $n \geq 2$, can be unscrambled using the operations of ld-excision and hi-excision/reinsertion if and only if $word(t)$ successfully reduces in S_n.

9   Graph Pointer Reduction Systems – Informal

A *graph* consists of *nodes* and *edges*, where each edge connects two nodes. In the standard pictorial representation of graphs, nodes are represented by circles and edges are represented as line segments between the circles (nodes) they connect. Thus Fig. 27 represents a graph with five nodes: v_1, v_2, v_3, v_4, v_5, and three edges: $\{v_1, v_3\}$, $\{v_2, v_3\}$, $\{v_2, v_5\}$. Here $\{v_1, v_3\}$ connects v_1 and v_3, $\{v_2, v_3\}$ connects v_2 and v_3, and $\{v_2, v_5\}$ connects v_2 and v_5.

Any two nodes connected by an edge are *adjacent* nodes, also referred to as *neighbors*. Hence in the graph from Fig. 26, v_5 has one neighbor (v_2), v_3

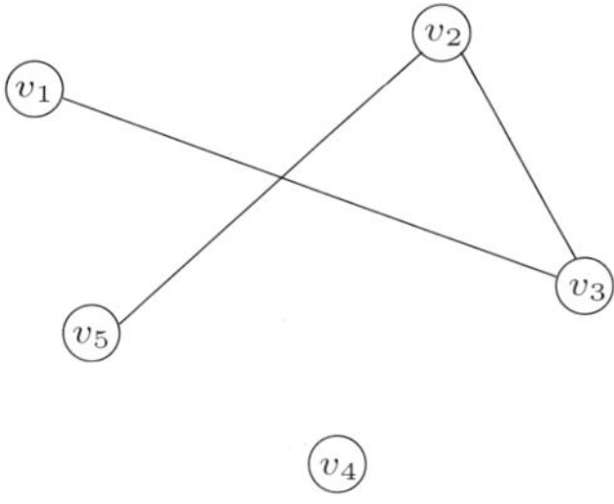

Fig. 27. A graph

has two neighbors (v_1 and v_2), and v_4 has no neighbors (and so is an *isolated node*). The *neighborhood* of a node consists of all the nodes adjacent to it; for a node v in a graph g its neighborhood is denoted by $ngh_g(v)$. Hence for the graph g from Fig. 26, $ngh_g(v_2) = \{v_3, v_5\}$, $ngh_g(v_1) = \{v_3\}$, and $ngh_g(v_4) = \emptyset$, the empty set (v_4 has no neighbors).

The *complement* of a graph g, denoted by $\bar{g}$, has the same set of nodes, but the "complementary" set of edges: two nodes are connected by an edge in $\bar{g}$ if and only if they are not connected in g. Thus, for the graph g from Fig. 27, its complement $\bar{g}$ is given in Fig. 28.

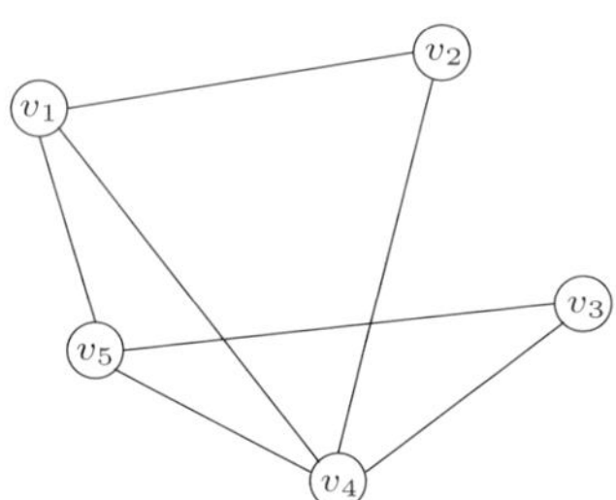

Fig. 28. The complement of the graph from Fig. 27

Given a graph g with a set of nodes V, we may be interested only in a subset of nodes U ($U \subseteq V$), and then only in this part of g that involves U – we call it the *subgraph of g induced by U*, and denote it by $sub_g(U)$. Thus, for the graph g from Fig. 27, and for the subset $U = \{v_1, v_2, v_3\}$ of V, $sub_g(U)$ is shown in Fig. 29.

We will consider the subgraphs induced by the neighborhoods of single nodes; hence for a graph g and a node v we are interested in $sub_g(ngh_g(v))$. In particular we will be especially interested in the complement of this graph – we will use the simple notation $< v >_g$ for it. Thus for the graph g from Fig. 27, and the node v_3, $< v_3 >_g$ is given in Fig. 30.

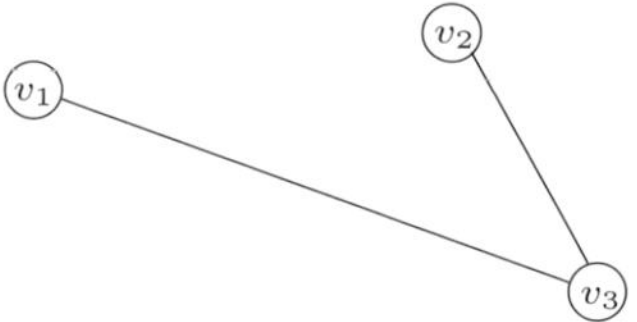

Fig. 29. The subgraph of the graph from Fig. 27 induced by $U = \{v_1, v_2, v_3\}$

Fig. 30. $< v_3 >_g$ for the graph g from Fig. 27

Labelled graphs are graphs in which each node has a label that is a letter from an alphabet of labels. Our labels will be very simple: $+$ and $-$. In our pictorial representation of labeled graphs, the label of each node will be given next to the circle representing this node. Fig. 31 gives an example of a labeled graph resulting from the graph in Fig. 27 by labelling its nodes.

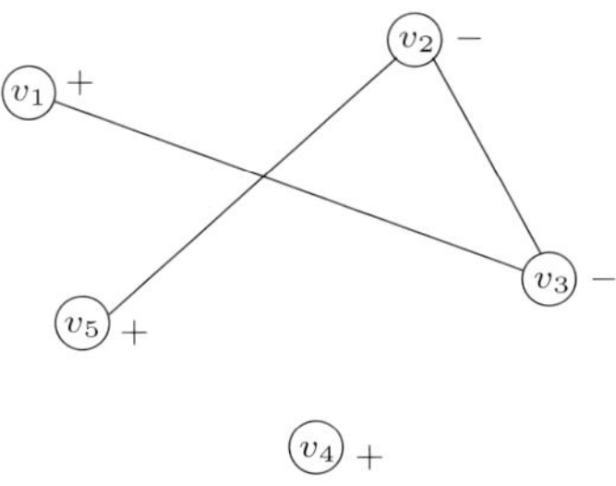

Fig. 31. A labelled graph

The notion of the complement of a graph gets modified to labelled graphs in the obvious way: if a node v has the label $+$ (respectively $-$) in a given graph, then it gets the label $-$ (respectively $+$) in the complement.

We are ready now to use graphs to represent reductions in string pointer reduction systems, and hence to represent the unscrambling processes.

To start with let's consider a word $x \in LW_n$. Let i be a pointer such that either i or $\bar{i}$ is present in x – let $point(x)$ be the set of all such pointers (*pointers of x*). For each pointer $i \in point(x)$, there are two occurrences from $\{i, \bar{i}\}$ in x; the subword beginning at one of these occurrences and ending at the other is called the *interval of i in x* and denoted by $int_x(i)$. Thus for the word $w = 346732545627$ that we considered in Sect. 6 we have: $int_w(3) = 34673$, $int_w(5) = \bar{5}4\bar{5}$, and $int_w(6) = 67\bar{3}2\bar{5}456$. These intervals are indicated in Fig. 32.

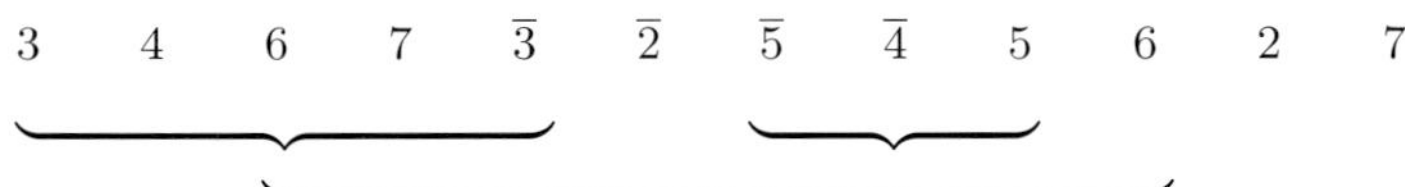

Fig. 32. Some intervals in w

Let $x \in LW_n$ and let i_1, i_2 be pointers of x. We say that $int_x(i_1)$ and $int_x(i_2)$ *overlap* in x if $int_x(i_1)$ and $int_x(i_2)$ contain common letters, and moreover $int_x(i_1)$ contains a letter not in $int_x(i_2)$ and $int_x(i_2)$ contains a letter not in $int_x(i_1)$. Then we say that i_1, i_2 *overlap* (*in* x) if $int_x(i_1)$ overlaps with $int_x(i_2)$. It is clearly seen from Fig. 32 that for the word x in this figure, 3 and 6 overlap, while 3 and 5 do not overlap. Note that 5 and 6 do not overlap, but rather $int_x(5)$ is included in $int_x(6)$.

Now, given a legal word x we translate it into a *legal labelled graph*, denoted g_x, as follows. The set V_x of nodes of g_x is the set of pointers of x, hence $V_x = point(x)$. A node $i \in V_x$ is labelled by $+$ if i is good in x, otherwise i is labelled by $-$ (recall that i is good in x if both i and $\bar{i}$ occur in x). Two nodes i_1, i_2 are connected by an edge if i_1, i_2 overlap in x, hence the set E_x of edges is the set $\{\{i_1, i_2\} : i_1, i_2 \text{ overlap in } x\}$. Thus for the word w considered above, the legal labelled graph g_w is given in Fig. 33.

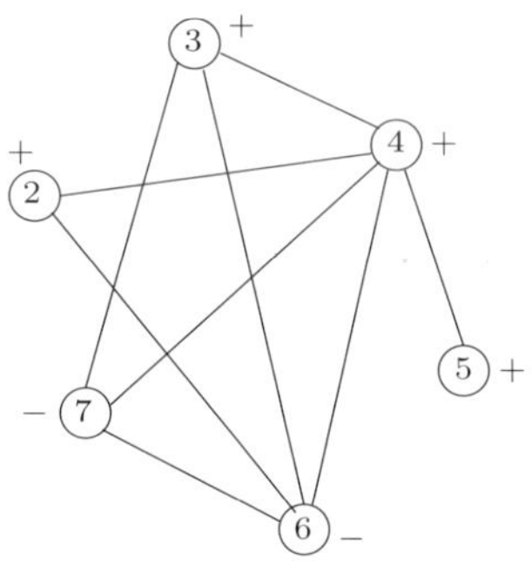

Fig. 33. g_w

A single reduction step for a legal labelled graph g can have one of two possible forms.

(1) A node v labelled by $+$ is removed together with all edges involving v, and moreover, the subgraph of g induced by the neighborhood of v (hence $sub_g(ngh_g(v))$) is changed into its complement (hence into $< v >_g$).

Thus for the graph g_w from Fig. 33 we may choose node 3 for removal. Since $ngb_{g_w}(3) = \{4, 6, 7\}$, we have to change the subgraph induced by $\{4, 6, 7\}$ into its complement $< 3 >_{g_w}$. This is illustrated in Fig. 34.

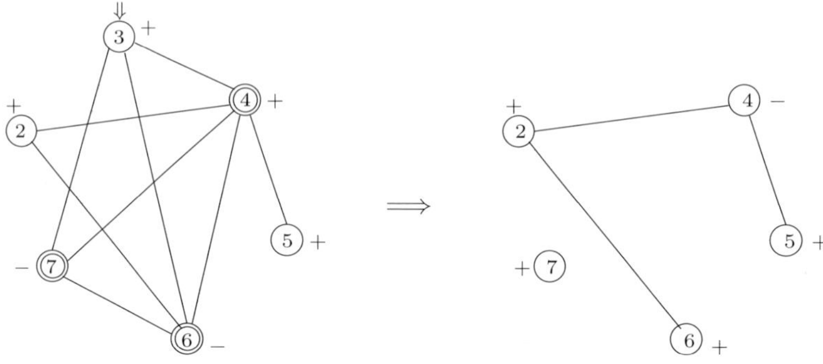

Fig. 34. Reducing node 3 in g_w

The left graph in Fig. 34 is g_w, where we have also indicated by a double arrow the node to be removed (hence 3), and we have doubly encircled the nodes from the neighborhood of the node to be removed (hence the nodes $4, 6, 7$). The right graph is the graph resulting from g_w by removing the node v, indicated by the double arrow, and changing $sub_g(ngh_g(v))$ into $< v >_g$ (hence in this case removing 3, and changing $sub_{g_w}(\{4, 6, 7\})$ into $< 3 >_{g_w}$).

(2) An isolated node labelled by $-$ is removed.

Note that it follows from (1) above that an isolated node labeled by $+$ can also be simply removed (with no other changes in the processed graph). Hence (1) and (2) imply that an isolated node can always be removed.

Let's analyze the first reduction step from Fig. 23, $w \Rightarrow u$, where $w = 346\overline{73}\overline{2}5\overline{4}5627$, and $u = \overline{7}6\overline{4}2545627$. Here u has resulted from w by removing pointer 3 and reverse switching subword 467. Obviously (reverse) switching 467 changes the goodness status of all involved pointers, $4, 6,$ and 7, meaning that 4 becomes bad in u (it was good in w), 6 becomes good in u (it was bad in w), and 7 becomes good in u (it was bad in w). That's why nodes $4, 6,$ and 7 in g_w have switched their $+/-$ labels.

The second observation is that reducing pointer 3 has an effect only on those pointers that overlap with 3 in w, hence pointers $4, 6, 7$. This means that only the goodness of these pointers may change, and the overlapping relationship may change only between these pointers. That's why the only edges that change are those in $sub_{g_w}(\{4, 6, 7\})$.

Finally, since the subword 467 becomes reversed (and switched), each overlapping between nodes in $\{4, 6, 7\}$ disappears. That's why we change $sub_{g_w}(\{4, 6, 7\})$ into its complement.

We continue now the rewriting process by removing nodes $7, 5, 6, 4$ and 2, in this order; these five steps are illustrated in Figs. 34, 35, 36, 37 and 38, respectively.

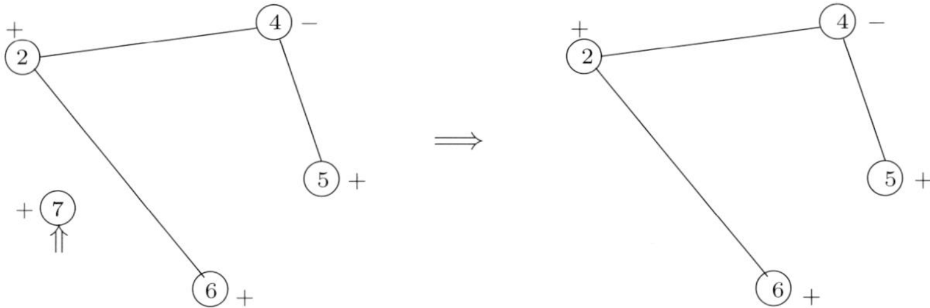

Fig. 35. Reducing node 7

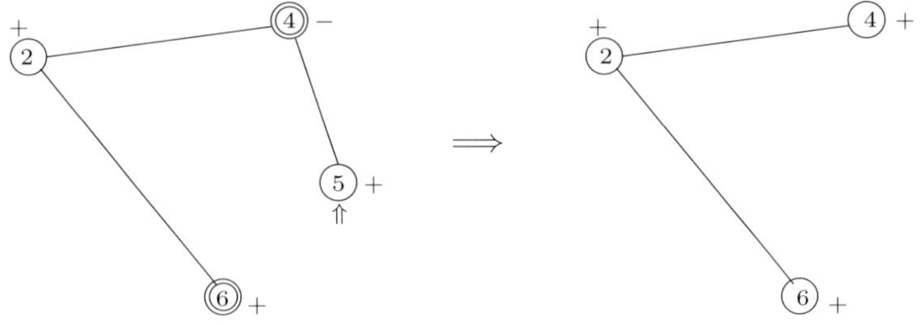

Fig. 36. Reducing node 5

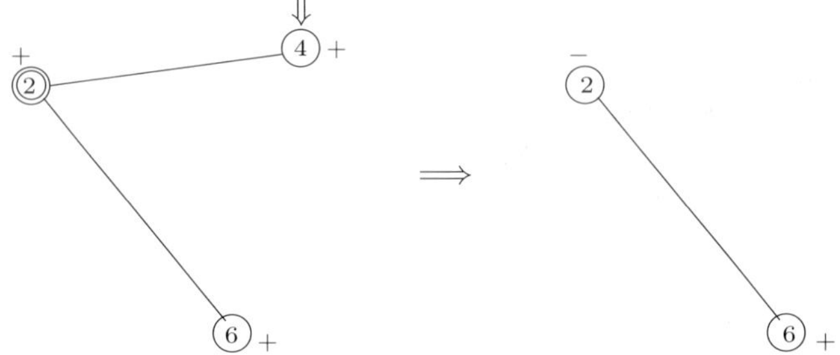

Fig. 37. Reducing node 4

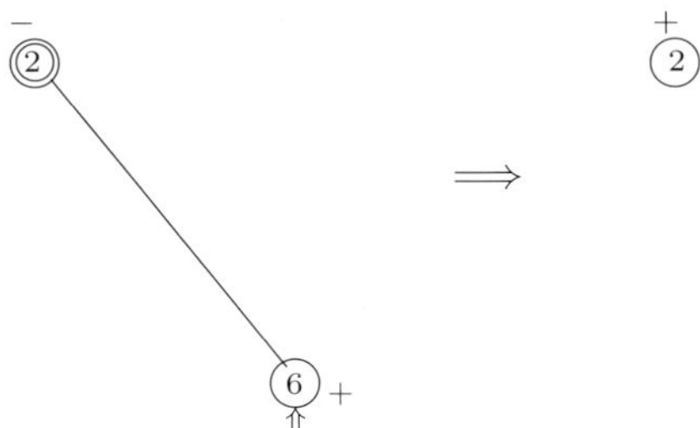

Fig. 38. Reducing node 6

Fig. 39. Reducing node 2

As can be seen from Fig. 39, the final result is the empty graph λ, the graph without nodes (it corresponds to the empty word, the word without letters).

10   Graph Pointer Reduction Systems – More Formal

A *graph* is an ordered pair $g = (V, E)$, where V is a finite set of *nodes*, and $E \subseteq \{\{u, v\} : u, v \in V \text{ and } u \neq v\}$ is a set of *edges*. Nodes $u, v \in V$ are *neighbors* in g if $\{u, v\} \in E$. For a node $v \in V$, the *neighborhood* of v in g is the set $\{u : \{u, v\} \in E\}$, denoted by $ngb_g(v)$. For a set $U \subseteq V$, the *subgraph of g induced by U*, denoted $sub_g(U)$, is the graph (U, E_U), where $E_U = \{\{u, v\} \in E : u, v \in U\}$. The graph $(\emptyset, \emptyset)$ is the *empty graph*, denoted by λ.

The *complement* of g, denoted by $\bar{g}$, is the graph $(V, \bar{E})$ where $\bar{E} = \{\{u, v\} : u, v \in V, u \neq v \text{ and } \{u, v\} \notin E\}$. For a node $v \in V$, the complement of $sub_g(ngb_g(v))$ is denoted by $< v >_g$.

A *labelled graph* is a triple $g = (V, E, \Theta, \psi)$ where (V, E) is a graph, called the *underlying graph of g*, Θ is an alphabet (of *labels*), and $\psi : V \to \Theta$ is the *labelling function* of g – it assigns to each node $v \in V$ a label from Θ. All the terminology and notation for graphs given above carries over to labelled graphs through their underlying graphs.

A *legal labelled graph* over Δ_n is a labelled graph $g = (V, E, \Theta, \psi)$ such that $V \subseteq \Delta_n$ and $\Theta = \{+, -\}$. Hence nodes of a legal graph come from Δ_n, and each of them is labelled either by $+$ or by $-$. Nodes labelled by $+$ are called *positive* and nodes labelled by $-$ are called *negative*. We use the notation LLG_n to denote the set of all legal labelled graphs over Δ_n.

The notion of complement extends in a natural way to legal graphs as follows. For a legal graph $g = (V, E, \{+, -\}, \gamma)$, its *complement* is the binary labelled graph $\bar{g} = (V, \bar{E}, \{+, -\}, \bar{\psi})$, where $(V, \bar{E})$ is the complement of (V, E), and $\bar{\psi}$ is defined by: for all $v \in V, \bar{\psi}(v) = +$ if and only if $\psi(v) = -$.

A *(legal) graph production over Δ_n*, for some $n \geq 2$, is of the form $(v, \ell) \to \lambda$, where $v \in \Delta_n$, and $\ell \in \{+, -\}$. We use K_n to denote the set of all graph productions.

For a graph production $\pi = (v, \ell) \to \lambda$ over Δ_n, and legal labelled graphs g, h over Δ_n, with $g = (V_g, E_g, \{+, -\}, \psi_g)$ and $h = (V_h, E_h, \{+, -\}, \psi_h)$, we say that g *directly reduces to h using π*, denoted by $g \Rightarrow_\pi h$, if one of the following two cases holds.

(1) $\ell = +$, $V_h = V_g - \{v\}$, $E_h = (E_g - (E_v \cup E_1)) \cup E_2$ with $E_v = \{\{u, z\} \in E_g :$ $u = v\}$, $E_1 = \{\{u, z\} \in E_g : u, z \in ngb_g(v)\}$ and $E_2 = \{\{u, z\} \notin E_g : u \neq z$ and $u, z \in ngb_g(v)\}$, and for each $u \in V_h$, $\psi_h(u) = \psi_g(u)$ for $u \notin ngb_g(v)$, while $\psi_h(u) = +$ if and only if $\psi_g(u) = -$ for $u \in ngb_g(v)$.

(2) $\ell = -$, v is isolated, $V_h = V_g - \{v\}$ and $E_h = E_g$.

Thus g directly reduces to h using π means that either $\ell = -$ and h results from g by removing v, or $\ell = +$ and h results from g by removing v together with all edges involving v, and replacing $sub_g(ngb_g(v))$ by its complement $<v>_g$.

Definition. For each $n \geq 2$, the *graph n-pointer reduction system, n-gprs* for short, is the ordered pair $G_n = (\Delta_n, K_n)$. We also say that G_n is a *graph pointer reduction system, gprs* for short.

The n-gprs G_n defines the direct reduction relation $\Rightarrow_{G_n}$, and the reduction relation $\Rightarrow^*_{G_n}$ on the set LLG_n of legal labelled graphs over Δ_n as follows. Let $g, h \in \mathrm{LLG}_n$. We say that g *directly reduces* to h in G_n, written $g \Rightarrow_{G_h} h$, if there exists a production $\pi \in K_n$ such that $g \Rightarrow_\pi h$ (note that since g is legal, so is h). Then we say that g *reduces* to h in G_n, written $g \Rightarrow^*_{G_n} h$, if either $g = h$ or there exists a sequence of labelled graphs $g_0, g_1, ..., g_m$ for some $m \geq 1$, such that $g_0 = g, g_m = h$, and $g_j \Rightarrow_{G_n} g_{j+1}$ for all $0 \leq j \leq m - 1$. Thus g reduces to h in G_n if g equals h or g directly reduces to h or g reduces to h through a finite sequence of direct reductions. We say that g *successfully reduces* in G_n (g is *successfully reducible in G_n*) if g reduces to λ in G_n (i.e., $g \Rightarrow^*_{G_n} \lambda$).

We return now to the problem of unscrambling the sequence of MDSs in the micronuclear version of a gene.

For a legal word x, the *graph of x*, denoted $graph(x)$, is the legal labelled graph $(V, E, \{+, -\}, \psi)$ such that $V = point(x)$, $E = \{\{i_1, i_2\} : i_1, i_2 \in point(x)$ and i_1, i_2 overlap in $x \}$, and for each $i \in V$, $\psi(i) = +$ if and only if i is good in x.

The usefulness of graph reduction for the study of unscrambling stems from the following result.

Theorem. Let t be a sequence of n MDSs, $n \geq 2$. Then $word(t)$ is successfully reducible in S_n if and only if $graph(word(t))$ is successfully reducible in G_n.

This result together with the theorem about successful reducibility in S_n from Section 7 yields the following corollary.

Corollary. A sequence t of n MDSs, $n \geq 2$, can be unscrambled using the operations of ld-excision or hi-excision/reinsertion if and only if $graph\,(word(t))$ is successfully reducible in G_n.

11   Linear vs Circular Unscrambling

In Sect. 8 and Sect. 10 we gave characterizations of successful unscrambling using the operations of ld-excision and hi-excision/reinsertion. However, successful unscrambling can be either circular or linear. In this section we give characterizations for each of these cases, thus we will be able to distinguish between them.

First we turn to the case of *linear unscrambling* – here mi_γ is unscrambled to the orthodox order placed on a linear molecule. We will give a characterization of these scrambled patterns of MDSs that can be linearly unscrambled; then the characterization of the circular unscrambling will be obtained by "duality" of these two cases.

We begin by coding sequences of MDSs into *marked legal words*. The difference with the coding into legal words is that we also include the begin and the end symbols (b and e), possibly barred, into the coding words. Thus, e.g., the sequence of MDSs from Fig. 21 is coded into the marked legal word $z = 3467\bar{3}\bar{2}5456b27e$ including b and e. Let *mword* be this coding function – hence $mword(mi_\gamma)$ is a marked legal word.

Here is how we proceed to determine whether z can be linearly unscrambled. Let x be the part of the molecule that directly preceeds (is directly upstream from) mi_γ, and let y be the part of the molecule that directly follows (is directly downstream from) mi_γ. What we know is that if mi_γ will be unscrambled, then the MDSs will fall into the orthodox order and will be assembled as in Fig. 20. What we don't know in advance is what will happen with IESs: which of them will get excised, and how the remaining ones will be positioned on the resulting molecule, before the assembly from Fig. 20 will be excised from the molecule to form the (macro)gene γ. By tracing this process we will be able to predict whether or not z can be linearly unscrambled. Again this will be accomplished manipulating the pointer representation of mi_γ - now however we consider marked legal words, hence we consider z.

Here is our procedure.

Step 1.

We move from left to right through z, letter by letter, and for each letter ℓ we encounter, we write down the ordered pair of letters (ℓ, ℓ') where ℓ' is the next letter of z. Then, if ℓ is the last letter of z, then we set ℓ' to be the first letter of z (we close z into a circle) - we call this pair the *closing pair* and we write it as $[\ell, \ell']$ rather than (ℓ, ℓ') as for all other pairs. Let $cpv(z)$ be the so obtained sequence of pairs (which is a word over the alphabet of all pairs of letters occurring in z).

Thus for our z, $cpv(z)$ is the following sequence:
$$(3, 4)(4, 6)(6, 7)(7, \bar{3})(\bar{3}, \bar{2})(\bar{2}, 5)(5, \bar{4})(\bar{4}, 5)(5, 6)(6, b)(b, 2)(2, 7)(7, e)[e, 3].$$

Step 2.

Note that each pair from $cpv(z)$ denotes uniquely either an MDS from mi_γ or an IES from mi_γ. For example, $(3, 4)$ denotes $M_3 = (3, m_3, 4)$ while $(7, \bar{3})$ denotes the IES flanked by pointers 7 and $\bar{3}$, hence the IES separating M_6 and the inverted M_2 in mi_γ. As a matter of fact, if we label the pairs in $cpv(z)$ by *odd* and *even* starting with the first one $(3, 4)$ being odd, and then alternating between odd and even, then each odd pair denotes an MDS and each even pair denotes an IES. The last pair, which is the closing pair and always even, will denote for us the pair (x, y), where the x will precede and y will follow whatever will result from the unscrambling of mi_γ (including possible combinations of IESs "hanging there").

Let then $op(z)$ be the set of all odd pairs from $cpv(z)$, and $ep(z)$ be the set of all even pairs from $cpv(z)$. As a matter of fact, we will extend the set of all pairs from $op(x)$ by adding for each pair (ℓ, ℓ') in it also its inverse $(bar(\ell'), bar(\ell))$ - let $OP(z)$ be the so obtained set of pairs. We do the same for the set $ep(z)$ obtaining the set $EP(z)$. We will need these two sets in the following step.

Step 3.

Begin with the letter e and find a pair beginning with e - it will be either odd (hence in $OP(z)$) or even (hence in $EP(z)$). Assume that such a pair is even – the reasoning is analogous if it is odd. As a matter of fact, in our case the pair is $[e, 3]$, and it is even. Now we have to continue choosing pairs only from $EP(z)$ in such a way that the first letter of the next pair equals the second letter of the given pair. We say that z is *closure successful* if we reach a pair involving b and the closing pair has been encountered on the way. In our case, already the first pair is the closing pair $[e, 3]$, so our z is closure successful if we reach b. Here is the *test sequence* of pairs from $EP(z)$ that we get following this procedure:

$$[e, 3](3, \bar{7})(\bar{7}, \bar{2})(\bar{2}, \bar{5})(\bar{5}, 4)(4, 6)(6, b).$$

Hence z is closure successful.

We can state now our characterization of linear unscrambling.

Theorem. A sequence t of n MDSs, $n \geq 2$, can be linearly unscrambled using the operations of ld-excision and hi-excision/reinsertion if and only if $word(t)$ successfully reduces in S_n, and $mword(t)$ is closure successful.

Restating this in terms of graph pointer reduction systems, we get the following result.

Theorem. A sequence t of n MDSs, $n \geq 2$, can be linearly unscrambled using the operations of ld-excision and hi-excision/reinsertion if and only if $graph(word(t))$ is successfully reducible in G_n, and $mword(t)$ is closure successful.

Clearly, we have the following duality: if a sequence of MDSs can be unscrambled, but not linearly, then it can be unscrambled circularly. Hence we have the following results for circular unscrambling.

Theorem. A sequence t of n MDSs, $n \geq 2$, can be circularly unscrambled using the operations of ld-excision and hi-excision/reinsertion if and only if $word(t)$ successfully reduces in S_n, and $mword(t)$ is not closure successful.

Theorem. A sequence t of n MDSs, $n \geq 2$, can be circularly unscrambled using the operations of ld-excision and hi-excision/reinsertion if and only if

$graph(word(t))$ is successfully reducible in G_n, and $mword(t)$ is not closure successful.

Since for mi_γ from Fig. 21, $word(mi_\gamma)$ successfully reduces in S_7 (see Fig. 24), and as shown above, $z = mword(mi_\gamma)$ is closure successful, we conclude that mi_γ can be linearly unscrambled using the operations of ld-excision and hi-excision/reinsertion.

Inspecting the test sequence of pairs $[e, 3](3, \bar{7})...(4, 6)(6, b)$ that we got for $z = mword(mi_\gamma)$ in Step 3 of the procedure above, we can infer how the linear molecule which includes γ (as given in Fig. 20) will look like. If we denote the consecutive IESs from Fig. 21 by $I_1, I_2, ..., I_6$, i.e., I_1 is the IES between M_3 and M_6,..., and I_6 is the IES between M_1 and M_7, then this linear molecule will begin with x followed by the sequence $\bar{I}_2$ $\bar{I}_6$ I_3 $\bar{I}_4$ I_1 I_5, followed by γ, and ending with y. As a matter of fact the sequence $\bar{I}_2.....I_5$ of IESs is "polluted" by (former) pointers, and actually this polluted sequence looks as follows: $\bar{3}$ I_2 $\bar{7}$ $\bar{I}_6$ $\bar{2}$ I_3 5 $\bar{I}_4$ 4 I_1 6 I_5. This pollution takes place because each time that hi-excision/reinsertion takes place, two IESs group together separated by a copy of the pointer that gets removed by this application of hi-excision/reinsertion (see Fig. 22 and Fig. 23). Hence the whole molecule looks now as follows: x 3 $\bar{I}_2$ $\bar{7}$ $\bar{I}_6$ $\bar{2}$ I_3 5 $\bar{I}_4$ 4 I_1 6 I_5 γ y. Here none of the IESs was lost because the unscrambling was achieved by using only hi-excision/reinsertion.

Now let us consider mi_γ given by

$(3, m_3, 4)(4, m_{41}, 5)(7, m_7, e)(\bar{7}, m_6, \bar{6})(\bar{6}, \bar{m}_5, \bar{5})(b, m_1, 2)(2, m_2, 3)$.

Here $z = mword(mi_\gamma) = 34457e\bar{7}\bar{6}\bar{6}\bar{5}b223$, and the test sequence is

$(e, \bar{7})(\bar{7}, \bar{6})(\bar{6}, \bar{6})(\bar{6}, \bar{5})(\bar{5}, b)$. Thus the test sequence does not include the closing pair $[3, 3]$, and so z is not closure successful. It can be easily checked that $word(mi_\gamma)$ successfully reduces in S_7, and so mi_γ can be circularly unscrambled. As a matter of fact, $(6, 4, 2, 7, 5, 3)$ is a successful reduction strategy here.

If we denote the consecutive IESs in mi_γ by $I_1, I_2, I_3, I_4, I_5, I_6$, then the circular molecule consists of γ followed by $I_3, \bar{I}_2$, and I_5 which joins the b end of γ. Again, the actual polluted sequence closing γ into a circle (reading from e to b, hence downstream from e) is: I_3 $\bar{7}$ $\bar{I}_2$ 5 I_5. This circular molecule containing γ was obtained by ld-excision aligned on pointer 3, which begins and ends mi_γ. Thus in terms of productions from S_7, the circular molecule is formed when we consider the last letter of a legal word to be consecutive to the first letter of this word, and then apply a square production to these two occurrences. As a matter of fact, if we map the successful reduction strategy $(6, 4, 2, 7, 5, 3)$ into the corresponding unscrambling chain for the sequence of MDSs (as we have done in Sect. 7 – see Fig. 24 and Fig. 25) up to the reduction of pointer 3, then we obtain $(3, m_3, 4, m_4, 5, m_5, 6, m_6, 7, m_7, e)(b, m_1, 2, m_2, 3)$ which is a cyclic conjugate of the (macro) gene

$\gamma = (b, m_1, 2, m_2, 3, m_3, 4, m_4, 5, m_5, 6, m_6, 7, m_7, e).$

Applying now ld-excision aligned on pointer 3 yields a circular molecule. This circular molecule containing γ is shown in Fig. 40.

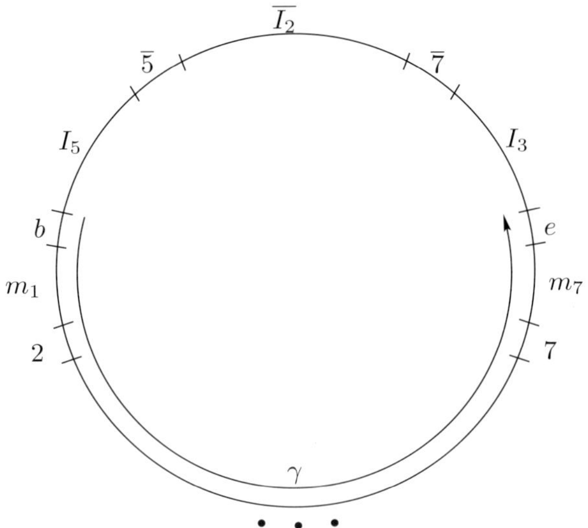

Fig. 40. The circular molecule containing γ

It is plausible that the same mechanism that can excise γ from the linear molecule from the first example above can also excise γ from the circular molecule in the second example. An important observation here is that the whole processing during the unscrambling remains intramolecular.

12   Back to Scrambling

We have by now a good understanding of the way that the two basic operations, the loop excision and the hairpin excision/reinsertion, play in the process of gene unscrambling. In our investigation of this process we have considered two types of alignment, the direct repeat and the inverted repeat.

We turn now to the process of scrambling itself, and in particular we postulate that scrambling can be originated by the operation of insertion and the operation of hairpin excision/reinsertion aligned on a *nonrepeat*. This hypothesis originates from our thinking about the computational nature of both unscrambling and scrambling, which leads us to believe that (from the computational point of view) scrambling may result from a "failed equality test".

Let us consider the situation depicted in Fig. 41, where we have two molecules with two *different* staggered cuts – one yielding the overhang α and the other yielding the overhang β.

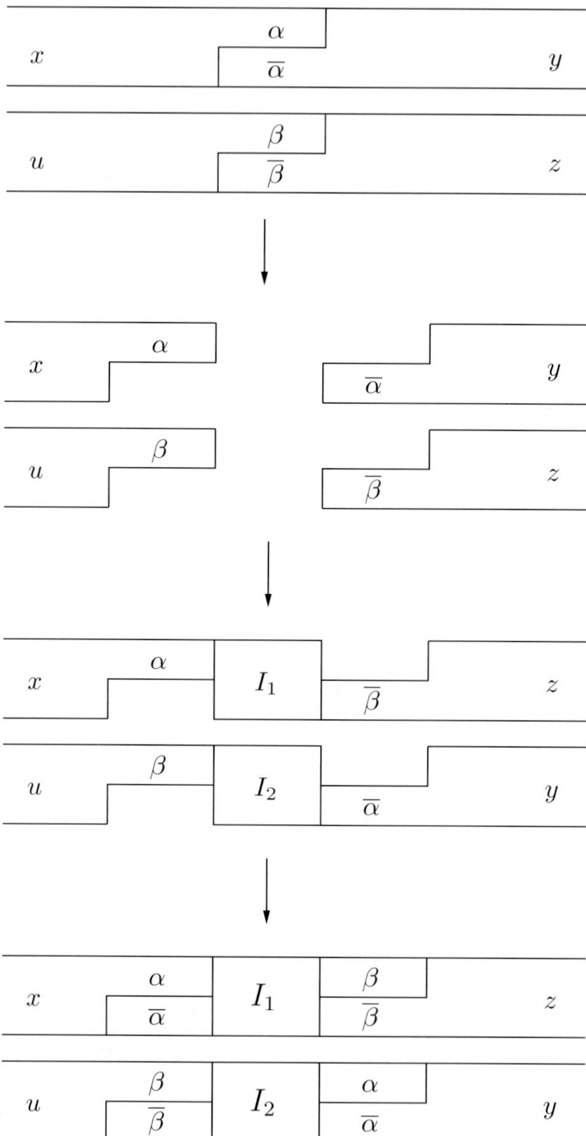

Fig. 41. Scrambling insertion

The recombinases doing these cuts do not "know" in advance whether these cuts are matching or not. Since the attempt to recombine these two molecules will fail, most of the time the molecules will be restored by annealing back α with $\bar{\alpha}$ and β with $\bar{\beta}$. But sometimes, the "persistent" recombinase may recombine the two molecules by slipping in IESs I_1 and I_2, then ligate them to single-stranded overhangs, and finally fill in the resulting gaps. This creates the nonrepeats (α, β) and (β, α) flanking the inserted IESs I_1 and I_2. When the same happens at some other place on a molecule, we get scrambling for direct repeats (α, α) and (β, β).

Thus, from the computational point of view, the nonrepeats were created because the equality test ($\alpha = \beta$?) has failed: the recombination could not take place, and so I_1 and I_2 were slipped in to achieve "pseudorecombination" anyhow. From this point of view, scrambling is the result of a failed equality test.

Also, when we consider the operation of hairpin excision but now aligned on a nonrepeat (α, β), we get scrambling for inverted repeats $(\alpha, \bar{\alpha})$ and $(\beta, \bar{\beta})$; this is illustrated in Fig. 42.

13   Discussion

After providing, in Sects. 2 through 5, the background information about the DNA processing during gene (un)scrambling in ciliates, this paper presents a systematic study of the computational nature of the unscrambling mechanism.

In particular, we study the use of two operations: ld-excision and hi-excision/ reinsertion, in the process of gene unscrambling – the latter operation is studied for the first time in this paper. To this aim we have introduced the formal framework of pointer reduction systems in both the string and the graph form. These systems have turned out to be convenient computational tools in the investigation of gene unscrambling. In particular, they provide computational characterizations of those scrambling patterns that can be unscrambled using the operations of the ld-excision and hi-excision/reinsertion.

In our final section, Sect. 12, we have indicated a possible origin of gene scrambling using the operation of insertion, and the operation of hairpin excision/reinsertion on nonrepeat.

In this paper we continue the study of computational aspects of gene (un)scrambling initiated in a series of papers by Landweber and Kari, see, e.g., [5,6]. However, the formal framework that we present here is very different, and also the questions we ask are quite different. Two important differences between our models are that we consider a new operation of hi-excision/reinsertion, and our model is based on intramolecular processing.

We see this paper as the first step in our attempt to obtain a systematic formal framework for the study of gene (un)scrambling in ciliates. There are still many problems to be investigated – we now indicate some of them.

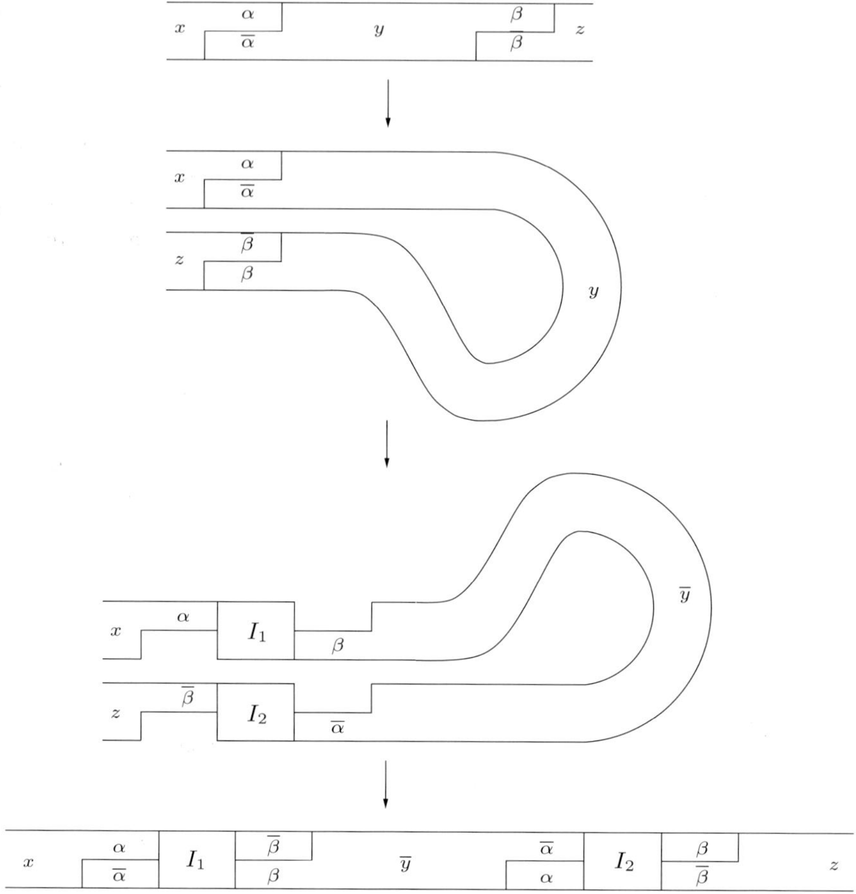

Fig. 42. Scrambling insertion by *hi*-excision/reinsertion

1. We have provided a computational characterization for scrambled patterns that can be unscrambled – this characterization is given in the form of successful reductions in a suitable string or graph pointer reduction system. What is still missing is a *combinatorial characterization* of such patterns.

2. The attractiveness of graph pointer reduction systems stems from the fact that there are less (legal) labelled graphs than (legal) strings in the sense that the same legal labelled graph may represent a whole family of legal strings: each string in this family has the same set of successful strategies for unscrambling! This opens a whole area of equivalences of scrambling patterns, which may turn out to be useful in determining evolutionary similarities (distance) between various species. An important formal problem here is a characterization of the set of all legal strings yielding the same legal labelled graph.

3. Unscrambling using ld-excision and hi-excision/reinsertion is limited: there are scrambled patterns that cannot be unscrambled in this way. A natural next step is to look for some more powerful (w.r.t. unscrambling) operations, and in particular to look for a *universal* set of such operations, i.e., a set of operations that can unscramble any pattern.

References

1. Adleman, L. (1994) Molecular computation of solutions to combinatorial problems, Science, 266:1021–1024.
2. Duharcourt, S., A-M. Keller, and E. Meyer (1998) Homology-dependent maternal inhibition of developmental excision of internal eliminated sequences in Paramecium tetraurelia. Mol. Cell. Biol. 18:7075–7085.
3. Hoffman, D.C. and D.M. Prescott (1996) A germline gene encoding DNA polymerase α in the hypotrichous ciliate Oxytricha nova is extremely scrambled. Nucl. Acids Res. 24:3337–3340.
4. Kari, L., Rubin, H., and D.H. Wood (1998) Proceedings of the 4th DIMACS meeting on DNA Based Computers, Philadelphia.
5. Landweber, L.F. and Kari, L. (1998) The evolution of cellular computing: nature's solution to a computational problem. In [4], pp. 3–15.
6. Landweber, L.F. and Kari, L. (2000) Universal molecular computation in ciliates. This volume.
7. Mitcham, J.L., D.M. Prescott, and M.K. Miller (1994) The micronuclear gene encoding β-telomere binding protein in Oxytricha nova. J. Euk. Microbiol. 41:478–480.
8. Mitcham, J.D., A.J. Lynn, and D.M. Prescott (1992) Analysis of a scrambled gene: the gene encoding α-telomere-binding protein in Oxytricha nova. Genes & Develop. 6:788–800.
9. Paun, G., Rozenberg, G., and A. Salomaa (1998) *DNA Computing* Springer-Verlag, Berlin, Heidelberg.
10. Prescott, D.M. (1999) The evolutionary scrambling and developmental unscrambling of germline genes in hypotrichous ciliates. Nucl. Acids Res., in press.
11. Prescott, D.M. and M.L. DuBois (1996) Internal eliminated segments (IESs) of Oxytrichidae. J. Euk. Microbiol. 43:432–441.
12. Prescott, D.M. (1992) The unusual organization and processing of genomic DNA in hypotrichous ciliates. Trends in Genet. 8:439–445.
13. Prescott, D.M. and A.F. Greslin (1992) Scrambled actin I gene in the micronucleus of Oxytricha nova. Develop. Genet. 13:66–74.
14. Wright, A-D.G. and D.H. Lynn (1997) Maximum ages of ciliate lineages estimated using a small subunit rRNA molecular clock: Crown eukaryotes date back to the Paleoproterozoic. Arch. Protistenkd. 148:329–341.

14   Acknowledgements

This work is supported by the NIGMS grant GM56161 to D.M. Prescott. The authors are indebted to L. Landweber and L. Kari for attracting our attention to computational aspects of gene (un)scrambling in ciliates. We are grateful to H.J. Hoogeboom and N. van Vugt for useful comments on the first version of this manuscript. We are also indebted to Mrs. Marloes Boon-van der Nat and to Mrs. Gayle Prescott for the expert typing of this manuscript, and to J. Hage and E. Winfree for their help in producing the figures.

Universal Molecular Computation in Ciliates

Laura F. Landweber and Lila Kari

Abstract. How do cells and nature "compute"? They read and "rewrite" DNA all the time, by processes that modify sequences at the DNA or RNA level. In 1994, Adleman's elegant solution to a seven-city Directed Hamiltonian Path problem using DNA [1] launched the new field of DNA computing, which in a few years has grown to international scope. However, unknown to this field, ciliated protozoans of genus *Oxytricha* and *Stylonychia* had solved a potentially harder problem using DNA several million years earlier. The solution to this "problem", which occurs during the process of gene unscrambling, represents one of nature's ingenious solutions to the problem of the creation of genes. Here we develop a model for the guided homologous recombinations that take place during gene rearrangement and prove that such a model has the computational power of a Turing machine, the accepted formal model of computation. This indicates that, in principle, these unicellular organisms may have the capacity to perform at least any computation carried out by an electronic computer.

1 Gene Unscrambling as Computation

Ciliates are a diverse group of 8000 or more unicellular eukaryotes (nucleated cells) named for their wisp-like covering of cilia. They possess two types of nuclei: an active *macronucleus* (soma) and a functionally inert *micronucleus* (germline) which contributes only to sexual reproduction. The somatically active macronucleus forms from the germline micronucleus after sexual reproduction, during the course of development. The genomic copies of some protein-coding genes in the micronucleus of hypotrichous ciliates are obscured by the presence of intervening non-protein-coding DNA sequence elements (internally eliminated sequences, or *IES*s). These must be removed before the assembly of a functional copy of the gene in the somatic macronucleus. Furthermore, the protein-coding DNA segments (macronuclear destined sequences, or *MDS*s) in species of *Oxytricha* and *Stylonychia* are sometimes present in a permuted order relative to their final position in the macronuclear copy. For example, in *O. nova*, the micronuclear copy of three genes (Actin I, α-telomere binding protein, and DNA polymerase α) must be reordered and intervening DNA sequences removed in order to construct functional macronuclear genes. Most impressively, the gene encoding DNA polymerase α (DNA pol α) in *O. trifallax* is apparently scrambled in 50 or more pieces in its germline nucleus [10]. Destined to unscramble its micronuclear genes by putting the pieces together again, *O. trifallax* routinely solves a potentially complicated computational problem when rewriting its genomic sequences to form the macronuclear copies.

This process of unscrambling bears a remarkable resemblance to the DNA algorithm Adleman [1] used to solve a seven-city instance of the Directed Hamiltonian Path problem. Adleman's algorithm involves the use of edge-encoding sequences as splints to connect city-encoding sequences, allowing the formation of all possible paths through the graph (Figure 1). Afterwards, a screening process eliminates the paths that are not Hamiltonian, i.e. ones which either skip a city, enter a city twice, or do not start and end in the correct origin and final destinations.

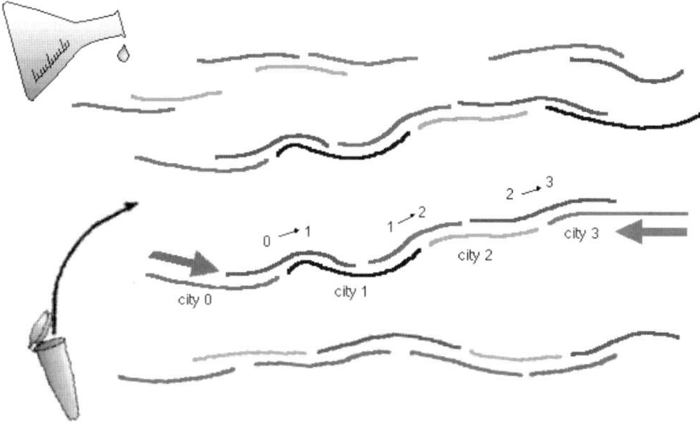

Fig. 1. DNA hybridization in a molecular computer. PCR primers are indicated by arrows.

The developing ciliate macronuclear "computer" (Figures 2-3) apparently relies on the information contained in short direct repeat sequences to act as minimal guides in a series of homologous recombination events. These guide-sequences provide the splints analogous to the edges in Adleman's graph, and the process of recombination results in linking the protein-encoding segments (MDSs, or "cities") that belong next to each other in the final protein-coding sequence ("Hamiltonian path"). As such, the unscrambling of sequences that encode DNA polymerase α accomplishes an astounding feat of cellular computation, especially as 50-city Hamiltonian path problems are sometimes considered hard problems in computer science and present a formidable challenge to a biological computer. Other structural components of the ciliate chromatin presumably play a significant role, but the exact details of the mechanism are still unknown.

2   The Path towards Unscrambling

Typical IES excision in ciliates involves the removal of short (14 - 600 bp) AT-rich sequences flanked by direct repeats of 2 to 14 bp. IESs are often

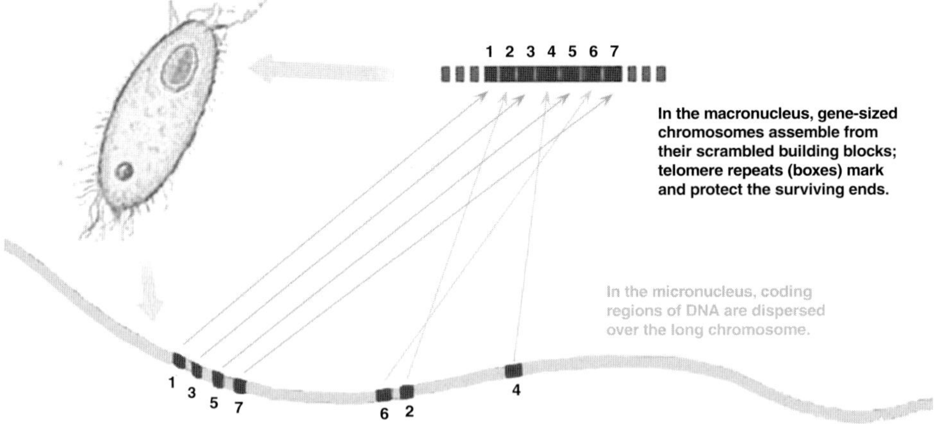

Fig. 2. Overview of gene unscrambling. Dispersed coding MDSs 1-7 reassemble during macronuclear development to form the functional gene copy (top), complete with telomere addition to mark and protect both ends of the gene.

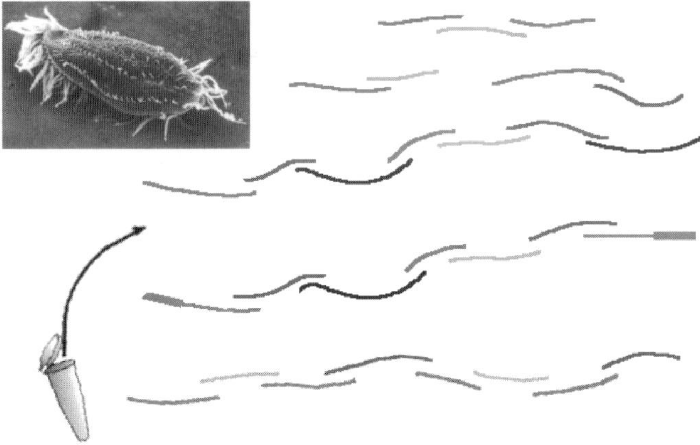

Fig. 3. A ciliate computer? Correct gene assembly in *Stylonychia* (inset) requires the joining of many segments of DNA, guided by short sequence repeats, only at the ends. Telomeres, indicated by thicker lines, mark the termini of correctly assembled gene-sized chromosomes. Note the similarities in principle to DNA computations that specifically rely on pairing of short repeats at the ends of DNA fragments, as in Adleman's experiment.

released as circular DNA molecules [21]. The choice of which sequences to remove appears to be minimally "guided" by recombination between direct repeats of only 2 to 14 base pairs.

Unscrambling is a particular type of IES removal in which the order of the MDSs in the micronucleus is often radically different from that in the macronucleus. For example, in the micronuclear genome of *Oxytricha nova*, the MDSs of α-telomere binding protein (α-TP) are arranged in the cryptic order 1-3-5-7-9-11-2-4-6-8-10-12-13-14 relative to their position in the "clear" macronuclear sequence 1-2-3-4-5-6-7-8-9-10-11-12-13-14. This particular arrangement predicts a spiral mechanism in the path of unscrambling that links odd and even segments in order (Figure 4; [15]).

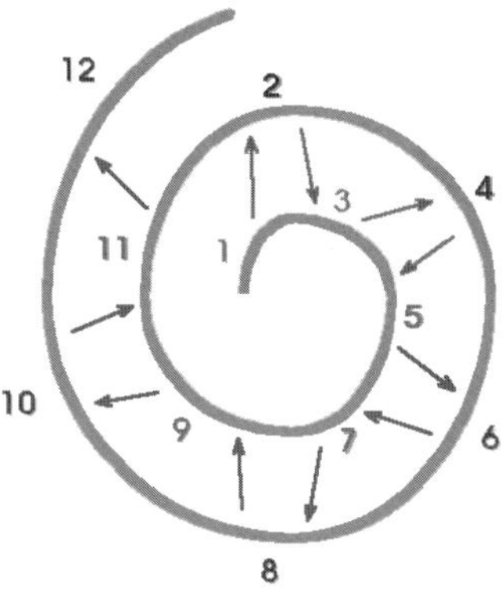

Fig. 4. Model for unscrambling in α-TP (adapted from [15]).

Homologous recombination between identical short sequences at appropriate MDS-IES junctions is implicated in the mechanism of gene unscrambling, as it could simultaneously remove the IESs and reorder the MDSs. For example, the DNA sequence present at the junction between MDS n and the downstream IES is generally the same as the sequence between MDS $n+1$ and its upstream IES, leading to correct ligation of MDS n to MDS $n+1$, over a distance. However, the presence of such short repeats (average length 4 bp between non-scrambled MDSs, 9 bp between scrambled MDSs [18]) implies that although these guides are necessary, they are certainly not sufficient to guide accurate splicing. Hence it is likely that the repeats satisfy more of a structural requirement for MDS splicing, and less of a role in substrate recognition. Otherwise, incorrectly spliced sequences (the results of promiscuous recombination) would dominate, especially in the case of very small (2-4 bp) repeats present thousands of times throughout the genome. This incorrect hybridization could be a driving force in the production of newly scrambled patterns in evolution. However, during macronuclear development only unscrambled molecules which contain 5' and 3' telomere addition sequences would be selectively retained in the macronucleus, ensuring that most promiscuously ordered genes would be lost.

3   Inversions as Catalysts of DNA Rearrangements

The gene encoding DNA polymerase α is broken into at least 44 MDSs in *O. nova* and 51 in *O. trifallax*, scrambled in a nonrandom order with an inversion in the middle, and some MDSs located at least several kilo-bases (kb) away from the main gene (in an unmapped PCR fragment). The resulting hairpin structural model predicted in Figure 5 could equip the ciliate with a dramatic shortcut to finding the correct solution to its DNA polymerase α unscrambling problem.

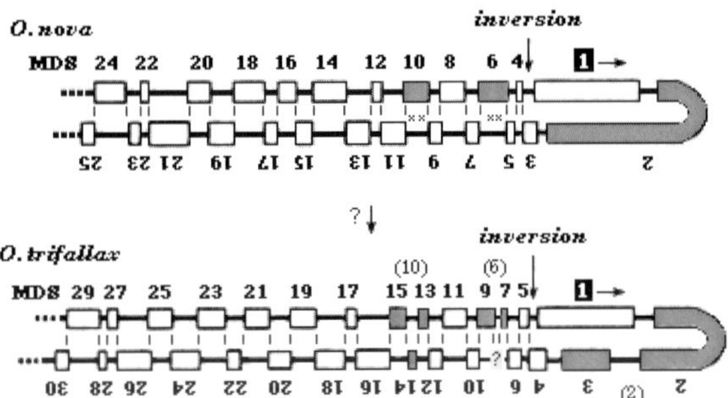

Fig. 5. Model for unscrambling of DNA pol α. Vertical lines indicate recombination junctions between scrambled MDSs, guided by direct repeats. MDS 10 in *O. nova* can also give rise to three new MDSs (13–15) in *O. trifallax*, one scrambled on the inverted strand, by two spontaneous intramolecular recombination events (x's) in the folded orientation shown. *O. nova* MDS 6 can give rise to *O. trifallax* MDSs 7-9 (MDS 8, shaded, is only 6 bp and was not identified in [10]). *O. trifallax* non-scrambled MDSs 2 and 3 could be generated by the insertion of an IES in *O. nova* MDS 2 (similar to a model suggested by M. Dubois in [10]).

Figures 5–6 outline a model for the origin and accumulation of scrambled MDSs. The appearance of an inversion is likely to encourage the formation of new MDSs in a nonrandomly scrambled pattern. By Muller's Ratchet, an inversion makes the addition of new MDSs much more likely, given that the hairpin structure, which juxtaposes coding and noncoding DNA sequences, would promote recombination, possibly between short arbitrary repeats. For example, the arrangement of MDSs 2, 6, and 10 in *O. nova* could have given rise to the arrangement of eight new MDSs in *O. trifallax* (Figure 5).

We have recently discovered scrambling in the gene encoding DNA polymerase α in the micronucleus of a different ciliate, *Stylonychia lemnae*, which

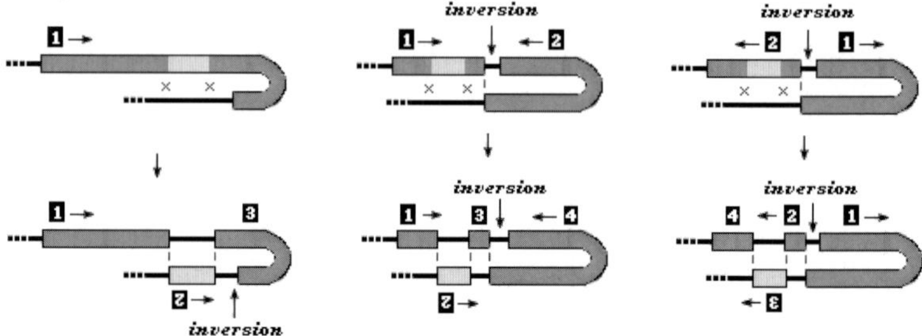

Fig. 6. Proposed model for the origin of a scrambled gene. Left: birth of a scrambled gene from a non-scrambled gene by a double recombination with an IES or any noncoding DNA (new MDS order 1-3-2 with an inversion between MDSs 3 and 2). Middle: generation of a scrambled gene with a non-random MDS order, from a non-scrambled gene with an inversion between two MDSs. Right: creation of new scrambled MDSs in a scrambled gene containing an inversion. Inversions may dramatically increase the production of scrambled MDSs, by stabilizing the folded conformation that allows reciprocal recombinations across the inversion.

enjoys the benefit of a working transformation system [22]. The scrambled gene in *S. lemnae* shares the presence of an inversion with the two *Oxytricha* species. These scrambled genes in ciliates thus offer a unique system in which to study the origin of a complex genetic mechanism and the role of inversions as catalysts of acrobatic DNA rearrangements during evolution (Figure 6). DNA polymerase α's complex scrambling pattern is possibly the best analog equivalent of a hard-path finding problem in nature. Alternate splicing at the RNA level, as well as other forms of programmed DNA rearrangements, could also be viewed as solutions to path-finding problems in nature. Dynamic processes, such as maturation of the immune response, provide examples of genuine evolutionary computation in cells, whereas the path-finding problems here may follow a more deterministic algorithm. Current effort is directed toward understanding how cells unscramble DNA, how this process has arisen, and how the "programs" are written and executed. Do they decode the message by following the shortest unscrambling path or by following a more circuitous but equally effective route, as in the case of RNA editing [12]? Also, how error prone is the unscrambling process? Does it actually search through several plausible unscrambled intermediates or follow a strictly deterministic pathway?

4 The Formal Model

Before introducing the formal model, we summarize our notation. An alphabet Σ is a finite, nonempty set. In our case $\Sigma = \{A, C, G, T\}$. A sequence of letters from Σ is called a string (word) over Σ and in our interpretation corresponds to a linear strand. The words are denoted by lowercase letters such as u, v, α_i, x_{ij}. The length of a word w is denoted by $|w|$ and represents the total number of occurrences of letters in the word. A word with 0 letters in it is called an empty word and is denoted by λ. The set of all possible words consisting of letters from Σ is denoted by Σ^*, and the set of all nonempty words by Σ^+. We also define circular words over Σ by declaring two words to be equivalent if and only if (iff) one is a cyclic permutation of the other. In other words, w is equivalent to w' iff they can be decomposed as $w = uv$ and $w' = vu$, respectively. A circular word $\bullet w$ refers to any of the circular permutations of the letters in w. $\Sigma^\bullet$ denotes the set of all circular words over Σ.

With this notation, we define intramolecular recombination using set theoretical notation as:

$$\{uxwxv\} \Longrightarrow \{uxv, \bullet wx\}$$

where u, w, x, and v are words in Σ^*, and x, the junction sequence that guides unscrambling, is nonempty.

Thus the defined operation models the process of intramolecular recombination. After x finds its second occurrence in $uxwxv$, the molecule undergoes a strand exchange in x that leads to the formation of two new molecules: uxv and a circular DNA molecule $\bullet wx$.

Intramolecular recombination also accomplishes the deletion of either sequence wx or xw from the original molecule $uxwxv$. The fact that $\bullet wx$ is circular implies that we can use any circular permutation of its sequence as an input for a subsequent operation.

In this model, the effects of intramolecular recombination can be reversed. Note that the operation in the forward direction is formally intramolecular recombination, whereas the operation in the reverse direction is intermolecular recombination. The intermolecular recombination

$$\{uxv, \bullet wx\} \Longrightarrow \{uxwxv\}$$

also accomplishes the insertion of the sequence wx or xw in the linear string uxv.

The above operations resemble the "splicing operation" introduced by Head in [7] and "circular splicing" [8,20,17]. [16], [3] and subsequently [23] showed that these models have the computational power of a universal Turing machine. (See [9] for a review.)

The process of gene unscrambling entails a series of successive or possibly simultaneous intra- and inter-molecular homologous recombinations. This is followed by excision of all sequences $\tau_s y \tau_e$, where the sequence y is marked by the presence of telomere addition sequences τ_s for telomere "start" (at its 5' end), and τ_e for telomere "end" (at its 3' end). Thus from a long sequence $u\tau_s y \tau_e v$, this step retains only $\tau_s y \tau_e$ in the macronucleus. Lastly, the enzyme telomerase extends the length of the telomeric sequences (usually double-stranded $TTTTGGGG_n$ repeats in these organisms) from τ_s and τ_e to protect the ends of the DNA molecule.

We now make the assumption that, by a clever structural alignment, such as the one depicted in Figure 4, or by other biological factors, the cell decides which sequences are non-protein-coding (IESs) and which are ultimately protein-coding (MDSs), as well as which sequences x guide homologous recombination. Moreover, such biological shortcuts are presumably essential to bring into proximity the guiding sequences x. Each of the n MDSs, denoted primarily by α_i, $1 \leq i \leq n$ is flanked by the guiding sequences $x_{i-1,i}$ and $x_{i,i+1}$. Each guiding sequence points to the MDS that should precede or follow α_i in the final sequence. The only exceptions are α_1, which is preceded by τ_s, and α_n, which is followed by τ_e in the input string or micronuclear molecule. Note that although present generally once in the final macronuclear copy, each $x_{i,i+1}$ occurs at least twice in the micronuclear copy – once after α_i and once before α_{i+1}.

We denote by ϵ_k an internal sequence that is deleted; ϵ_k does not occur in the final sequence. Thus, since unscrambling leaves one copy of each $x_{i,i+1}$ between α_i and α_{i+1}, an IES is nondeterministically either $\epsilon_k x_{i,i+1}$ or $x_{i-1,i}\epsilon_k$, depending on which guiding sequence $x_{i,i+1}$ is eliminated. Similarly, an MDS is technically either $\alpha_i x_{i+1}$ or $x_{i-1,i}\alpha_i$. For this model, either choice is equivalent.

Removal of nonscrambled IESs in *Euplotes crassus* actually leaves extra sequences (including a duplication of x_{ij}) at the junctions between ϵ_k's in the resulting non-protein-coding products. This may result when the x_{ij}'s are as short as two nucleotides [11]. It is unknown whether unscrambling also introduces extra sequences, since it uses considerably longer x_{ij}'s on average. However, since the extra sequences have always been found at junctions between ϵ_k's, this would not affect our unscrambling model.

The following example models unscrambling of a micronuclear gene that contains MDSs in the scrambled order 2-4-1-3:

$$\{u\ x_{12}\ \alpha_2\ x_{23}\ \epsilon_1\ x_{34}\ \alpha_4\ \tau_e\ \epsilon_2\ \tau_s\ \alpha_1\ x_{12}\ \epsilon_3\ x_{23}\ \alpha_3\ x_{34}\ v\} \Longrightarrow$$

$$\{u\ x_{12}\ \epsilon_3\ x_{23}\ \alpha_3\ x_{34}\ v\ ,\quad \bullet\alpha_2\ x_{23}\ \epsilon_1\ x_{34}\ \alpha_4\ \tau_e\ \epsilon_2\ \tau_s\ \alpha_1\ x_{12}\ \} =$$

$$\{u\ x_{12}\ \epsilon_3\ x_{23}\ \alpha_3\ x_{34}\ v,\quad \bullet\epsilon_1\ x_{34}\ \alpha_4\ \tau_e\ \epsilon_2\ \tau_s\ \alpha_1\ x_{12}\ \alpha_2\ x_{23}\} \Longrightarrow$$

$$\{u\ x_{12}\ \epsilon_3\ x_{23}\ \epsilon_1\ x_{34}\ \alpha_4\ \tau_e\ \epsilon_2\ \tau_s\ \alpha_1\ x_{12}\ \alpha_2\ x_{23}\ \alpha_3\ x_{34}\ v\} \Longrightarrow$$

$$\{u\ x_{12}\ \epsilon_3\ x_{23}\ \epsilon_1\ x_{34}\ v,\quad \bullet\alpha_4\ \tau_e\ \epsilon_2\ \tau_s\ \alpha_1\ x_{12}\ \alpha_2\ x_{23}\ \alpha_3\ x_{34}\} =$$

$$\{u\ x_{12}\ \epsilon_3\ x_{23}\ \epsilon_1\ x_{34}\ v\ ,\quad \bullet\tau_s\ \alpha_1\ x_{12}\ \alpha_2\ x_{23}\ \alpha_3\ x_{34}\ \alpha_4\ \tau_e\ \epsilon_2\} \Longrightarrow$$

$$\{\tau_s\ \alpha_1\ x_{12}\ \alpha_2\ x_{23}\ \alpha_3\ x_{34}\ \alpha_4\ \tau_e\ ,\quad \epsilon_2\ ,\quad u\ x_{12}\ \epsilon_3\ x_{23}\ \epsilon_1\ x_{34}\ v\}$$

Note that the process is nondeterministic in that, for example, one could start by replacing the first step, between homologous sequences x_{12}, by recombination between the homologous sequences x_{34} instead, obtaining the same result in the same number of steps.

Once the cell has "decided" which are the α_i's, $x_{i,i+1}$'s and ϵ_i's, the process that follows is simply sorting, requiring a linear number of steps (possibly fewer than n if some of the recombination events take place simultaneously). Part of this "decision" process entails finding the correct "path" linking the pieces of protein-coding regions in the correct order, with the occurrence of $\alpha_i x_{i,i+1}$ and $x_{i,i+1}\alpha_{i+1}$ in the micronuclear sequence providing the link between α_i and α_{i+1} in the macronuclear sequence. The junction sequences $x_{i,i+1}$ thus serve the role of the "edge" sequences in Adleman's graph.

A computational difficulty is the presence of multiple copies of the sequences $x_{i,i+1}$ which may direct the formation of incorrect "paths". Indeed, throughout the genome, such simple sequences may be present in extreme redundancy. Some of the $x_{i,i+1}$ even overlap with each other. For example, in the *O. trifallax* gene encoding DNA polymerase α, $x_{24,25} = \text{GAGAGATAGA}$ contains $x_{1,2} = \text{AGATA}$ as a subsequence. The search for the proper junction sequences thus amounts to finding the correct "path" and is potentially the most costly part of the computation. Production of incorrect paths will not necessarily lead to the production of incorrect proteins unless the path sequences start and end with the correct telomere addition sites (τ_s and τ_e), since these ensure survival of the genes in the macronucleus. Analogous to the PCR primers in Adleman's experiment, the role of telomeres here is thus to preserve those strands that start and end with the correct origin and final destinations.

5   Computational Power of Gene Rearrangement

In this section we define the notion of a *guided recombination system* that models the process taking place during gene rearrangement, and prove that such systems have the computational power of a Turing machine, the most widely used theoretical model of electronic computers.

The following strand operations generalize the intra- and intermolecular recombinations defined in the preceding section by assuming that homologous recombination is influenced by the presence of certain contexts, i.e., the presence of either an IES or an MDS flanking a junction sequence x_{ij}. The observed dependence on the old macronuclear sequence for correct IES removal in *Paramecium* suggests that this is the case [14]. This restriction captures the fact that the guide sequences do not contain all the information for accurate splicing during gene unscrambling.

We define the contexts that restrict the use of recombinations by a *splicing scheme*, [7,8], a pair $(\Sigma, \sim)$ where Σ is the alphabet and $\sim$, the pairing relation of the scheme, is a binary relation between triplets of nonempty words satisfying the following condition: If $(p, x, q) \sim (p', y, q')$ then $x = y$.

In the splicing scheme $(\Sigma, \sim)$ pairs $(p, x, q) \sim (p', x, q')$ now define the contexts necessary for a recombination between the repeats x. Then we define *contextual intramolecular recombination* as

$$\{uxwxv\} \Longrightarrow \{uxv, \bullet wx\}, \text{ where } u = u'p, w = qw' = w''p', v = q'v'.$$

This constrains intramolecular recombination within $uxwxv$ to occur only if the restrictions of the splicing scheme concerning x are fulfilled; i.e., the first occurrence of x is preceded by p and followed by q, and its second occurrence is preceded by p' and followed by q'.

Similarly, if $(p, x, q) \sim (p', x, q')$, then we define *contextual intermolecular recombination* as

$$\{uxv, \bullet wx\} \Longrightarrow \{uxwxv\} \text{ where } u = u'p, v = qv', w = w'p' = q'w''.$$

Informally, intermolecular recombination between the linear strand uxv and the circular strand $\bullet wx$ may take place only if the occurrence of x in the linear strand is flanked by p and q and its occurrence in the circular strand is flanked by p' and q'. Note that sequences p, x, q, p', q' are nonempty, and that both contextual intra- and intermolecular recombinations are reversible by introducing pairs $(p, x, q') \sim (p', x, q)$ in $\sim$.

The operations defined in the preceding section are particular cases of contextual recombinations, where all the contexts are empty, i.e, $(\lambda, x, \lambda) \sim (\lambda, x, \lambda)$ for all $x \in \Sigma^+$. This would correspond to the case where recombination may occur between every repeat sequence, regardless of the contexts.

If we use the classical notion of a set, we can assume that the strings entering a recombination are available for multiple operations. Similarly, there would be no restriction on the number of copies of each strand produced by recombination. However, we can also assume some strings are only available in a limited number of copies. Mathematically this translates into using *multisets*, where one keeps track of the number of copies of a string at each moment. In the style of [6], if $\mathbf{N}$ is the set of natural numbers, a multiset of Σ^* is a mapping $M : \Sigma^* \longrightarrow \mathbf{N} \cup \{\infty\}$, where, for a word $w \in \Sigma^*$, $M(w)$ represents the number of occurrences of w. Here, $M(w) = \infty$ means that there are unboundedly many copies of the string w. The set $\text{supp}(M) = \{w \in \Sigma^* \mid M(w) \neq 0\}$, the *support of M*, consists of the strings that are present at least once in the multiset M.

We now define a *guided recombination system* that captures the series of dispersed homologous recombination events that take place during these gene rearrangements in ciliates.

Definition. *A guided recombination system is a quadruple $R = (\Sigma, \sim, A)$ where $(\Sigma, \sim)$ is a splicing scheme, and $A \in \Sigma^+$ is a linear string called the axiom.*

A guided recombination system R defines a *derivation relation* that produces a new multiset from a given multiset of linear and circular strands, as follows. Starting from a "collection" (multiset) of strings with a certain number of available copies of each string, the next multiset is *derived from* the first one by an intra- or inter-molecular recombination between existing strings. The strands participating in the recombination are "consumed" (their multiplicity decreases by 1), whereas the products of the recombination are added to the multiset (their multiplicity increases by 1).

For two multisets S and S' in $\Sigma^* \cup \Sigma^\bullet$, we say that S derives S' and we write $S \Longrightarrow_R S'$, iff one of the following two cases hold:

(1) there exist $\alpha \in \mathrm{supp}(S)$, $\beta, \bullet\gamma \in \mathrm{supp}(S')$ such that
- $\{\alpha\} \Longrightarrow \{\beta, \bullet\gamma\}$ according to an intramolecular recombination step in R,
- $S'(\alpha) = S(\alpha) - 1$, $S'(\beta) = S(\beta) + 1$, $S'(\bullet\gamma) = S(\bullet\gamma) + 1$;

(2) there exist $\alpha', \bullet\beta' \in \mathrm{supp}(S)$, $\gamma' \in \mathrm{supp}(S')$ such that
- $\{\alpha', \bullet\beta'\} \Longrightarrow \{\gamma'\}$ according to an intermolecular recombination step in R,
- $S'(\alpha') = S(\alpha') - 1$, $S'(\bullet\beta') = S(\bullet\beta') - 1$, $S'(\gamma') = S(\gamma') + 1$.

Those strands which, by repeated recombinations with initial and intermediate strands eventually produce the axiom, form the language of the guided recombination system. Formally,

$$L_a^k(R) = \{w \in \Sigma^* \mid \{w\} \Longrightarrow_R^* S \text{ and } A \in \mathrm{supp}(S)\},$$

where the multiplicity of w equals k. Note that $L_a^k(R) \subseteq L_a^{k+1}(R)$ for any $k \geq 1$.

In a Turing machine (TM), a read/write head scans an infinite tape composed of discrete "squares", one square at a time. The read/write head communicates with a control mechanism under which it can read the symbol in the current square or replace it by another. The read/write head is also able to move on the tape, one square at a time, to the right and to the left (note the analogy to the action of RNA or DNA polymerase). The set of words that make a Turing machine finally halt is considered its language.

Formally [19] a rewriting system $TM = (S, \Sigma \cup \{\#\}, P)$ is called a *Turing machine* iff:

(i) S and $\Sigma \cup \{\#\}$ (with $\# \notin \Sigma$ and $\Sigma \neq \emptyset$) are two disjoint alphabets referred to as the *state* and the *tape* alphabets.

(ii) Elements s_0 and s_f of S, and B of Σ are the *initial* and *final* state, and the *blank symbol*, respectively. In addition, a subset T of Σ is specified and referred to as the *terminal* alphabet. It is assumed that T is not empty.

(iii) The productions (rewriting rules) of P are of the forms

$$\begin{aligned}
&(1)\ s_i a \longrightarrow s_j b \quad \text{(overprint)}\\
&(2)\ s_i ac \longrightarrow a s_j c \quad \text{(move right)}\\
&(3)\ s_i a\# \longrightarrow a s_j B\# \quad \text{(move right and extend workspace)}\\
&(4)\ c s_i a \longrightarrow s_j ca \quad \text{(move left)}\\
&(5)\ \# s_i a \longrightarrow \# s_j Ba \quad \text{(move left and extend workspace)}\\
&(6)\ s_f\, a \longrightarrow s_f\\
&(7)\ a\, s_f \longrightarrow s_f
\end{aligned}$$

where s_i and s_j are in S, $s_i \neq s_f$, $s_j \neq s_f$, and a, b, c are in Σ. For each pair (s_i, a), where s_i and a are in the appropriate ranges, P either contains no productions (2) and (3) (respectively, (4) and (5)) or else contains both (3) and (2) for every c (respectively, contains both (5) and (4) for every c). There is no pair (s_i, a) such that the word $s_i a$ is a subword of the left side in two productions of the forms (1), (3), (5).

A configuration of the TM is of the form $\#w_1 s_i w_2 \#$, where $w_1 w_2$ represents the contents of the tape, #s are the boundary markers, and the position of the state symbol s_i indicates the position of the read/write head on the tape: if s_i is positioned at the left of a letter a, this indicates that the read/write head is placed over the cell containing a. The TM changes from one configuration to another according to its rules. For example, if the current configuration is $\#w s_i a w' \#$ and the TM has the rule $s_i a \longrightarrow s_j b$, this means that the read/write head positioned over the letter a will write b over it, and change its state from s_i to s_j. The next configuration in the derivation will thus be $\#w s_j b w' \#$.

The Turing machine TM *halts* with a word w iff there exists a derivation that, when started with the read/write head positioned at the beginning of w, eventually reaches the final state; i.e., if $\#s_0 w \#$ derives $\#s_f \#$ by successive applications of the rewriting rules (1)-(7) The language $L(TM)$ *accepted* by TM consists of all words over the terminal alphabet T for which the TM halts. Note that TM is *deterministic*: at each step of the rewriting process, the application of at most one production is possible.

Theorem. *Let L be a language over T^* accepted by a Turing machine $TM = (S, \Sigma \cup \{\#\}, P)$ as above. Then there exist an alphabet Σ', a sequence $\pi \in \Sigma'^*$, depending on L, and a recombination system R such that a word w over T^* is in L if and only if $\#^6 s_0 w \#^6 \pi$ belongs to $L_a^k(R)$ for some $k \geq 1$.*

Proof. Consider that the rules of P are ordered in an arbitrary fashion and numbered. Thus, if TM has m rules, a rule is of the form $i : u_i \longrightarrow v_i$ where $1 \leq i \leq m$.

We construct a guided recombination system $R = (\Sigma', \sim, A)$ and a sequence $\pi \in \Sigma'^*$ with the required properties. The alphabet is $\Sigma' = S \cup \Sigma \cup \{\#\} \cup \{\$_i \mid 0 \leq i \leq m + 1\}$. The axiom, i.e., the target string to be achieved at the end of the computation, consists of the final state of the TM bounded by markers:

$$A = \#^{n+2} s_f \, \#^{n+2} \$_0 \$_1 \ldots \$_m \$_{m+1},$$

where n is the maximum length of the left-side or right-side words of any of the rules of the Turing machine.

The sequence π consists of the catenation of the right-hand sides of the TM rules bounded by markers, as follows:

$$\pi = \$_0 \, \$_1 e_1 v_1 f_1 \$_1 \, \$_2 e_2 v_2 f_2 \$_2 \ldots \$_m e_m v_m f_m \$_m \, \$_{m+1},$$

where $i:\ u_i \longrightarrow v_i$, $1 \le i \le m+1$ are the rules of TM and $e_i, v_i \in \Sigma \cup \{\#\}$.

If a word $w \in T^*$ is accepted by the TM, a computation starts then from a strand of the form $\#^{n+2}s_0 w \#^{n+2}\pi$, where we will refer to the subsequence starting with $\$_0$ as the "program", and to the subsequence at the left of $\$_0$ as the "data".

We construct the relation $\sim$ so that

(i) The right-hand sides of rules of TM can be excised from the program as circular strands which then interact with the data.

(ii) When the left-hand side of a TM rule appears in the data, the application of the rule can be simulated by the insertion of the circular strand encoding the right-hand side, followed by the deletion of the left-hand side.

To accomplish (i), for each rule $i : u \longrightarrow v$ of the TM, we introduce in $\sim$ the pairs

$$(C) \qquad (\$_{i-1}, \$_i, evf) \sim (evf, \$_i, \$_{i+1}),$$

for all $e, f \in \Sigma \cup \{\#\}$.

To accomplish (ii) for each rule $i : u \longrightarrow v$ of the TM, we add to the relation $\sim$ the pairs

$$(A) \qquad (ceu, f, d) \sim (\$_i ev, f, \$_i ev),$$

$$(B) \qquad (c, e, uf\$_i) \sim (uf\$_i, e, vfd),$$

for all $c \in \{\#\}^* \Sigma^*$, $d \in \Sigma^* \{\#\}^*$, $|c| = |d| = n$, $e, f \in \Sigma \cup \{\#\}$.

Following the above construction of the alphabet Σ', sequence π, and recombination system R, for any $x, y \in \Sigma'$ we can simulate a derivation step of the TM as follows:

$$\{xceufdy\$_0 \ldots \$_{i-1}\$_i evf\$_i\$_{i+1} \ldots \$_{m+1}\} \Longrightarrow_R$$

$$\{xceufdy\$_0 \ldots \$_{i-1}\$_i\$_{i+1} \ldots \$_{m+1}, \bullet\$_i evf\} \Longrightarrow_R$$

$$\{xceuf\$_i evfdy\$_0 \ldots \$_{i-1}\$_i\$_{i+1} \ldots \$_{m+1}\} \Longrightarrow_R$$

$$\{xcevfdy\$_0 \ldots \$_{i-1}\$_i\$_{i+1} \ldots \$_{m+1}, \bullet\$_i euf\}.$$

The first step is an intramolecular recombination using contexts (C) around the repeat $\$_i$ to excise $\bullet\$_i evf$. Note that if the current strand does not contain a subword $\$_i evf\$_i$, this can be obtained from another copy of the original linear strand, which is initially present in k copies. The second step is an intermolecular recombination using contexts (A) around the repeat f, to insert $\$_i evf$ after $ceuf$. The third step is an intramolecular recombination using contexts (B) around the direct repeat e to delete $\$_i euf$ from the linear strand. Thus, the "legal" insertion/deletion succession that simulates one TM derivation step claims that any u in the data that is surrounded by at least $n+1$ letters on both sides may be replaced by v. This explains why in our choice of axiom we needed $n+1$ extra symbols $\#$ to provide the contexts allowing recombinations to simulate all TM rules, including (3) and (5).

From the fact that a TM derivation step can be simulated by recombination steps we deduce that, if the TM accepts a word w, then we can start a derivation in R from

$$\#^{n+2} s_0 w \#^{n+2} \pi = \#^{n+2} s_0 w \#^{n+2} \$_0 \$_1 \ldots \$_i e_i v_i f_i \$_i \ldots \$_m \$_{m+1}$$

and reach the axiom by only using recombinations according to R. This means that our word is accepted by R; that is, it belongs to $L_a^k(R)$ for some k. Note that if some rules of the TM have not been used in the derivation then they can be excised in the end, and that k should be large enough so that we do not exhaust the set of rewriting rules.

For the converse implication, it suffices to prove that starting from the strand $\#^{n+2} s_0 w \#^{n+2} \pi$, no other recombinations except those that excise rules of TM from the program and those that simulate steps of the TM in the data are possible in R.

In the beginning of the derivation we start with no circular strands and k copies of the linear strand

$$\#^{n+2} s_0 w \#^{n+2} \$_0 \ldots \$_i e_i v_i f_i \$_i \ldots \$_{m+1}, w \in T^*,$$

where $i : u_i \longrightarrow v_i$ are TM rules, $e_i, f_i \in \Sigma \cup \{\#\}$, $1 \leq i \leq m$.

Assume now that the current multiset contains linear strands of the form $\delta_0 \pi$, where $\delta_0 \in \Sigma'^*$ contains only one state symbol and no $\$_i$ symbols and

$$\pi = \$_0 r_1 r_2 \ldots r_m \$_{m+1},$$

with r_i either encoding the right-hand side of a rule or being the remnant of a rule, i.e., $r_i \in \{\$_i e_i v_i f_i \$_i\} \cup \{\$_i\}$, $1 \leq i \leq m$. Moreover, assume that the circular strands present in the multiset are of the form $\bullet \$_i e_i v_i f_i$, with e_i, v_i, f_i as before.

Then:

(i) We cannot use (A) or (B) to insert or delete in the program because that would require the presence of strands $ceufd$ or $\$_i evf \$_i ev$ (if we want to use (A)) or $ceuf\$_i$ or $uf\$_i evfd$ (if we want to use (B)). However, none of these strands can appear in the program. Indeed, the 1st, 3rd, and 4th word all contain subwords over $\Sigma \cup \{\#\}$ of length at least $n + 3$, and this is more than the length of the longest subword over $\Sigma \cup \{\#\}$ present in the program. The 2nd word cannot appear in the program because no marker $\$_i$ appears alone in p, as p always contains at least two consecutive markers.

(ii) We cannot use (C) to insert or delete in the data because that would require the presence in δ_0 of two consecutive markers $\$_{i-1} \$_i$ or $\$_i \$_{i+1}$, which contradicts our assumptions.

(iii) We cannot use (C) to insert in the program because that would require the presence of a circular strand with two markers - a contradiction with our assumptions.

Arguments (i) - (iii) show that the only possible recombinations are either deletions in the program using (C), which result in the release of circular strands $\bullet \$_i e v f$, or insertions/deletions in the data using (A) and (B).

Assuming that the data contains as a subword the left-hand side of a TM rule $i : u \longrightarrow v$, and assuming that the necessary circular strand $\bullet \$_i e v f$ has already been excised from the program, the next step is to show that the only possible insertions/deletions in the data are those simulating a rewriting step of TM using rule i.

Indeed, in this situation:

(1) It is not possible to delete in δ_0 using (A), or to insert or delete using (B), as all these operations would require a $\$_i$ in δ_0. Therefore only an insertion in δ_0 using (A) is possible. An insertion according to (A) may take place only between a sequence $ceuf$ and a sequence d, where u contains a state symbol, i.e., the read/write head, c and d have length n, and e and f are letters. This means that, for the insertion to take place, the linear word has to be of the form

$$\delta_0 \ \pi = xceufdy \ \pi$$

and the intermolecular recombination with the circular strand $\bullet \$_i e v f$ inserts $\$_i e v f$ between u and f, producing the linear strand

$$\delta_1 \ \pi = xceuf\$_i evfdy \ \pi.$$

Note that as δ_0 contains only one state symbol and no marker $\$_i$, the newly formed word δ_1 contains only two state symbols (read/write heads), one in u and one in v, and only one marker $\$_i$. (Here we use the fact that every rule $u \longrightarrow v$ of the TM has exactly one state symbol on each side.)

(2) Starting now from $\delta_1 \pi$:

(2a) We can delete in δ_1 using (B) and, as there is only one $\$_i$ in δ_1, there is only one position where the deletion can happen. After the release of the strand $\bullet \$_i e u f$ as a circular strand, the linear strand produced is

$$\delta_2 \ \pi = xcevfdy \ \pi.$$

(2b) No insertion in δ_1 using (A) may take place, as the marker $\$_i$ "breaks" the contexts necessary for further insertions.

Indeed, the occurrence of another insertion according to (A) requires that the read/write head symbol be both followed and preceded by at least $(n+1)$ letters different from $\$_i$. In δ_1, the first read/write head is in u and the number of letters following it is at most $|u| - 1 + |f| \leq n - 1 + 1 = n$, which is not enough as a right context for insertion using (A). The second read/write head is in v and the number of letters preceding it is at most $|e| + |v| - 1 \leq 1 + n - 1 = n$, which is not enough as a left context for insertion using (A).

(2c) No deletion in δ_1 using (A) may occur, as this would require the presence of a repeat f bordered by a $\$_i ev$ on each side. This would imply that the current strand δ_1 contains two markers $\$_i$, which is not true.

(2d) No insertion in δ_1 using (B) is possible, as that would require the presence of a circular strand containing $\$_i evfd$. The length of such a strand would be at least $1 + |e| + |v| + |f| + |d|$; that is, at least $n + 4$, which is more than the length of any initial or intermediate circular strand. Indeed, all the circular strands produced from the program have length $n + 3$, and the only circular strands that are released are of the form $\bullet\$_i euf$, as seen in (2a), and thus also have lengths of at most $n + 3$.

The arguments above imply that the only possible operations on the data simulate legal rewritings of the TM by tandem recombination steps that necessarily follow each other.

Together with the arguments that the only operations affecting the program are excisions of circular strands encoding TM rules, and that the circular TM rules do not interact with each other, this proves the converse implication.

From the definition of the Turing machine we see that n, the maximum length of a word occurring in a TM rule, equals 4, which completes the proof of the theorem.

Q.E.D.

The preceding theorem implies that if a word $w \in T^*$ is in $L(TM)$, then $\#^6 s_0 w \#^6 \pi$ belongs to $L_a^k(R)$ for some k and therefore belongs to $L_a^i(R)$ for any $i \geq k$. This means that in order to simulate a computation of the Turing machine on w, any sufficiently large number of copies of the initial strand will do. The assumption that sufficiently many copies of the input strand are present at the beginning of the computation is in accordance with the fact that there are multiple copies of each strand available during the (polytene chromosome) stage where unscrambling occurs. Note that the preceding result is valid even if we allow interactions between circular strands or within a circular strand, formally defined in [13] as circular contextual intra- and intermolecular recombinations.

The proof that a guided recombination system can simulate the computation of a Turing machine suggests that the micronuclear gene, present in multiple copies, consists of a sequence encoding the input data combined with a sequence encoding a program, i.e., a list of encoded "computation instructions". The computation instructions can be excised from the micronuclear gene and become circular "rules" that can recombine with the data. The process continues then by multiple intermolecular recombination steps involving the linear strand and circular rules, as well as intramolecular recombinations within the linear strand itself. The resulting linear strand, which is the functional macronuclear copy of the gene, can then be viewed as the output of the computation performed on the input data following the computation instructions excised as circular strands.

The last step, telomere addition and the excision of the strands between the telomere addition sites, can easily be added to our model as a final step consisting of the deletion of all the markers, rule delimiters and remaining rules from the output of the computation. This would result in a strand that contains only the output of the computation (macronuclear copy of the gene) flanked by end markers (telomere repeats).

In conclusion, we have developed a model for the acrobatic process of gene unscrambling in hypotrichous ciliates. While the model is consistent with our limited knowledge of this biological process, it needs to be rigorously tested using molecular genetics. We have shown, however, that the model is capable of universal computation. This both hints at future avenues for exploring biological computation and opens our eyes to the range of complex behaviors that may be possible in ciliates, and potentially available to other evolving genetic systems.

Acknowledgements. We thank Jarkko Kari for essential contribution to the proof of the theorem in its present form; Rani Siromoney and Gilles Brassard for suggestions; Erik Winfree and Gheorghe Păun for comments; Grzegorz Rozenberg, Richard Lipton, David Prescott, Hans Lipps, and Ed Curtis for discussions. This research was partially supported by the Natural Sciences and Engineering Research Council of Canada (to L.K.) and NIGMS grant GM59708 (to L.F.L.).

References

1. Adleman, L.M. 1994. Molecular computation of solutions to combinatorial problems. *Science* 266: 1021–1024.
2. Bartel, D.P. and J.W. Szostak. 1993. Isolation of New Ribozymes from a Large Pool of Random Sequences. *Science* 261: 1411–1418.
3. Csuhaj-Varju, E., R. Freund, L. Kari, and G. Paun. 1996. DNA computing based on splicing: universality results. In Hunter, L. and T. Klein (editors). *Proceedings of 1st Pacific Symposium on Biocomputing.* World Scientific Publisher, Singapore, pp. 179–190.
4. DuBois, M. and D.M. Prescott. 1995. Scrambling of the actin I gene in two *Oxytricha* species. *Proc. Natl. Acad. Sci., U.S.A.* 92: 3888–3892.
5. Denninghoff, R.W and R.W. Gatterdam, 1989. On the undecidability of splicing systems. *International Journal of Computer Mathematics* 27: 133–145.
6. Eilenberg, S. 1984. *Automata, Languages and Machines.* Academic Press, New York.
7. Head, T. 1987. Formal language theory and DNA: an analysis of the generative capacity of specific recombinant behaviors. *Bull. Math. Biology* 49: 737–759.
8. Head, T. (1991). Splicing schemes and DNA. In Rozenberg, G. and A. Salomaa (editors). *Lindenmayer systems*, Springer Verlag, Berlin. pp. 371–383.
9. Head, T., G. Paun, and D. Pixton. 1997. Language theory and molecular genetics. In Rozenberg, G. and A. Salomaa (editors). *Handbook of Formal Languages*, vol 2., Springer Verlag, Berlin. pp. 295–358.
10. Hoffman, D.C. and D.M. Prescott. 1997. Evolution of internal eliminated segments and scrambling in the micronuclear gene encoding DNA polymerase α in two *Oxytricha* species. *Nucl. Acids Res.* 25: 1883–1889.

11. Klobutcher, L.A., L.R. Turner and J. LaPlante. 1993. Circular forms of developmentally excised DNA in *Euplotes crassus* have a heteroduplex junction. *Genes Dev.* 7: 84–94.

12. Landweber, L. F., A.G. Fiks and W. Gilbert. 1993. The boundaries of partially edited cytochrome c oxidase III transcripts are not conserved in kinetoplastids: implications for the guide RNA model of editing. *Proc. Natl. Acad. Sci. USA* 90: 9242–9246.

13. Landweber, L.F. and L. Kari. 1999. The evolution of cellular computing: nature's solution to a computational problem. *Biosystems* 52: 3–15.

14. Meyer, E. and S. Duharcourt. 1996. Epigenetic Programming of Developmental Genome Rearrangements in Ciliates. *Cell* 87: 9–12.

15. Mitcham, J.L., A.J. Lynn and D.M. Prescott. 1992. Analysis of a scrambled gene: The gene encoding α-telomere-binding protein in *Oxytricha nova*. *Genes Dev.* 6: 788–800.

16. Paun, G. 1995. On the power of the splicing operation. *Int. J. Comp. Math* 59: 27–35.

17. Pixton, D., 1995. Linear and circular splicing systems. In *Proceedings of the First International Symposium on Intelligence in Neural and Biological Systems*. IEEE Computer Society Press, Los Alamos. pp. 181–188.

18. Prescott, D.M. and M.L. Dubois. 1996. Internal Eliminated Segments (IESs) of Oxytrichidae. *J. Euk. Microbiol.* 43: 432–441.

19. Salomaa, A. 1973. *Formal Languages*. Academic Press, New York.

20. Siromoney, R., K.G. Subramanian and V. Rajkumar Dare. 1992. Circular DNA and splicing systems. In Nakamura, A. (editor). *Parallel Image Analysis. Lecture Notes in Computer Science 654*, Springer Verlag, Berlin. pp. 260–273.

21. Tausta, S.L. and L.A. Klobutcher. 1989. Detection of circular forms of eliminated DNA during macronuclear development in *E. crassus*. *Cell* 59: 1019–1026.

22. Wen, J., C. Maercker and H.J. Lipps. 1996. Sequential excision of internal eliminated DNA sequences in the differentiating macronucleus of the hypotrichous ciliate *Stylonychia lemnae*. *Nucl. Acids Res.* 24: 4415–4419.

23. Yokomori, T., S. Kobayashi and C. Ferretti. 1997. Circular Splicing Systems and DNA Computability. In *Proc. of IEEE International Conference on Evolutionary Computation'97*. IEEE Computer Society Press, Los Alamos. pp. 219–224.

Toward *in vivo* Digital Circuits

Ron Weiss, George E. Homsy, and Thomas F. Knight, Jr.

Abstract. We propose a mapping from digital logic circuits into genetic regulatory networks with the following property: the chemical activity of such a genetic network *in vivo* implements the computation specified by the corresponding digital circuit. Logic signals are represented by the synthesis rates of cytoplasmic DNA binding proteins. Gates consist of structural genes for output proteins, fused to promoter/operator regions that are regulated by input proteins. The modular approach for building gates allows a free choice of signal proteins and thus enables the construction of complex circuits. This paper presents simulation results that demonstrate the feasibility of this approach. Furthermore, a technique for measuring gate input/output characteristics is introduced. We will use this technique to evaluate gates constructed in our laboratory. Finally, this paper outlines automated logic design and presents BioSpice, a prototype system for the design and verification of genetic digital circuits.

1 Introduction

We seek to design and build biochemical reaction networks *in vivo* that implement the digital logic abstraction, and are thus capable of carrying out computational functions. This would allow us to fit biological cells with digital "prostheses" that enable the cells to perform user-specified computational processes. Programmable computation in living cells would be an enabling technology for a host of applications such as drug and biomaterial manufacturing, nanomachine assembly, sensor/effector arrays, programmed therapeutics, and as a tool for studying genetic regulatory networks.

Our approach is to use synthesis rates of DNA binding proteins as logic signals. Since DNA binding proteins can function as transcriptional repressors, the effect of one protein on the transcription rate of another can represent the flow of logical information. The simplest logic gate, the inverter, is built from a single operator/promoter region that can be bound by an (input) repressor, fused to a structural gene coding for the output protein. Since the input protein represses transcription of the gene for the output protein, this system implements a digital inverter, provided the steady-state input–output transfer function is sufficiently sigmoidal. Since transcription rates are additive, we can build combinatorial gates from inverters with the same output protein.

In the remainder of this paper we describe related work (Sect. 2), present and discuss the general approach for implementing gene expression based digital logic (Sect. 3), describe an example of a chemical reaction model for the digital abstraction and show simulation results (Sect. 4), introduce a mechanism for quantifying the steady-state behavior of gates *in vivo* (Sect. 5), discuss some issues in microbial circuit design (Sect. 6), describe BioSpice, a prototype simulator for designing and verifying genetic digital circuits (Sect. 7), and offer conclusions and avenues for future work (Sect. 8).

2   Related Work

At least as early as 1974, Roessler and others [19–22] noted the possibility of building universal automata by coupling bistable chemical reactions, and that chemical reaction kinetics share a formal relationship with electronic circuit action. Okamoto *et al.* studied a cyclic enzyme system and showed that it had some properties of a McCulloch–Pitts neuron. In 1991, Hjelmfelt *et al.* [7] showed in principle how to construct neural networks from coupled chemical reactions, and determined specific connections for the construction of chemical logic gates. Later, Arkin and Ross [1] refined this method to allow use of enzymes with lower binding cooperativity, and applied their model to an in-depth analysis of a portion of the glycolytic cycle.

Recently, McAdams and others [11–13] have constructed mathematical models of various genetic regulatory networks *in vivo*.

Neidhardt and Savageau [15] have noted the need for useful high-level logical abstractions to improve our understanding of the integrative molecular biology of the cell.

Monod and Jacob [14], Sugita [23], Kauffman [9], and Thomas [24] have all made various and partially successful attempts at describing the global qualitative dynamics of genetic regulatory systems, by simplifying those systems to binary signal levels and pursuing a treatment in terms of boolean networks.

3   Gene Expression Based Logic

3.1   General Approach

Based on the model proposed by Knight and Sussman [10], we are developing an engineering discipline for designing and implementing digital logic *in vivo*. We seek a mapping from digital logic circuits into genetic regulatory networks with the following property: the chemical activity of such a genetic network *in vivo* expresses the computation specified by the corresponding digital circuit. Our approach uses the translation rates of repressor proteins as signals, and constructs genetic regulatory elements that constrain the signals to realize the desired logic function.

To build an inverter, select an existing promoter with an operator (repressor binding site), and fuse it to a structural gene for a distinct repressor

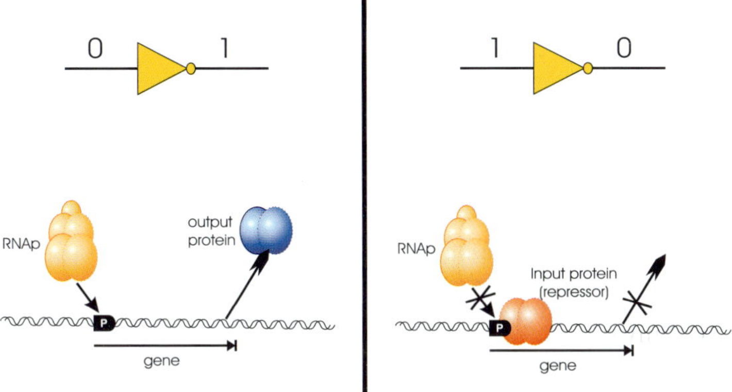

Fig. 1. The two idealized cases for a biological inverter. If input repressor is absent, RNA_P (RNA polymerase) transcribes the gene for the output protein and enables its synthesis. If input repressor is present, no output protein is synthesized.

protein. The steady-state translation rate of the "output" protein will decrease monotonically with increasing concentration of the "input" protein. Given the assumptions in Sect. 3.2, the concentration of the input is linear in its translation rate. Then the two signals, defined by translation rates, are related by a monotonic decreasing function. Figure 1 shows the two ideal cases in the truth table of a biological inverter.

NAND gates are built by combining inverters with common output genes. These NAND gates can serve as building blocks for any desired finite state machine, within practical limitations such as the number of distinct signal proteins available.

3.2   The Module Abstraction

Consider a logic element consisting of an input protein, A, acting on an operator, O_A, associated with a promoter P. Let P be fused to a structural gene G_Z coding for the output protein Z.

For abstraction purposes, decouple the transfer function of this logic element into *synthesis* and *decay* stages. The synthesis stage, denoted by S, is the mapping from the input protein concentration, π_A, to the reference translation flux ϕ_P, when the system is in steady-state:

$$S : \pi_A \longrightarrow \phi_P \tag{1}$$

The reference translation flux is the rate at which a protein coded for by a single structural gene would be synthesized from promoter P. The decay stage that follows, denoted by D, is the mapping from the reference translation flux ϕ_P to the output protein concentration π_Z:

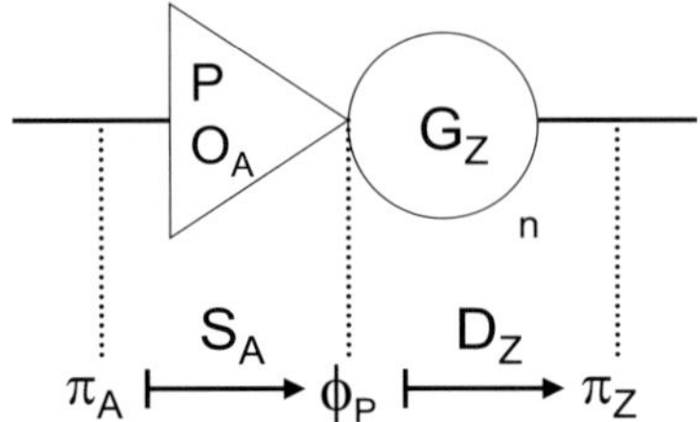

Fig. 2. Synthesis and decay components of an inverter. Concentrations are denoted by π and translation fluxes by ϕ. S_A is the synthesis mapping determined by operator O_A, and D_Z is the decay mapping of Z determined by G_Z.

$$D : \phi_P \longrightarrow \pi_Z \tag{2}$$

The characteristics of S depend on the thermodynamics of A binding to O_A, the promoter strength of P, and the effectiveness of the ribosome binding site. These are all properties of O_A and P. In general, S will be nonlinear.

In turn, D depends on the degradation rate of the mRNA for Z and the degradation rate of Z itself. These are properties of G_Z. If we assume that mRNA degradation and protein degradation are first-order kinetic processes, and that one of them is rate limiting, we can conclude that D is linear. This will be useful in our subsequent analysis.

Figure 2 shows the components of this abstraction. The structural gene G_Z carries a subscript, n, denoting the *cistron count* of G_Z. This count represents the number of distinct ribosome binding sites and copies of the structural gene for Z that are fused to the same promoter. S is subscripted by A to indicate that the translation rate is a function of π_A, the concentration of A. And respectively, D is subscripted by Z to indicate that the concentration of Z is a function of ϕ_Z, the (aggregate) translation rate of structural genes for Z.

Since there are n cistrons coding for Z, the aggregate translation rate of structural genes for Z (the output signal) is just n times the reference translation rate ϕ_P. This yields the transfer function of the inverter:

$$\phi_Z = n \cdot \phi_P = n \cdot D(S(\phi_A)) \tag{3}$$

Figure 2 clarifies several points:

- The distinction between protein concentrations and translation fluxes
- Fluxes determine concentrations, and concentrations determine fluxes
- S depends only on O_A and P
- D depends only on G_Z

The second fact above illustrates "flux/concentration duality": the logical action of the circuit can be characterized either by fluxes or by concentrations. Either forms a complete description of the system, since each determines the other. This notation allows for the fusion of distinct structural genes to a single promoter by connecting the promoter/operator stage to multiple gene stages, one for each output protein of interest.

3.3   Gates: Implementation of Combinatorial Logic

The approach to combinatorial logic is to "wire-OR" the outputs of multiple inverters by assigning them the same output gene. Since the output protein will be expressed in the absence of either input protein, this configuration implements a NAND gate (Fig. 3). Since the performance of a NAND gate relies solely on that of its constituent inverters, well-engineered inverters will yield well-engineered combinatorial gates.

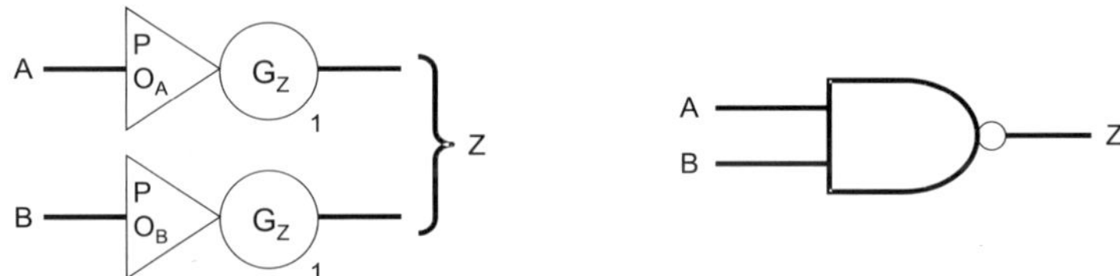

Fig. 3. Wire-OR-ing the outputs of two inverters yields a NAND gate.

3.4   Advantages

Modularity of the network design affords a free choice of signal proteins. Any suitable repressor protein can be used for any signal; the issue of "suitability" is discussed in Sect. 6.3. This modularity is necessary for implementing a "bio-compiler": a program that consults a library of repressor proteins and their associated operators and generates genetic logic circuits directly from gate-level descriptions. Contrast this modularity with the method of Hjelmfelt *et al.*, which requires proteins that modify other proteins, and where all signals are protein concentrations. The resulting physico-chemical interdependence of successive logic stages makes simple modularity almost impossible.

In addition, the library of naturally available signal proteins is large. Any repressor protein with sufficiently cooperative DNA binding and that does not interfere with normal cell operation should be appropriate. In our first set of experiments, CI proteins from lambdoid phages will serve as signals. The protein signal library will soon be as large as the family of lambdoid phages. In the future, combinatorial chemistry techniques, combined with a method such as phage display, should yield large libraries of novel DNA binding proteins and corresponding operators.

$$A + A \xrightarrow[\text{dimerization}]{k_{dim(a)}} A_2 \qquad (4) \qquad\qquad A \xrightarrow[\text{decay}]{k_{dec(a)}} \qquad (12)$$

$$A_2 \xrightarrow[\text{single}]{k_{sngl(a)}} A + A \qquad (5) \qquad\qquad A_2 \xrightarrow[\text{decay}]{k_{dec(a2)}} \qquad (13)$$

$$Z + Z \xrightarrow[\text{dimerization}]{k_{dim(z)}} Z_2 \qquad (6) \qquad\qquad Z \xrightarrow[\text{decay}]{k_{dec(z)}} \qquad (14)$$

$$Z_2 \xrightarrow[\text{single}]{k_{sngl(z)}} Z + Z \qquad (7) \qquad\qquad Z_2 \xrightarrow[\text{decay}]{k_{dec(z2)}} \qquad (15)$$

$$G_Z + A_2 \xrightarrow[\text{repress 1}]{k_{rprs(a2)}} G_Z A_2 \qquad (8) \qquad\qquad G_Z A_2 \xrightarrow[\text{decay}]{k_{dec(ga2)}} G_Z \qquad (16)$$

$$G_Z A_2 \xrightarrow[\text{dissociation}]{k_{dis(a2)}} G_Z + A_2 \qquad (9) \qquad\qquad G_Z A_4 \xrightarrow[\text{decay}]{k_{dec(ga4)}} G_Z A_2 \qquad (17)$$

$$G_Z A_2 + A_2 \xrightarrow[\text{repress 2}]{k_{rprs(a4)}} G_Z A_4 \qquad (10)$$

$$G_Z A_4 \xrightarrow[\text{dissociation}]{k_{dis(a4)}} G_Z A_2 + A_2 \qquad (11) \qquad\qquad mRNA_Z \xrightarrow[\text{decay}]{k_{dec(mrna)}} \qquad (18)$$

$$G_Z + RNA_p \xrightarrow[\text{transcribe}]{k_{xscribe}} G_Z + RNA_p + mRNA_Z \qquad (19)$$

$$mRNA_Z + rRNA \xrightarrow[\text{translate}]{k_{xlate}} mRNA_Z + rRNA + Z \qquad (20)$$

Table 1. Chemical reactions that implement an inverter. A is the input protein and Z the output.

4   Modeling and Simulation

This section presents a chemical model of a reaction system implementing an inverter, and provides a simulation of its dynamic behavior. The feasibility of the model is explored by testing whether non-trivial circuits composed of such inverters exhibit the desired logical behavior.

4.1   Chemical Reactions Implementing an Inverter

Natural gene regulation systems exhibit characteristics useful for implementing *in vivo* logic circuits. These include transcriptional control, repression through cooperative binding, and degradation of proteins and mRNA transcripts. Table 1 presents one possible chemical model of the reactions involved in such a system. In particular, this model reflects the characteristics of the lambda CI repressor operating on the lambda $O_R 1$ and $O_R 2$ operators.

$k_{dim(a)}$	8.333	$k_{rprs(a2)}$	66.67	$k_{dec(a)}$	.5775	$k_{dec(ga2)}$	.2887
$k_{sngl(a)}$	.1667	$k_{dis(a2)}$	.2	$k_{dec(a2)}$	.5775	$k_{dec(ga4)}$	.2887
$k_{dim(z)}$	8.333	$k_{rprs(a4)}$	333.3	$k_{dec(z)}$	.5775	$k_{dec(mrna)}$	2.0
$k_{sngl(z)}$	.1667	$k_{dis(a4)}$	.25	$k_{dec(z2)}$	.5775	$k_{xscribe}$	.0001
						k_{xlate}	.03

Table 2. Kinetic constants used in the simulations. The units for the first order reactions are 100 sec^{-1} and the units for the second order reactions are $\mu M^{-1} \cdot 100$ sec^{-1}.

The repressor protein A represents the input signal, and protein Z represents the output signal. In contrast to Sect. 3.2, G_Z now denotes the concentration of the *active* form of the structural gene for Z. A structural gene is active only when its associated operator is *unbound* by a repressor.

A_2 and Z_2 denote the dimeric forms of A and Z, respectively, and $G_Z A_2$ and $G_Z A_4$ represent the repressed (i.e., inactive) forms of the gene. $mRNA_Z$ is the gene transcript coding for Z. RNA_p is RNA polymerase, and $rRNA$ is ribosomal RNA that participates in translation of the $mRNA$[1]. Important aspects of the model include dimerization of the proteins (reactions 4–7), cooperative binding (reactions 8–11), transcription and translation (reactions 19, 20), and degradation of proteins and mRNA (reactions 12–18).

This relates to the abstraction explained in Sect. 3.2 as follows: If π_A (the *total* amount of A in the system) is fixed, then equations 4, 5, 8–11, 16, and 17 determine G_Z. By reactions 19 and 20, G_Z determines the reference translation rate ϕ_P from the promoter. This results in the synthesis mapping S. Respectively, ϕ_P determines π_Z (the total amount of Z in the system), given equations 6, 7, 14, 15, and 18. This results in the decay mapping D.

The kinetic constants used in this simulation (Table 2) are based on the literature describing the phage λ promoter P_R and repressor (cI) mechanism [5,18].

Figure 4 shows the dynamic behavior of the inverter as modeled with the above chemical reactions. The three graphs show the concentrations of the input protein A, the active gene G_Z, and the output protein Z. The concentrations include both the monomeric and dimeric forms.

The reactions proceed as follows: First, π_Z increases until it stabilizes when the expression and decay reactions reach a balance. Then, an externally imposed drive increases π_A. As a result, the concentration of G_Z decreases as A binds free operator. Then π_Z decreases as no additional Z is synthesized and existing Z is degraded. Finally the drive π_A decreases, A degrades, G_Z

---

[1] The simulations in this section assume that the concentrations of RNA_p and $rRNA$ are fixed. Section 5 discusses how to measure the effect of fluctuations in these concentrations, as well as other factors, on the inverter's behavior. Once these effects have been quantified, robust gates can be designed.

recovers, and π_Z once again reaches the HIGH signal range. Note that the gate switching time (measured in minutes for this mechanism) is governed by the rate of recovery of G_Z.

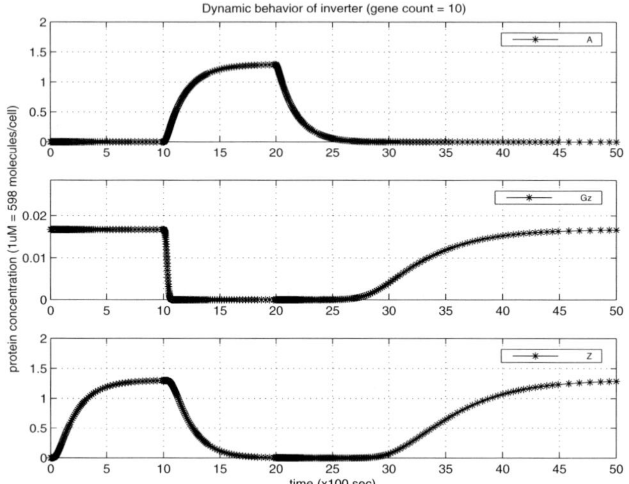

Fig. 4. The dynamic behavior of the inverter. The top is the input protein, the middle is the active (unrepressed) form of the output gene, and the bottom is the output protein.

4.2   Connections: Analysis of a Ring Oscillator

A ring oscillator is a simple circuit that can help determine the utility of our inverters for building complex logic circuits. The oscillator consists of three inverters connected in a series loop. The simulation results in Fig. 5 depict the expected oscillation in protein concentrations, as well as a phase shift between the values. Note, however, that oscillation occurs close to the LOW end of the signal values. This results from the skewed transfer curve that describes the steady-state characteristics of the inverter. Sections 5 and 6 discuss this issue in depth.

4.3   Storage: Analysis of an RS Latch

Another good test circuit is the RS latch, a component for persistently maintaining a data bit. It consists of two cross-coupled NAND gates, with inputs $\overline{S}$ and $\overline{R}$ for setting and resetting the complementary output values A and B (Fig. 6). The inverters with inputs $\overline{R}$ and B and common output A constitute one of the NAND gates, while the inverters with inputs $\overline{S}$ and A and common output B constitute the other NAND gate. Figure 7 shows the dynamic

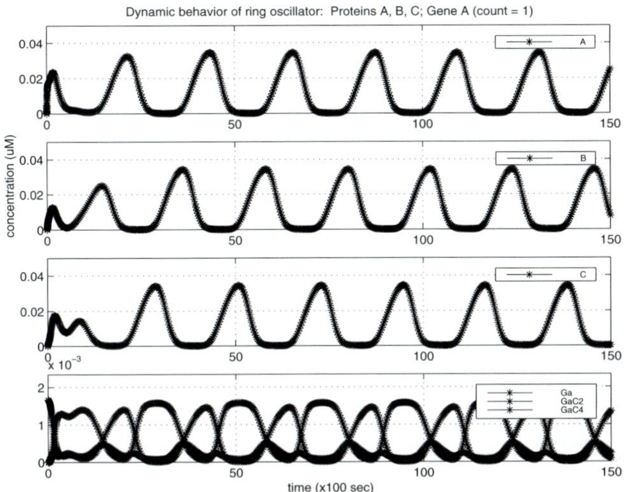

Fig. 5. Dynamic behavior of a ring oscillator. The top three curves are the outputs of the three inverters. Note the 120° phase shift between successive stages. The bottom shows various repression states of the first inverter's output gene.

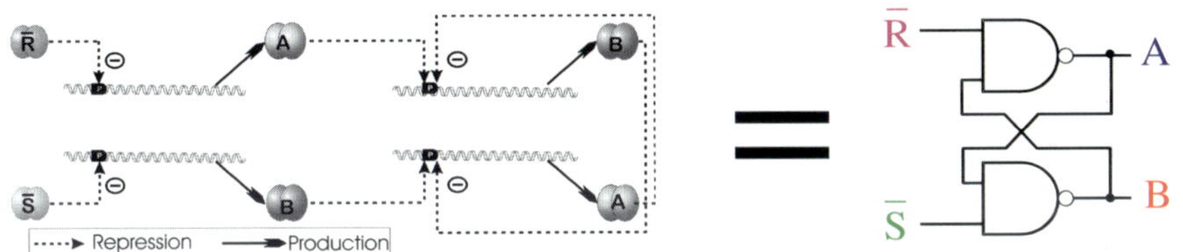

Fig. 6. Gene logic and schematic representations of an RS latch, used for storage.

behavior of this RS latch. As expected, both long and short pulses effectively set and reset the latch.

4.4   Caveats

Published data on kinetic constants is scarce and often imprecise. In several cases, the constants were guessed from published equilibrium constants. This situation is rapidly getting better, and we expect to have more accurate and complete data in the near future.

In cells, typical promoter copy counts correspond to very low concentrations. Therefore, the stochastic noise in concentrations resulting from the discreteness of the transcription reactions can be significant (see Arkin and Ross [1]). To decrease this stochastic variance, we will use medium to high promoter copy numbers in our experiments.

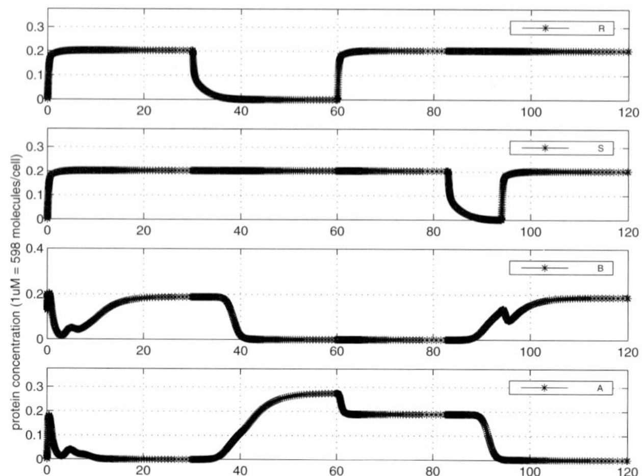

Fig. 7. Dynamic behavior of the RS latch. The top two curves represent the $\overline{\text{RESET}}$ and $\overline{\text{SET}}$ inputs, respectively. The bottom two curves are the complementary outputs. The initial behavior shows the system settling into a steady-state.

5 Measuring Transfer Functions

A transfer function is the relation between the input signal and the output signal of a gate or circuit in steady-state. Sect. 5.1 describes how to measure an individual signal in a genetic circuit by constructing a probe that measures expression activity *in vivo*. Sect. 5.2 introduces a mechanism for estimating the transfer function by measuring many different points of the transfer curve. Finally, Sect. 5.3 discusses how to account for systematic fluctuations and noise inherent in biological systems by generalizing the transfer function to a *transfer band*. We will use the techniques outlined in this section to measure and verify the characteristics of biological logic circuits *in vivo*.

5.1 Measuring a Signal

Recall that the translation flux ϕ_Z of all genes coding for a protein Z represents a logic signal. Thus the relevant chemical reaction for a signal is the rate of translation of the $mRNA$ product of G_Z into the protein product Z:

$$mRNA_Z + rRNA \xrightarrow[\text{translation}]{k_{xlate}} mRNA_Z + rRNA + n \cdot Z \qquad (21)$$

where G_z is the active (unrepressed) form of the gene, k_{xlate} represents the rate of translation from $mRNA$ into the protein product, and n denotes the cistron count. Then, assuming this is the only production of Z in the system,

$$\phi_Z \equiv n \cdot k_{xlate} \cdot [mRNA_Z] \cdot [rRNA] \tag{22}$$

To measure the signal, insert a reporter protein RP as an additional structural gene, and assume that for the concentrations of interest, RP remains mostly in monomeric form:

$$mRNA_Z + rRNA \xrightarrow[\text{translation}]{k_{xlate}} mRNA_Z + rRNA + n \cdot Z + RP \tag{23}$$

$$RP \xrightarrow[\text{decay}]{k_{dec(rp)}} \tag{24}$$

Then, the time derivative for the reporter concentration is:

$$\frac{d[RP]}{dt} = k_{xlate}[mRNA_Z][rRNA] - k_{dec(rp)}[RP] \tag{25}$$

At equilibrium:

$$0 = k_{xlate}[mRNA_Z][rRNA] - k_{dec(rp)}[RP] \tag{26}$$

Since $k_{xlate}[mRNA_Z][rRNA] = k_{dec(rp)}[RP]$, by substitution into 22,

$$\phi_Z = n \cdot k_{dec(rp)} \cdot [RP] \tag{27}$$

We know n, and can measure $[RP]$ (up to an unknown multiplicative factor) by picking, for example, a fluorescent protein for RP and measuring its fluorescence. By using the same reporter for each measured signal in the circuit, we obtain approximations of signals all scaled by the same factor.

5.2   Measuring the Transfer Curve of an Inverter

Once an individual signal can be measured, the transfer function of a gate is estimated by measuring many points on the curve. A point on the transfer curve is a steady-state relation between the translation flux ϕ_A of the input protein and the translation flux ϕ_Z of the output protein. A point is measured by constructing a system with an unknown but fixed ϕ_A and measuring ϕ_A and ϕ_Z.

To obtain many points, construct multiple systems yielding various fixed values of ϕ_A and observe the corresponding values of ϕ_Z. Let $P_D{}^j$ represent a constitutive promoter (i.e., "drive") resulting in a fixed value of ϕ_A, say $\phi_A{}^j$. Let $\mathcal{I}$ denote the transfer function of inverter I. Then, for each drive $P_D{}^j$, the value pair $\left(\frac{\phi_A{}^j}{k_{dec(rp)}}, \frac{\mathcal{I}(\phi_A{}^j)}{k_{dec(rp)}} \right)$ can be measured with the reporter RP as

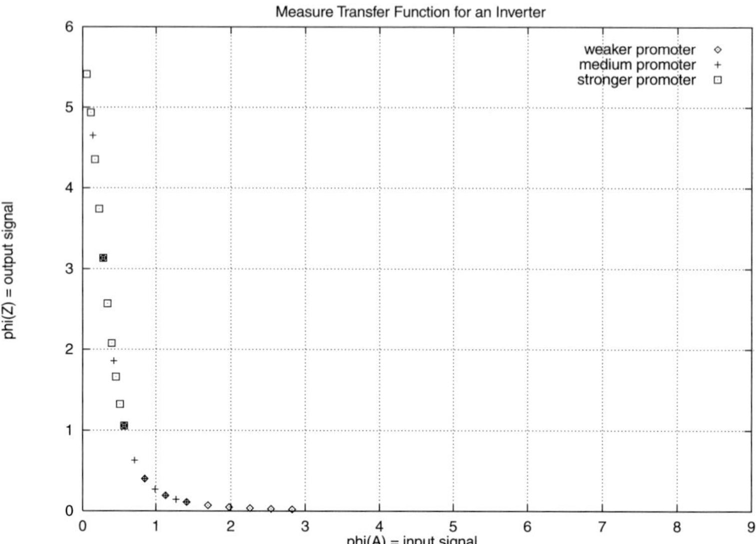

Fig. 8. Measuring points on the transfer function of an inverter using three promoters, each with ten different cistron counts.

described above. This requires two separate experiments, one to measure the drive and one to measure the output. With a set of these points, we obtain the transfer curve of a gate, where all points are expressed in the same (albeit unknown) units.

There are at least three mechanisms for obtaining a variety of drives. First, one can choose different promoters. Second, one can modify the strength of a given promoter region through base-pair substitutions, resulting in different transcription initiation rates. Third, for a given promoter, increasing the cistron count of the gene for the drive yields a multiplicative increase in the drive. Figure 8 illustrates points on an inverter's transfer curve obtained by simulating thirty different drives. The simulation computes ten points for each of three different promoters with different RNA_p affinities. For each promoter, the different points indicate the effect of including between one and ten cistrons.

To measure a complete transfer curve, the range of inputs must cover both the LOW and HIGH input ranges. This will require drives with both strong and weak promoters. One does not need to know the characteristics of $P_D{}^j$ a priori to use it for measuring points on the transfer curve. Also, drives with similar characteristics simply add redundancy to the measurements.

To measure more complex circuits, measure the values of the relevant signals by inserting the structural gene for the reporter at the appropriate promoters.

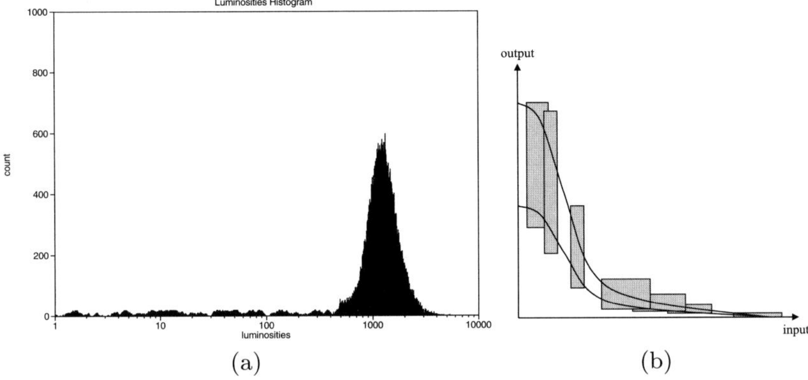

Fig. 9. (a) A typical flow cytometry histogram of scaled luminosities, showing one dominant peak. (b) Approximation of a transfer band from several distinct drives. The measurement of each drive yields a shaded rectangle. The band lies between the two bold curves.

5.3   The Transfer Band: Models vs. Reality

While this simple model yields idealized discrete points on the transfer function, actual biological systems exhibit both systematic fluctuations and noise that are not modeled in Sect. 4. The *transfer band* is a concept capturing these fluctuations. Specifically, it is the mapping from individual input values to the output value ranges exhibited by viable cells.

Flow cytometry [4] is a technology useful for quantifying gene expression activities of individual cells. First, a gene coding for a fluorescent reporter (e.g., GFP) is fused to the same promoter as the gene of interest. Then luminosity values are measured for individual cells as they flow through a cytometer's capillary, yielding a histogram of luminosities. Figure 9(a) shows such a histogram for a large number of "identical" cells (i.e., with the same promoter/reporter construct). The variance results from systematic fluctuations in protein expression rates (due to cell growth cycle and environmental factors, for example), idiosyncratic fluctuations due to stochastic noise in gene expression, measurement error, and non-viable or damaged cells.

We expect our experiments to yield histograms with one clearly dominant peak. For a drive $P_D{}^j$, $\phi_A{}^j$ is now a distribution. Then, let $\underline{\phi}_A{}^j$ be the minimum value of the drive distribution's peak, and $\overline{\phi}_A{}^j$ be the maximum value of that peak. Cells with values in this range are said to be *operational*. In the same manner, let $\phi_Z{}^j$ represent the corresponding distribution of output values, and let $\underline{\phi}_Z{}^j$ and $\overline{\phi}_Z{}^j$ be the minimum and maximum value of the output distribution's peak.

Then the measurement of the input and output distributions for each drive yields a rectangular region with the lower left corner at $\left(\underline{\phi}_A{}^j, \underline{\phi}_Z{}^j\right)$

and the upper right corner at $\left(\overline{\phi}_A{}^j, \overline{\phi}_Z{}^j\right)$. If the sample density is high, then the transfer band lies within the union of all such rectangles. Figure 9(b) illustrates the approximation of a transfer band from several flow cytometry measurements.

6   Microbial Circuit Design

The objective of *microbial circuit design* is to take a desired logic circuit and a database of kinetic rates as input, and produce a genetic network that implements the circuit. The design process requires searching the database and assigning suitable proteins to each gate, where the dynamic behavior of the gate depends on these choices. The gates must be robust enough to use a wide variety of proteins with different reaction kinetics.

This section outlines some of the key implementation choices in the design process, defines how to match gate input/output threshold levels, and describes mechanisms to modify the steady-state characteristics of an inverter in order to achieve these levels.

6.1   Implementation Choices

In subsequent publications, we will present detailed analysis of how to design microbial circuits. Some of the key issues are:

- *Global gene copy number:* The circuits will be implemented on one or several plasmids. Since high copy number plasmids place a metabolic burden on the cell, while low copy number plasmids may result in large stochastic noise, we intend to use medium copy number plasmids such as *pBR322*.
- *Output proteins:* An output protein must be soluble, bind some known operator site(s), and be inessential for normal cell function. To ensure sufficient gain and noise margins, binding should be highly cooperative (e.g., Lambda CI repression uses two dimers).
- *Promoter/operator regions:* Operators should bind repressors cooperatively, and promoters should be weak enough to not saturate subsequent gate inputs.
- *Signal threshold levels:* The gate input thresholds must be chosen to provide high gain near the switching threshold, adequate noise margins at the HIGH and LOW signal levels, and balanced transition times.
- *Per-gate cistron count:* The cistron count can be adjusted for each output protein to match threshold levels.

6.2   Matching Thresholds

Transfer functions suitable for implementing digital gates must have LOW and HIGH ranges such that signals in the LOW range map strictly into the

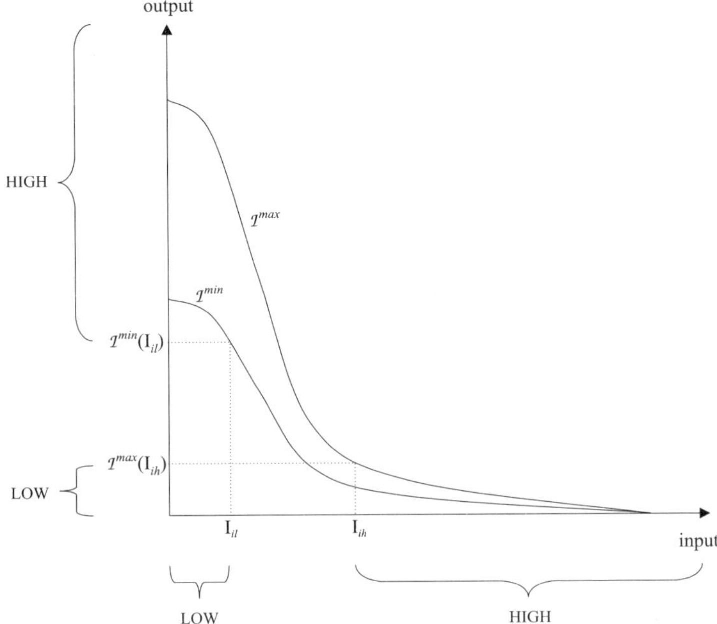

Fig. 10. HIGH and LOW input ranges for a hypothetical inverter. The transfer band is defined by the two curves.

HIGH range, and *vice versa*. The strictness of the inclusion reduces noise from input to output. For electronic digital circuits, the LOW and HIGH signal ranges are the same for all gates because the circuit is composed of transistors with identical threshold voltages, spatially arranged. However, in biological digital circuits, the gate components (proteins) have different characteristics depending on their reaction kinetics. Therefore, the designer of biological digital circuits must take explicit steps to ensure that the signal ranges for coupled gates are appropriately matched.

A given transfer band can be defined by a pair of functions. As shown in Fig. 10, let $\mathcal{I}^{min}$ be the function that maps an input to the minimum corresponding operational output, and let $\mathcal{I}^{max}$ be the function that maps an input to the maximum corresponding operational output.

If I_{il} and I_{ih} are the input thresholds, then the suitability condition given above can be written as:

$$[\text{in LOW}] \qquad \langle 0, \mathrm{I}_{il} \rangle \xrightarrow{\text{into}} \langle \mathcal{I}^{min}(\mathrm{I}_{il}), \mathcal{I}^{max}(0) \rangle \qquad [\text{out HIGH}]$$
$$[\text{in HIGH}] \qquad \langle \mathrm{I}_{ih}, \infty \rangle \xrightarrow{\text{into}} \langle 0, \mathcal{I}^{max}(\mathrm{I}_{ih}) \rangle \qquad [\text{out LOW}]$$

Consider the case of two inverters, I and J, with J's output coupled to I's input. Then, the coupling is correct *iff*:

$$\langle \mathcal{J}^{min}(\mathbf{J}_{il}), \mathcal{J}^{max}(0)\rangle \subset \langle \mathbf{I}_{ih}, \infty\rangle$$
$$\langle 0, \mathcal{J}^{max}(\mathbf{J}_{ih})\rangle \subset \langle 0, \mathbf{I}_{il}\rangle$$

Then the following conditions are necessary and sufficient for correct coupling:

$$\mathcal{J}^{min}(\mathbf{J}_{il}) > \mathbf{I}_{ih}$$
$$\mathcal{J}^{max}(\mathbf{J}_{ih}) < \mathbf{I}_{il}$$

6.3   Modifying the Inverter Characteristics

The first step in developing the microbial circuit design process is to design, build, and characterize several inverters. It is likely that these inverters will not match correctly according to the definitions above. Fortunately, there are techniques to adjust them so they are matched for use in complex circuits. These include:

- Modifying the strength of the promoter or the ribosome binding site (RBS) changes the output scaling of an inverter. DNA sequence determinants of promoter and RBS strengths have been studied extensively [3,6,8]. Figure 11 shows the effect of hypothetical reductions in promoter strength on the transfer functions of an inverter and two inverters in series.
- Modifying the repressor/operator binding affinity changes the input scaling of an inverter and the shape of its transfer function. This is also accomplished via base-pair substitutions, although the effects of these substitutions are different for each repressor/operator pair. Figure 12 shows the effect of hypothetical reductions in this affinity on the transfer function of an inverter.
- Altering the degradation rate of a protein changes the steady-state relation between its synthesis rate and its concentration. This can be done on a per-protein basis by changing the few amino acid residues on the C terminus [2,16,17].
- The simulated transfer functions shown above are not ideal due to the lack of noise margin at the LOW signal level. Autorepression could improve this by limiting the steady-state concentration of the output protein to a much lower maximum value. An operator that binds the output protein may be added to the promoter/operator region to accomplish autorepression.

7   BioSpice

BioSpice is a prototype system for simulating and verifying genetic digital circuits. It takes as inputs a network of gene expression systems (including

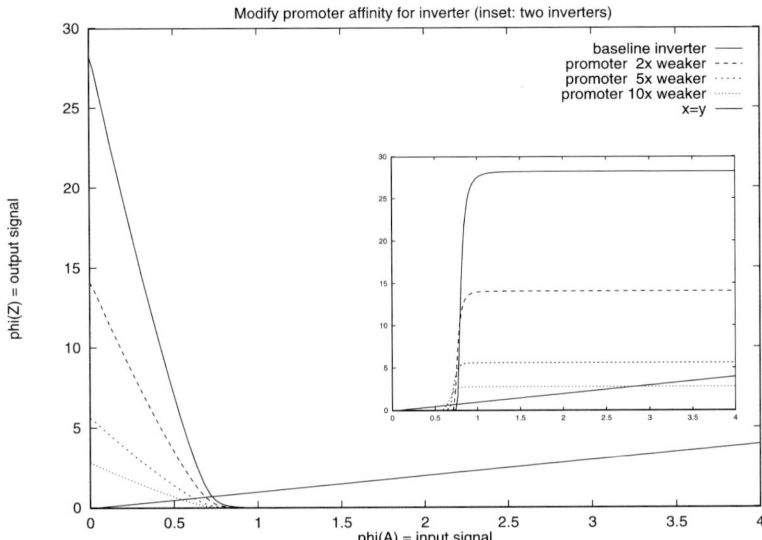

Fig. 11. The effects of reducing the binding affinity of the RNA polymerase on the transfer functions of an inverter. Inset shows the effects on the transfer functions of two inverters in series. The diagonal lines correspond to input equals output.

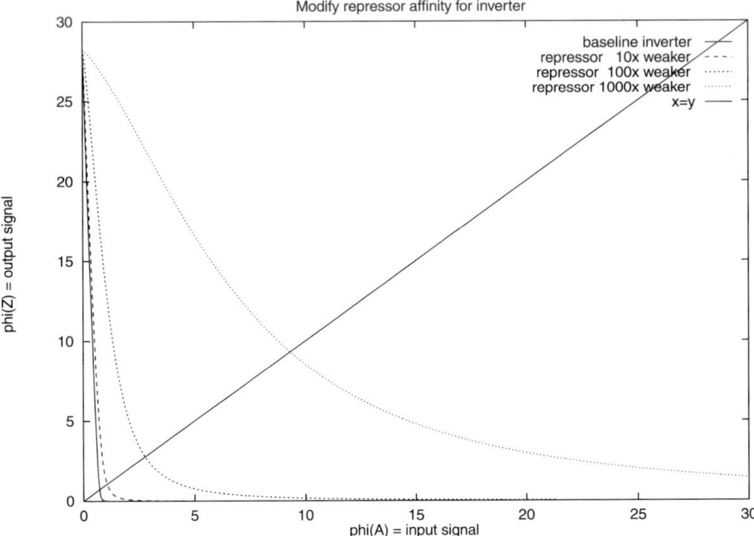

Fig. 12. The effects of reducing the binding affinity of the repressor on the transfer function of an inverter.

the relevant protein products) and a small layout of cells on some medium. BioSpice then consults a database of reaction kinetics and diffusion rates in order to simulate the dynamic characteristics of the target system. The simulation computes the time-domain behavior of concentration of intracellular proteins and intercellular message passing chemicals.

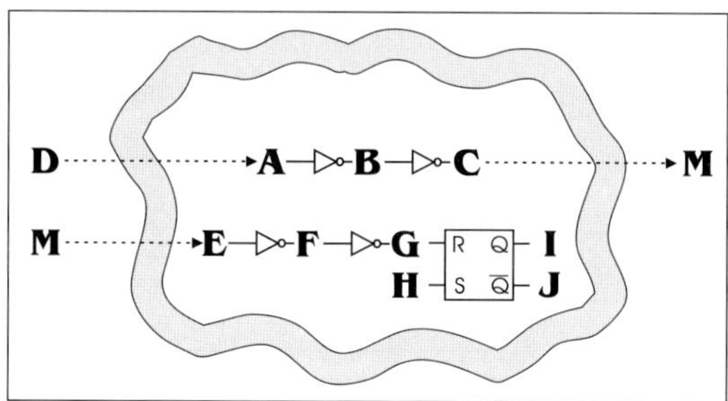

Fig. 13. Gate level representation of a genetic circuit to accomplish a simple bacterial task.

Consider a simple bacterial task, where upon receipt of a message (represented by inward diffusion of a message passing chemical), a cell communicates to its neighbors and instructs them to set a state bit. Figure 13 represents a genetic digital circuit designed to perform this task. The initiating signal D is a chemical that traverses the cell membrane and results in the presence of protein A in the cytoplasm. This can be achieved with certain signal transduction pathways. The presence of A results in controlled synthesis of C. Notice that the gate with input A can be chosen or adjusted to be sensitive to even small quantities of A. Once a sufficient concentration of A accumulates, C is synthesized and secreted into the surrounding environment as protein M. M diffuses through the medium and serves as a message to neighboring cells. In response to M, the neighbors each set their RS latch, whose output is I.

Figure 14 shows a BioSpice simulation of the above system on a 4×4 grid (representing the medium) with two bacterial cells (heavily shaded squares). The initial condition, depicted in the top-left snapshot, shows that the output of the RS latch (represented by I) is LOW. Then, a drive D is introduced into the environment next to one of the cells, as illustrated in the top-right snapshot. This causes the cell to transmit a message M. Once the other cell receives M (recognizable by the presence of E) it uses G to set the RS latch. Finally, when the drive is removed and the message M decays, the value of I remains latched at HIGH.

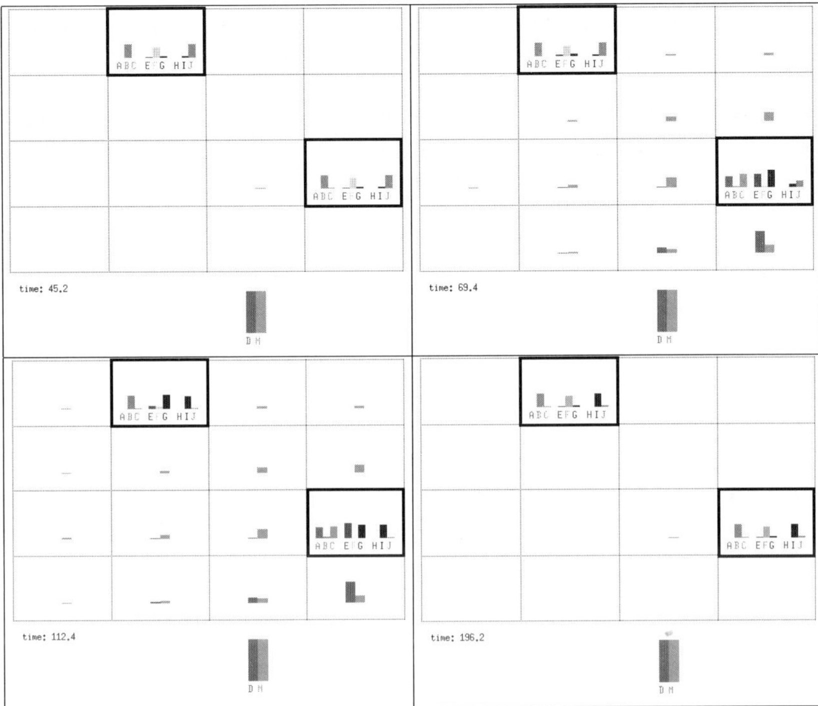

Fig. 14. BioSpice simulation snapshots of intracellular protein and intercellular message chemical concentrations.

8   Conclusions

This paper presents a design paradigm for gene expression based digital logic implemented *in vivo*. The proposed modular abstraction enables the construction of complex circuits using a library of interchangeable components. Simulation results indicate the feasibility of this paradigm. This paper also presents a measurement technique for characterizing the steady-state behavior of the system components. This technique accommodates systematic fluctuations and noise. Microbial circuit design uses these measurements in matching gates for correct function in complex logic circuits. BioSpice is a prototype tool for simulation and verification of distributed genetic digital systems.

If the initial experiments are successful, future work will concentrate on developing the technology from a simple laboratory model with one or two flip flops to genetic circuits of several hundred or thousand gates. An important component of this effort is designing new repressors and matching operator sequences, either *de novo* or by altering existing systems. Other related problems include harnessing signal transduction pathways to accomplish environmental sensing and intercellular communication.

Acknowledgements. Ron Weiss and Thomas F. Knight, Jr. are supported in part by DARPA and ONR. George E. Homsy is supported in part by an NSF Graduate Research Fellowship, and by a Merck/MIT Graduate Research Fellowship.

References

1. A. Arkin and J. Ross. Computational functions in biochemical reaction networks. *Biophysical Journal*, 67:560–578, August 1994.
2. James U. Bowie and Robert T. Sauer. Identification of C-terminal extensions that protect proteins from intracellular proteolysis. *Journal of Biological Chemistry*, 264(13):7596–7602, 1989.
3. David E. Draper. Translational initiation. In Frederick C. Neidhardt, editor, *Escherichia Coli and Salmonella*, pages 902–908. ASM Press, Washington, D.C., 2nd edition, 1992.
4. Alice Longobardi Givan. *Flow Cytometry: First Principles*. Wiley-Liss, New York, 1992.
5. Roger W. Hendrix. *Lambda II*. Cold Spring Harbor Press, Cold Spring Harbor, New York, 1983.
6. Peter H. von Hippel, Thomas D. Yager, and Stanley C. Gill. Quantitative aspects of the transcription cycle in Escherichia coli. In *Transcriptional Regulation*, pages 179–201. Cold Spring Harbor Laboratory Press, Cold Spring Harbor, New York, 1992.
7. A. Hjelmfelt, E. D. Weinberger, and J. Ross. Chemical implementation of neural networks and Turing machines. *Proc. Natl. Acad. Sci.*, 88:10983–10987, December 1991.
8. M. Thomas Record Jr., William S. Reznikoff, Maria L. Craig, Kristi L. McQuade, and Paula J. Schlax. Escherichia coli RNA polymerase ($e\sigma^{70}$), promoters, and the kinetics of the steps of transcription initiation. In Frederick C. Neidhardt, editor, *Escherichia coli and Salmonella*, pages 792–821. ASM Press, Washington, D.C., 2nd edition, 1992.
9. S. A. Kauffman. Gene regulation networks: a theory for their global structure and bahviors. In A. Moscona and A. Monroy, editors, *Current Topics in Developmental Biology*, volume 6, pages 145–182. Academic Press, New York, 1971.
10. Thomas F. Knight, Jr. and Gerald Jay Sussman. Cellular gate technology. In C.S. Calude, J. Casti, and M.J. Dinneen, editors, *First International Conference on Unconventional Models of Computation*, pages 257–272, Auckland, NZ, 1998. Springer-Verlag.
11. Harley H. McAdams and Adam Arkin. Stochastic mechanisms in gene expression. *Proc. Natl. Acad. Sci.*, 94:814–819, February 1997.
12. Harley H. McAdams and Adam Arkin. Simulation of prokaryotic genetic circuits. *Annu. Rev. Biophys. Biomol. Struc.*, 27:199–224, 1998.
13. Harley H. McAdams and Lucy Shapiro. Circuit simulation of genetic networks. *Science*, 269(5224):650–656, August 1995.

14. J. Monod and F. Jacob. *Cellular Regulatory Mechanisms*, pages 389–401. Cold Spring Harbor Press, New York, 1961.
15. Frederick C. Neidhardt and Michael A. Savageau. Regulation beyond the operon. In Frederick C. Neidhardt, editor, *Escherichia coli and Salmonella*, pages 1310–1324. ASM Press, Washington, D.C., 2nd edition, 1992.
16. Andrew A. Pakula and Robert T. Sauer. Genetic analysis of protein stability and function. *Annual Review of Genetics*, 23:289–310, 1989.
17. Dawn A. Parsell, Karen R. Silber, and Robert T. Sauer. Carboxy-terminal determinants of intracellular protein degradation. *Genes and Development*, 4:277–286, 1990.
18. Mark Ptashne. *A Genetic Switch: Phage lambda and Higher Organisms*. Cell Press and Blackwell Scientific Publications, Cambridge, MA, 2nd edition, 1986.
19. O. Roessler. *J. Theor. Biol.*, 36:413–417, 1972.
20. O. Roessler. In M. Conrad, W. Guettinger, and M. Dal Cin, editors, *Lecture Notes in Biomathematics 4*, pages 399–418. Springer, Berlin, 1974.
21. O. Roessler. In M. Conrad, W. Guettinger, and M. Dal Cin, editors, *Lecture Notes in Biomathematics 4*, pages 546–582. Springer, Berlin, 1974.
22. F. Seelig and O. Roessler. *Z. Naturforsch*, 27:1441–1444, 1972.
23. M. Sugita. Functional analysis of chemical systems in vivo using a logic circuit equivalent. *J. Theor. Biol.*, 4:179–192, 1963.
24. R. Thomas. Boolean formalization of genetic control circuits. *J. Theor. Biol.*, 42:563–585, 1973.

Evolution of Genetic Organization in Digital Organisms

Charles Ofria and Christoph Adami

Abstract. We examine the evolution of expression patterns and the organization of genetic information in populations of self-replicating digital organisms. Seeding the experiments with a linearly expressed ancestor, we witness the development of complex, parallel secondary expression patterns. Using principles from information theory, we demonstrate an evolutionary pressure towards overlapping expressions causing variation (and hence further evolution) to sharply drop. Finally, we compare the overlapping sections of dominant genomes to those portions which are singly expressed and observe a significant difference in the entropy of their encoding.

1 Introduction

Life on Earth is the product of approximately four billion years of evolution, with the vast majority of beginning and intermediate states lost to us forever. The exact details of how we evolved to become what we are may be impossible to ascertain for sure, but we may still be able to better understand the evolutionary pressures exerted on life, and from that reconstruct sections of the path our evolution is likely to have taken.

Here we look at a fundamental issue to life as we know it: the organization of the genetic code and the differentiation in its expression. DNA is structured into many distinct genes which can be concurrently active, transcribed and expressed in an asynchronous (i.e., differentiated) manner. Extant living systems have evolved to a state in which multiple genes influence each other, typically without sharing genetic material. It appears that in all higher life forms each gene has its own unique position on the genome, while the transcription products often interact with unique positions "downstream". Those organisms which do exhibit *overlapping* expression patterns are mostly virii and bacteriophages [9]. This suggests that genomes containing only purely localized, non-overlapping genes must have evolved later on [6].

Upon initial inspection, the reason for a spatially separated layout appears uncertain. A modular design may be quite common in artificially created coding schemes such as computer programs, but, in fact, only reflects a designer's quest to create humanly understandable structures. Evolution has no such incentive, and will always exert pressure towards the most immediate solution given the current circumstances. A more compressed coding scheme, perhaps with overlapping genes, would allow a sufficiently shorter code that would minimize the mutational load and hence be able to preserve its information with a higher degree of accuracy. Furthermore, such overlapping regions might be used for gene regulation. Why this is not much more common becomes clearer when we observe those examples from nature where these overlapping reading frames do exist, such as DNA phages [9] and eukaryotic viruses [14]. Even in these organisms only some sections of code overlap, but examination of those sections reveals that they contain little variation—almost all of the nucleotides are effectively frozen in their current state from one generation to the next [7,8]. This occurs because for any mutation to be neutral in such a section of genetic code, it must be neutral to *both* of the genes which it would affect. Further, most neutral mutations in DNA occur in the third nucleotide of a codon, as substitutions in that position are often synonymous. When overlapping genes have offset (out-of-phase) reading frames, however, the position of the third nucleotide in one gene maps to the first or second in the other, leaving no redundancy.

We have investigated the development of genome organization and differentiation in *digital organisms*: populations of self-replicating computer code living in a computer's memory. Such "Artificial Life" systems have proven to be useful test cases to investigate the biochemical paradigm because the computational chemistry the digital organisms are based on share *Turing universality* with their biochemical cousins, i.e., just as any type of organism appears to be implementable in biochemistry, the digital organisms can in principle compute any (partially-recursive) function [2]. Due to the ease with which experiments can be prepared, data can be gathered, and trials can be repeated, digital organisms present an important tool for studying universal traits in the evolution and development of symbolic sequences. Differentiation in digital organisms was first investigated within the tierra architecture [12,15,13], and we comment on those results below.

For the present study, we have extended our avida system [2] to allow for the expression of a *second* gene to occur in parallel. We then processed the evolution of 600 populations from a seed program to complex information-processing sequences for an average of over 9000 generations each. The 600 trials were divided into four sets which differ in the length of the seed program, constraints on size evolution, and their ability to express multiple portions of code in parallel. All populations with a genetic basis allowing for the development of multiple *threads* learn to use them almost immediately (each thread is an instruction pointer which executes the code independently), but the methods by which this happens are quite distinct and varied. In the next section, we outline the most important design characteristics of the avida system, focusing mostly on the particular experimental setup needed for this study. Also, we outline the kind of observables which we record and discuss measures of differentiation. In Sect. 3 we present results obtained with our multiple-expression digital chemistry and compare them to controls in which no secondary expression was allowed. In Sect. 4 we study the evolution of differentiation for different experimental boundary conditions, while Sect. 5 explores in more detail the organization and development of genes at the hand of an example. We close in Sect. 6 with a discussion of the evidence and conclusions, and issue caveats about applying the lessons learned directly to biochemistry.

2   Experimental Details

2.1   The Avida Platform

The computer program avida is an auto-adaptive genetic system [1] designed primarily for use as a platform in Artificial Life research. The system consists of a population of self-reproducing strings of instructions with a Turing-complete genetic basis subjected to Poisson-random mutations during reproduction. The population adapts to a combination of an intrinsic fitness criterion (self-reproduction) and an externally imposed (extrinsic) fitness landscape provided by the researcher by creating an information-rich environment.

A normal avida organism is a computer program written in a very simple assembly language, with 28 possible commands for each line (Table 1).

Table 1. Standard (single expression) avida instruction set

Instruction type	Mnemonic
flow control	jump-b, jump-f, call, return
conditionals	if-n-eq, if-less, if-bit-1
self-analysis	search-f, search-b
computation	shift-l, shift-r, inc, dec, swap
	swap-stk, push, pop, add, sub, nand
metabolic	alloc, divide, copy
I/O	get, put
labels	nop-A, nop-B, nop-C

These programs exist on a two-dimensional lattice with toroidal boundary conditions and are executed on simple virtual CPUs, residing at the lattice-sites, which process their code and allow them to interact with their environment and perform functions such as self-replication, as well computations on numbers which are found in the external environment. For more details on the virtual CPUs in avida, see [10].

In order to study the evolution of code expression, we have extended the instruction set of Table 1 to allow for more than one instruction pointer to execute a program's code. Within the biochemical metaphor, the *simultaneous* execution of code is viewed as the concurrent expression of two genes, i.e., the chemical action of two proteins. The first new instruction allows a program to initiate a new expression: fork-th. Its execution creates a new instruction pointer ("forking off a thread") which immediately executes the next instruction, while the original thread skips it. Thus, `fork-th` is the rough equivalent of a promoter sequence in biochemistry. In a sense, this secondary expression is rather trivial and leads to redundancy; if the second thread is not sufficiently altered by the instruction following the `fork-th`, it simply executes the identical code as the first thread in lock-step. Of course, we are interested in how the organisms use this redundancy as a starting point to *diversify* the expression.

The second new instruction *inhibits* an expression: `kill-th` removes the instruction pointer which executed it, while the third addition `id-th` *identifies* which pointer is currently executing the code, i.e., which pattern is currently being expressed. We expect the three commands together to be useful in the *regulation* of expression. In principle, more than two instruction pointers can be generated by repeated issuing of the `fork-th` command, but here we restrict ourselves to a maximum of two threads in order not to complicate the analysis. In nature, of course, complex genomes express hundreds of proteins simultaneously.

As our experiments begin with a self-replicating program which does not use any of the multiple expression commands, the first question might be whether or not multiple expression will develop at all. In fact, it does almost instantly, as secondary expression (typically in the trivial mode mentioned earlier) appears to be immediately beneficial, perhaps in the same manner as simple gene doubling or a second promoter sequence. From here on, differentiation evolves, i.e., the two instruction pointers begin to adapt independently, to express more and more different code. Ultimately one might expect that each pointer executes an entirely different section of code, achieving local separation of genes and fully parallelized execution. The mode and manner in which this separation occurs is the subject of this investigation.

Several hundred independent experimental trials and controls were obtained in this study, testing different experimental conditions. For each of these trials we keep a record of a variety of statistics, including the dominant genotype at each time step, from which we can track the progression of evolution of the population, in particular by studying the details of its expression patterns.

2.2   Basic Analysis Metrics

In order to track the differentiation of the threads, we need to develop a means to monitor the divergence between the two instruction pointers roaming the genome. Also, to study the evolutionary pressures such as the *mutational load*, we need to introduce some standard (and some less standard) observables which allow us to track the *adaptability* of the population. This is one of the major advantages of digital chemistries—some of the data that we collect is impossible to accurately obtain in biochemical systems, and even less practical to analyze.

Fitness is measured as the number of offspring a genome produces per unit time, normalized to the replication rate of the ancestor. Thus, in all experiments the fitness of the dominant genotype starts at one and increases. Fitness improvements are due to two effects: the optimization of the gene for replication (the "copy-loop") leading to a smaller gestation time, as well as the development of new genes which accomplish computations on externally provided random numbers. These computations are viewed as the equivalent of exothermic catalytic reactions mediated by the expression products. We reward the accomplishment of all bit-wise logical operations performed on up to three numbers by speeding up the successful organism's CPU at a rate commensurate to the difficulty of the computation.

Fidelity is the probability for an organism to produce an offspring perfectly identical to itself, i.e., the probability that the offspring is unaffected by mutations during the copy process. For pure copy-mutations (each instruction copied is mutated with a probability R_c),

$$F = (1 - R_c)^\ell \tag{1}$$

where ℓ is the organism's sequence length. In an adapting population, other factors can affect the fidelity and lead to low-fidelity organisms even while the theoretical fidelity is high. On the other hand, the development of error-correction schemes could increase the actual fidelity.

Neutrality ν is the probability that an organism's fitness is unaffected by a single point mutation in its genome. This is calculated by obtaining all possible one-point mutations of the examined genome, and processing each of them in isolation to determine fitness. The neutrality is then the number of neutral mutations divided by the total tested:

$$\nu = \frac{N_{\text{neut}}}{\ell(\boldsymbol{D} - 1)} \, , \tag{2}$$

where $\boldsymbol{D}$ is the number of different instructions in the digital chemistry, i.e., the size of the instruction set.

The preceding three indicators are key in determining the ability of an organism to thrive in an **avida** environment. Fitness, fidelity, and neutrality correspond respectively to an organism's ability to create offspring, for those offspring to have a minimum mutational load, and for them to survive those mutations which they do bear. Apart from this, however, there is another aspect which is necessary for a phylogenetic branch to be successful, and that is its ability to further adapt to its environment. To characterize this, we define two more genomic attributes.

Neutral Fidelity is a measure which can be calculated once an organism's neutrality is known. It is the probability that an organism will give birth to an identical *or equivalent* offspring. Taking $f_c = R_c(1 - \nu)$ to be the probability for a line to be mutated *and* be non-neutral to the organism, we obtain the neutral fidelity as:

$$F_{\text{neut}} = (1 - f_c)^\ell \, . \tag{3}$$

Genomic Diffusion Rate is the probability for an offspring to have a genome *different* from its parent, but to be otherwise equivalent (i.e., neutral). This is obtained by subtracting the genome's fidelity from its neutral fidelity:

$$D_g = F_{\text{neut}} - F \, . \qquad (4)$$

This is a particularly important indicator as it is the rate at which new, viable genotypes are being created, which in turn is the pace at which genetic space is being explored, and therefore directly proportional to the rate of adaptation.

2.3   Differentiation Measures

The following measures and indicators keep track of code differentiation. In biochemistry, the differentiation of expression can be very varied, and includes overlapping reading frames (in phase and out of phase), overlapping operons and promoter sequences, and gene regulation. Obviously, there are no reading frames in our digital chemistry, but it is possible for a sequence of instructions to give rise to a different computation depending on which thread is executing it, in particular if one gene contains another (as is very common in overlapping biochemical genes [16]). Also, thread identification may lead one thread to execute instructions which are skipped by the other thread, and threads may interact to turn each other on and off—a case of digital gene regulation. All such differentiation, however, has to evolve from the trivial secondary expression discussed earlier, and we consequently need to monitor the divergence of thread execution with suitable measures.

Expression Distance is a metric we use to determine the divergence of the two instruction pointers. Simply put, this measurement is the average *distance* (in units of instructions) between the sections of the genome actively being expressed by the individual threads. At the initial point leading to secondary expression, this distance is zero as the two threads execute the same code in lock-step. If this value is high relative to the length of the genome, it is a strong indication that the instruction pointers are expressing different sections of the genetic code at any one time, while if it is low, they most likely move together with identical (or nearly so) execution patterns. However, this measure only indicates the differentiation between execution at a particular point in *time*, implying that if the execution is simply time-offset, this metric may be misleading.

Expression Differentiation distinguishes execution patterns with characteristically differing *behavior*. Each execution thread is recorded with time, and a count is kept of how *often* each portion of the genome is expressed. The expression differentiation is the fraction of the genome in which those counts differ. Thus, the ordering of execution (time-delay) is irrelevant for this metric; only whether the code ends up getting expressed *differently* by one thread vs. the other is important.

2.4   Information Theoretic Measures

We use information theory in order to distinguish sequences which do or do not code for genes. In our digital chemistry, regions which do not code for a gene are either *unexecuted,* i.e., the instruction pointer skips over them, or else *neutral,* i.e. their execution will typically not affect the behavior of the program. Trivial neutral instructions often involve the **nop** instructions (see Table 1), which perform no function on their own when executed but do act to modify other instructions. Thus, even though their execution is neutral, their particular value can still severely affect the functioning of the organism. A perfectly neutral position sports any of the D instructions with equal probability among a population of sequences, while a maximally fixed position can only have one of the D instructions there. To distinguish these, we define per-site entropy and per-genome entropy.

Per-Site Entropy of a locus is determined by trying out each of the D instructions at that position and evaluating the fitness of the resulting organisms. All neutral positions are assigned an equal probability to be expected at that site, while deleterious mutations are assigned a vanishing probability (as they would be selected against). Due to the uniform assignment of probabilities, the per-site entropy of locus x_i (normalized to the maximum entropy $\log(D)$) is

$$H(x_i) = \frac{\log N_{\mathrm{neut}}(x_i)}{\log(D)} \; . \tag{5}$$

In an equilibrated population, this theoretical value of the per-site entropy is a good indicator for the actual per-site entropy, measured across the population (if the population is large enough). As positive mutations are extremely rare and we are only interested in the diversity of the population when it is in equilibrium, for the purposes of this measurement they are treated as if they were neutral. An indicator for the randomness within a sequence is per-genome entropy.

Per-Genome Entropy is approximated by the sum of the per-site entropies

$$H = \sum_{i}^{\ell} H(x_i) \; . \tag{6}$$

The actual per-genome entropy is in fact smaller, as the above expression neglects *epistatic* effects which lead to correlations between sites. For most purposes, however, the sum of the per-site entropies is a good approximation for the randomness. Measuring the entropy of the population by recording the individual genomic abundances is fruitless, as the sampling error is of the order of the entropy [4].

3   Single Expression vs. Multiple Expression

Let us first examine adaptability as measured by the average increase in fitness for both single and multiple expression chemistries. In Fig. 1A, the fitness is averaged for the 200 trials[1] which were seeded with small ($\ell = 20$) seed sequences and no size constraint (set I), for each of the chemistries. While the average increases relatively smoothly in time[2], it should be noted that each individual fitness history is marked by periods of stasis interrupted by sharp jumps, giving rise to a "staircase" picture reminiscent of the adaptation of *E. coli* [5]. During adaptation, the sequence length increases commensurately with the acquired information, as shown in Fig. 1B.

Clearly, the trials in which multiple expression is possible adapt more *slowly* than the single-expression controls, a behavior that may appear at first glance to be paradoxical as the only difference in the underlying coding of the multiple expression trials is an *increased* functionality. However, as we have noted previously, the neutral fidelity of an organism directly determines the fraction of its offspring which are viable. As this value is inversely correlated to the length of the genome, there is a pressure for the genomes to evolve towards shorter length. Normally, this pressure is counteracted by the adaptive forces which require the organism to store more information in its genome, requiring increased length. Overlapping expression patterns (here, multiple parallelized execution) allows this adaptation to occur while minimizing the length requirement. Hence, multiple-expression genomes adapt more slowly.

The pitfalls of compacting so much information into the same portion of the genome are illustrated in Fig. 1C where we plot the average genomic diffusion rate D_g for both chemistries. It is evident in this graph that initially both sets of experiments explore genetic space at a comparable rate, but at approximately 5000 updates (on average) the diffusion rates diverge markedly, followed by a corresponding divergence in the fitness of the organisms (that a higher diffusion rate leads directly to higher fitness in an information-rich environment is shown in [3].) Investigating the course of evolution further, we see that it is precisely at this point that the differentiated, yet overlapping, use of multiple threads is typically established.

To further implicate overlapping expression in reduced adaptation for the populations, we consider (as was done in [7] for the bacteriophage $\Phi X174$)) the substitution rate of instructions for overlapping versus non-overlapping

---

[1] Each trial is seeded with a single ancestor, which quickly multiplies to reach the maximum number of programs in the population, set to 3,600 for these trials. The population was subjected to copy mutations at a rate of 7.5×10^{-3} per instruction copied, and a rate of 0.5% of single insert or delete mutations per gestation period.

[2] Time is measured in arbitrary units called *updates*. Every update represents the execution of an average of 30 instructions per program in the population.

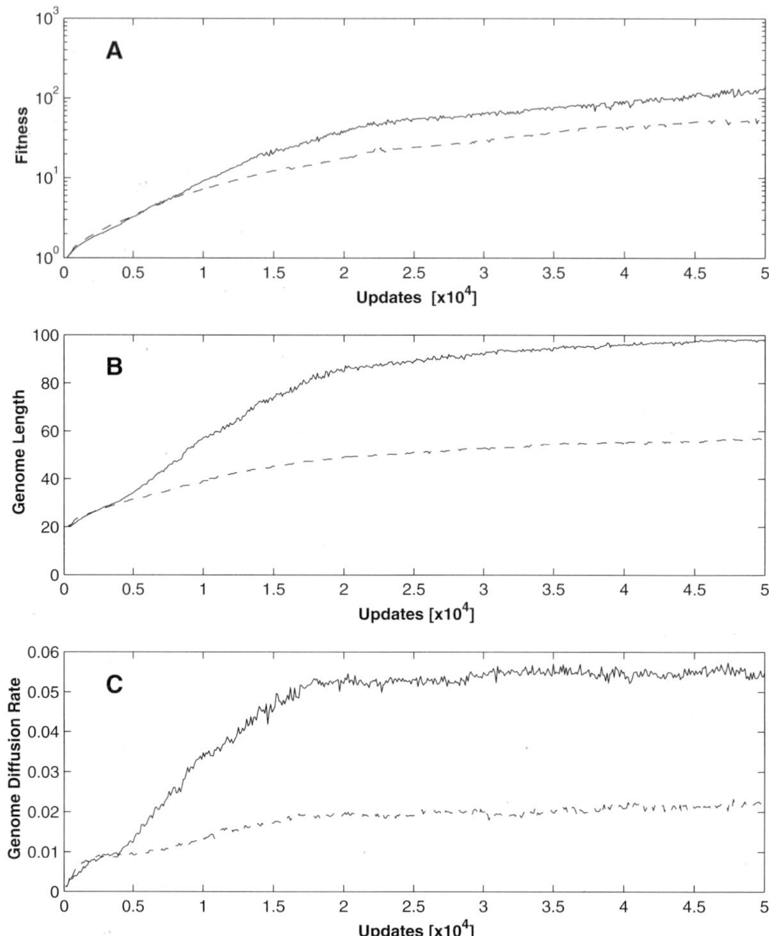

Fig. 1. (A); Average fitness as a function of time (in updates) for 200 populations evolved from $\ell = 20$ ancestors; their average sequence length (B); and the average genomic diffusion rate (C) for the single expression chemistry controls (solid line) and the multiple expression chemistry (dashed line).

genes. The substitution rate in avida is equal to the neutrality (at equilibrium). We find the *substitution suppression* (the neutrality in multiply expressed code divided by the neutrality in singly expressed code) to be between 0.53 and 0.56 for the three sets of trials (Table 2), similar to (but not quite as severe as) the suppression ratio of between 0.4 and 0.5 observed in the bacteriophages [7]. This was to be expected, as there are no reading frames in avida which implies that two non-differentiated threads do not constrain the evolution any more than a single thread. When the instruction pointers do adapt independently and the threads differentiate, neutrality is compromised. Consequently, the instructions within sections of overlapping code are comparatively "frozen" into their state.

Table 2. Average neutrality of the final dominant genotype: multiply-expressed code (column 1), singly expressed code (column 2), and their ratio (column 3), for 200 populations grown from $\ell = 20$ ancestors (variable length) [set I], 100 populations grown from $\ell = 80$ ancestors (variable length) [set II], and 100 populations grown from $\ell = 80$ ancestors (constant length) [set III].

Set	ν_{mult}	ν_{single}	ratio
I	0.109	0.202	0.539
II	0.197	0.346	0.569
III	0.082	0.145	0.566

4   Evolution of Differentiation

Let us now track the evolution of differentiation in more detail. We first address the *de novo* evolution of multiple expression, i.e., the development of multi-threading from linear execution. This question has previously been addressed within tierra [12], a population of self-replicating computer programs that served as the inspiration to our avida. In initial experiments, usage of multiple threads would not evolve spontaneously, but hand-written programs that had secondary expressions would evolve towards multiple expression [15]. More recently, experiments were carried out within a network version of the tierra architecture, which showed that a program which used different instruction pointers to execute different genes would not lose this ability [13]. The failure of multiple expression to evolve spontaneously in this system can be tracked back to problems with tierra's digital chemistry and the lack of an information-rich environment [11].

Within avida, the ability to use more than a single thread begins to develop within the first 5000 updates and is very common after about 10,000 updates, depending on the experimental boundary conditions. Fig. 2A shows the (averaged) percentage of a program's lifetime in which more than one thread is active, for the populations of set I (solid line), set II (dashed line), and set III (dotted line). It is apparent that multiple expression develops much more readily in smaller genomes, due to the fact that the logistics are less daunting.

In panels B and C of Fig. 2 we display two indicators of differentiation (defined earlier), the expression distance and the expression differentiation, respectively. The expression distance appears to be sensitive to the experimental starting condition, as set II and set III show a value over twice that of set I. We observe that this is due to the small size of the ancestor used in set

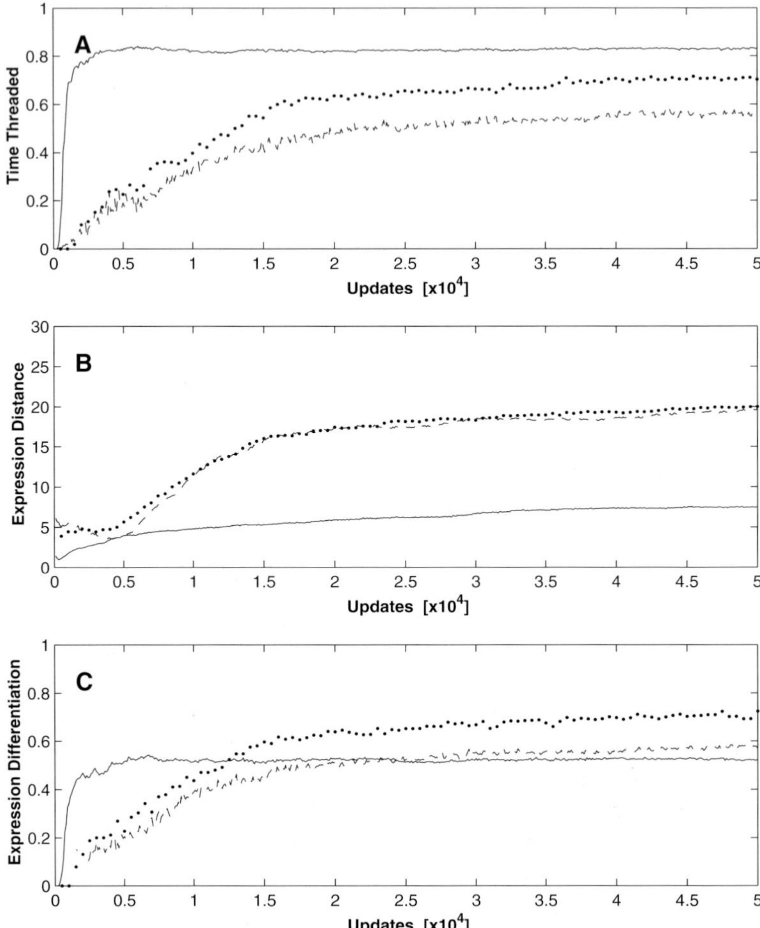

Fig. 2. Differentiation measures. (A) Average fraction of lifetime spent with secondary expression, as a function of time (in updates), (B) average expression distance, and (C) average expression differentiation. Set I (solid line), set II (dashed line), and set III (dotted line).

I: because that ancestor develops threading very quickly, it loses adaptability earlier and lags both in average fitness and average sequence length. In fact, those averages are dragged down by a significant percentage of the trials in set I, which were stuck in an evolutionary dead-end. Sets II and III were seeded with an ancestor of length $\ell = 80$ and did not suffer from this lot. Fig. 2C shows the *expression differentiation*, i.e., the fraction of code that is executed differently by the two threads. This fraction is less dependent on experimental conditions, and the genomes appear to develop towards 0.5. Note, however, that this measure cannot accurately reflect differentiation which is

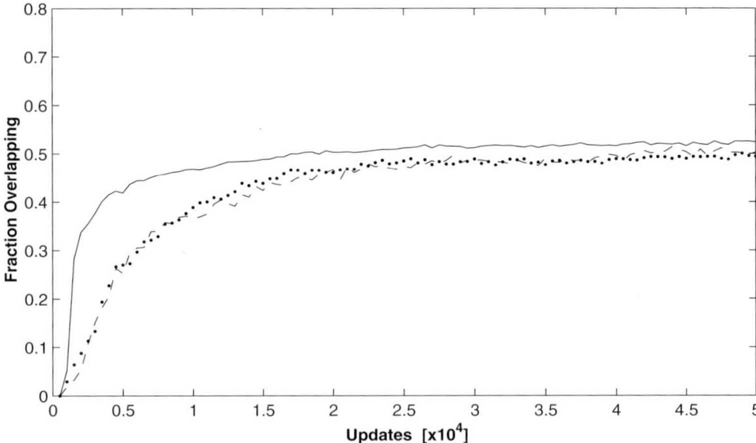

Fig. 3. Average fraction of doubly expressed code for the three experimental sets. Set I (solid line), set II (dashed line), set III (dotted line).

more subtle than threads executing particular instructions a different number of times. For example, two threads which execute a stretch of code in an identical manner but that start execution at different points "upstream" may end up calculating very different functions, and thus have quite different behaviors. This difference will thus be underestimated. While the preceding graphs seem to indicate that differentiation stops about halfway through the duration we recorded, this is actually not so, as the more microscopic analysis of the following section reveals. Finally, Fig. 3 shows the evolution of the fraction of code that is executed by multiple threads.

We anticipate that this fraction rises swiftly at first, but then levels off, as it is not advantageous to multiply express all genes (see below). However, we might anticipate that the fraction would start to decline at some point, when the organism develops the ability to localize its genes and use independent instruction pointers for each of them. We do not witness this trend in Fig. 3, presumably because there is no cost associated with the development of secondary expression. This should be viewed as a peculiarity of the digital environment rather than as a universal feature, which we hope to eliminate with future refinements of the avida world.

5   Evolution of Genetic Locality

To get a better idea of how evolution acts upon programs harboring multiple threads, we must look at exactly what is being expressed. We can loosely characterize all organisms by tracking three separate genes. They are "self-analysis" (*slf*), "replication" (*rpl*), and "computation" (*cmp*). To follow the progression of these genes through time, we examine a sample experiment

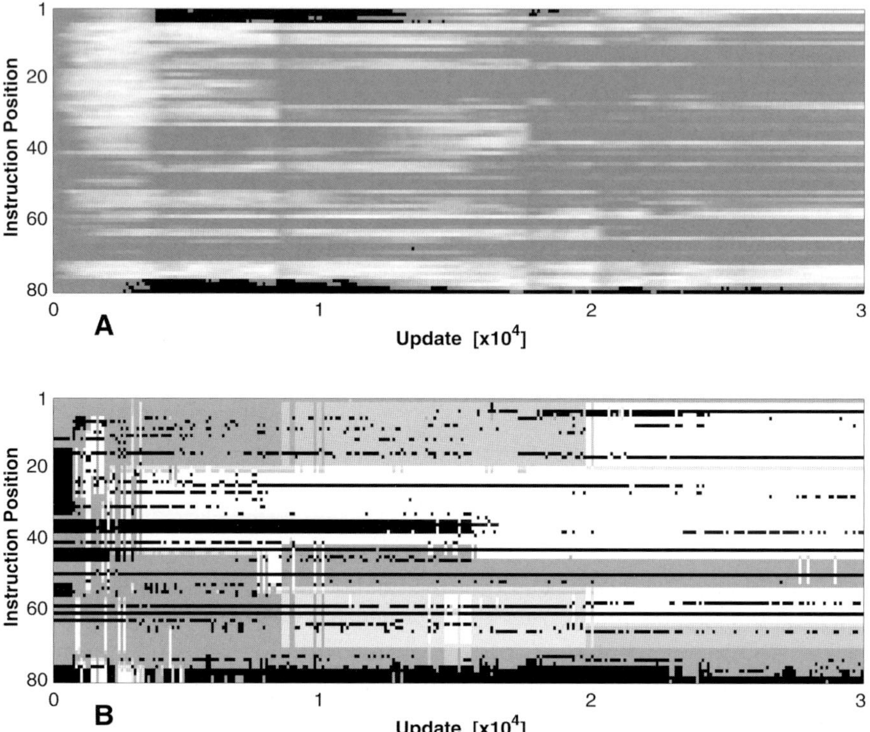

Fig. 4. (A) Per-site entropy for each locus as a function of time for a standard (set III) trial. Random (variable) positions with near-unit per-site entropy are bright, while "fixed" instructions with per-site entropy near zero are dark. (B) Thread identification within a genome. Black indicates instructions which are never directly executed, dark grey denotes instructions executed by a single thread when no other thread is active, while sections which are executed by a single thread while another thread is executing a different section are colored in lighter shades of grey. Sections with overlapping expression are in white.

seeded with an ancestor of size 80 (as before, capable only of self-replication), in an environment in which size-altering mutations are strictly forbidden (a trial from set III). This limitation was enforced in order to better study the functionality of the organism and the location of its genes. Similar studies, done with all 400 trials used to collect the bulk of the data for this report, show comparable behavior.

In Fig. 4A we follow the per-site entropies for each locus as a function of time. Positions are labeled by 1 to 80 on the vertical axis, while time proceeds horizontally. A grey-scale coding has been employed to denote the variability of each locus, where the white end denotes more variable positions and the dark end more fixed positions. Because the per-site entropies have been calculated by obtaining the frequency with which each instruction appears at that locus within the population (as opposed to the theoretical estimate based on neutrality), major evolutionary transitions are identifiable by dark vertical bands. Fig. 4B shows which portion of the code is expressed by which pointer, by two pointers simultaneously, or not at all.

The first gene *slf* uses pattern matching on `nop` instructions in order to find the limits of its genome and from that calculate its length. This value is used for *elongation* (via the command `alloc`), which adds empty memory to the genome and prepares it for the "execution" of the replication gene. Note that avidian genomes are circular. There are two interesting points to note about the evolution of *slf*: First, there are many methods by which the organism can determine its own genomic length, so this gene tends to vary widely. Most of the time the organism keeps pattern-matching techniques, but matches different portions of the code. However, often an organism shifts to purely numerical methods, performing mathematical operations upon itself which yield the genome length "by accident". The other evolutionary characteristic of this gene is that there is no benefit in expressing it multiple times as it has a fixed result which needs to be applied only once during the gestation cycle. Looking at Fig. 4, the *slf* gene initially spans from lines 44 to 61 plus the first four lines and last four lines of the genome which are boundary markers fashioned from `nop` instructions. The first major modification to the *slf* gene occurs around update 3000. The pattern used to mark the limits of the genome is a series of four `nop-A` instructions. As a newly allocated genome has all of its sites initialized to `nop-A`, the genome is re-organized such that these lines are no longer copied. This reduces the possibility of variation in these sections of code to zero. This is apparent in Fig. 4A as the positions of these limit patterns become completely black, indicating vanishing entropy.

The *slf* gene is continuously undergoing minor changes as it becomes more optimized and thus requires fewer lines of code to perform its function. Near update 13,000 it shifts dramatically and is replaced by one in which size is calculated using only the final boundary markers. The distance from the gene to the final marker is determined and then manipulated numerically in order obtain the number which is the size of the organism. Looking at the first four lines of Fig. 4A around this update, we see that they are slowly phased out and increase in entropy as they are no longer as critical to the organism's survival. Finally, the size of the pattern marking the end boundary of the organism is shortened until it becomes only a single line. By the end of the evolution shown, the *slf* gene only occupies lines 48 through 56. Note that all of these lines are only expressed a single time.

The next gene under consideration is the actual replication gene *rpl*. This sequence of instructions uses the blank memory allocated in the self-analysis phase and enters a "copy-loop" which moves line by line through the genome, copying each instruction into the newly available space. When this process is finished, it severs the newly created copy of itself which is then placed in an adjacent lattice site. These dynamics spawn a new organism which, if the copy process were free of mutations, would be identical to the parent. In Fig. 4, the organism being tracked has its replication gene on lines 65 to 71 until update 24,000 at which time this gene actually grows an additional line becoming much more efficient by "unrolling" its copy-loop. What this means is that it is now able to copy *two* lines each time through the loop. From the dark color of these lines, it is obvious that they have very low entropy, and are therefore very difficult to mutate. The copy-loop is a very fragile portion of code, critical to the self-replication of the organism, yet we do see some evolution occurring here when multiple threads are in use. Often the secondary thread will simply "fall through" the copy-loop (not actually looping through to copy the genome) and move on to the next gene, while the other thread performs the replication. However, sometimes the two threads will actually evolve together to use the copy-loop in different ways, with each thread copying *part* of the genome. In Fig. 4, most of the *rpl* gene is executed by only one thread. The *rpl* gene is followed by junk code which, while executed sporadically, does not affect the fitness in any way (as evidenced by the light shading in Fig. 4A for these lines).

The most interesting of the genes is the computation gene *cmp*. The ancestor does not possess this gene at all, so it evolves spontaneously during the adaptive process. There are 78 different computations rewarded in this environment, all of which are based on bit-wise logical operations. The organisms have three main commands which they use to accomplish the computations: a `get` instruction which retrieves numbers from the environment, a `put` instruction to return the processed result, and a `nand` instruction which computes the logical operation not-and (see Table 1). Any logical operation can be computed with a properly arranged collection of `nand` instructions.

The *cmp* gene(s) evolve uniquely in each trial, enabling the organisms to perform differing sets of tasks. There are, however, certain themes which we see used repeatedly whereby the same section of code is used by both threads, but their initial values (i.e., the processing performed thus far on the inputs) differ. Consequently, this section of code performs radically different tasks, actually encouraging this overlapping. Portions of this algorithm which might have some neutrality for a single thread of execution will now be frozen due to the added constraints imposed by a secondary execution. The size of *cmp* grows during adaptation as a number of computations are performed, and the gene is almost always expressed by both threads as this is always advantageous. In Fig. 4, the *cmp* gene stretches from line 1 to line 42 (at update 30,000), while it is considerably smaller earlier. Furthermore, the genome manages to execute the entire gene by both threads (the transition from single expression of part of *cmp* to double expression is visible around update 20,000). This gene ends up being expressed many times (as the instruction pointers return to this section many times during execution). All in all, 17 different logical operations are being performed by this gene.

By the end of the evolution tracked in Fig. 4, most of the genes appear to occupy localized positions on the genome. The *cmp* gene (white sections in Fig. 4) is revisited many times by both threads with differing initial conditions for the registers, allowing the genome to maximize the computational output. In the meantime, those sections have become fixed (their variability is strongly reduced) as witnessed by their dark shading in Fig. 4A.

6   Discussion and Conclusions

The path taken by evolution from simple organisms with few genes towards the expression of multiple genes via overlapping and interacting gene products in complex organisms is difficult to retrace in biochemistry. Artificial Life, the creation of living systems based on a different chemistry but using the same universal principles at work in biochemical life, may help to understand some key principles in the development of gene regulation and the organization of the genetic code. We have examined the emergence and differentiation of code expression *in parallel* within a digital chemistry, and found some of the same constraints affecting multiply expressed code as those observed in the overlapping genes of simple biochemical organisms. For example, multiply expressed code is more fragile with respect to mutations than code that is "transcribed" by only one instruction pointer, and as a result evolves more slowly. During most stages of evolution, two constraints are most notable: the pressure to reduce sequence length in order to lessen the mutational load, and the pressure to *increase* sequence length in order to be able to store more information. Simple organisms can give in to both pressures by using overlapping genes, gaining in the short term but mortgaging the future: the reduced evolvability condemns such organisms to a slower pace of adaptation and exposes them to the risk of extinction in periods of changing environmental conditions.

This trend is clearly visible in the evolution of digital organisms, as is a trend towards multiple expression of as much of the code as possible. We believe that this latter feature is *not* universal, but rather is due to the fact that multiple expression in avida is cheap, i.e., no resources are being used in order to express more code. In a more realistic chemistry, this would not be the case: adding an instruction pointer should put some strain on the organism and use up energy; in such circumstances multiple expression would only emerge if the advantage of the secondary expression outweighs the cost of it. We also expect more complex gene regulation in such an environment, as genes would be turned on only when needed.

Still, under extreme conditions we believe that multiple overlapping genes are a standard path that any chemistry might follow. Even though evolution slows down, such organisms can be rescued by either the development of error-correction algorithms or an external change in the error rate. In either case, a drastic reduction of the mutational load would enable the sequence length to grow and the overlapping genes to be "laid out" (for example by gene-duplication). The corresponding easing of the coding constraints might give rise to an explosion of diversity and possibly the emergence of multi-cellularity.

Acknowledgements. We would like to thank Grace Hsu and Travis Collier for collaboration in the initial stages of this work. Access to a Beowulf system was provided by the Center for Advanced Computing Research at the California Institute of Technology. This work was supported by the National Science Foundation.

References

1. Adami, C. (1995) Learning and complexity in genetic auto-adaptive systems, Physica **D 80**, 154.
2. Adami, C. (1998) *Introduction to Artificial Life*, Telos Springer-Verlag, New York.
3. Adami, C., Collier, T. C., and Ofria, C. (1999) Robustness and evolvability of computer languages, to be published.
4. Basharin, G. P. (1959) On a statistical estimate of the entropy of a sequence of independent random variables, Theory Probability Appl. **4**, 333.
5. Elena, S. F., Cooper, V. S., and Lenski, R. E. (1996) Punctuated evolution caused by selection of rare beneficial mutations, Science **272**, 1802.
6. Keese, P. and Gibbs, A. (1992) Origins of genes: "Big bang" or continuous creation? Proc. Natl. Acad. Sci. **89**, 9489–9493.
7. Miyata, T. and Yasunaga, T. (1978) Evolution of overlapping genes, Nature **272**, 532.
8. Mizokami, M., Orito, E., Ohba, K., Lau, J. Y. N., and Gojobori, T. (1997) Constrained evolution with respect to gene overlap of Hepatitis B virus, J. Mol. Evol. **44** (Suppl. 1), S83–S90.
9. Normark, S., Bergström, S., Edlund, T., Grundström, T., Jaurin, B., Lindberg, F. P., and Olsson, O. (1983) Overlapping genes, Ann. Rev. Gen. **17**, 499–525.
10. Ofria, C., Brown, C. T., and Adami, C. (1998) Avida User's Manual, in [2].
11. Ofria, C., Collier, T. C., Hsu, G., and Adami, C. (1999) Evolution of differentiated expression patterns in digital organisms, KRL preprint MAP-250 (February 1999).
12. Ray, T. S. (1992) An approach to the synthesis of life, in *Proc. of Artificial Life II*, C. G. Langton, C. Taylor, J. D. Farmer, and S. Rasmussen, Eds., Addison-Wesley, Redwood City, p. 371.
13. Ray, T. S. and Hart, J. (1998) Evolution of differentiated multi-threaded digital organisms, in *Proc. of Artificial Life VI*, C. Adami, R. K. Belew, H. Kitano, and C. E. Taylor, Eds., MIT Press, Cambridge, MA, p. 295.
14. Samuel, C. E. (1989) Polycistronic animal virus messenger RNAs, Prog. Nucleic Acids Res. Mol. Biol. **37**, 127–153.
15. Thearling, K. and Ray, T. S. (1994) Evolving multi-cellular artificial life, in *Proc. of Artificial Life IV*, R. A. Brooks and P. Maes, Eds., MIT Press, Cambridge, MA, p. 283.

16. Watson, J. D., Hopkins, N. H., Roberts, J. W., Steitz, J. A., and Weiner, A. M. (1987) *Molecular Biology of the Gene*, Fourth Edn., Benjamin Cummings, Menlo Park, CA, p. 457.

Toward Code Evolution
By Artificial Economies

Eric B. Baum and Igor Durdanovic

Abstract. We have begun exploring code evolution by artificial economies. We implemented a reinforcement learning machine called Hayek2 consisting of agents, written in a machine language inspired by Ray's Tierra, that interact economically. The economic structure of Hayek2 addresses credit assignment at both the agent and meta levels. Hayek2 succeeds in evolving code to solve Blocks World problems, and has been more effective at this than our hillclimbing program and our genetic program (GP). Our hillclimber and our GP also performed well, learning algorithms as strong as a simple search program that incorporates hand-coded domain knowledge. We made efforts to optimize our hillclimbing program and it has features that may be of independent interest. Our GP using crossover performed far better than a version utilizing other macro-mutations or our hillclimber, bearing on a controversy in the genetic programming literature.

1   Introduction

We address the reinforcement learning problem of learning to interact with a system (that we call the "world") where one has sensations, may take actions, and that will give rewards in response to certain sequences of actions. We would like to understand how one might set up a machine that would learn to extract rewards efficiently. Our hope is to work toward methods that might ultimately be capable of addressing very complex worlds, involving huge state spaces such as are apparently addressed by natural intelligences. This involves automated programming: the system must output a complex program for interacting with the world. We believe the critical problem here is credit assignment: a large number of computational and/or physical actions are taken, and some reward or penalty results. What in the learner or hypothesis should be modified?

There has been considerable work on using artificial evolution techniques patterned on biological evolution to optimize functions and attempt automated programming, c.f. [14,12]. Biological evolution has created intelligence, but we believe it is a poor model to emulate because it does not address credit assignment. Economies differ from ecologies in that rules imposed on economies protect property rights and avoid the tragedy of the commons [16]. The evolution of economies is radically different from the evolution of ecologies for this reason. Although this is relevant to the topic of evolution as computation, we, will not discuss this much in this paper because we have written about it at some length elsewhere, c.f. [5].

Here we will report experiments on three approaches. The first is a system we call Hayek2 that maintains an artificial economy of agents. The agents can earn money in the world, can pass money to other agents, and can create other agents in which they become "investors". The rules of the economy are set up so that each agent is motivated to increase the performance of the whole system. This addresses the credit assignment problem explicitly and dynamically: the agents learn to assign credit. There are two issues involved in creating such an economy.

- What rules should the economy operate under, so that assignment of credit motivates the agents correctly? We address this in Sect. 2.1.
- What representation language should the agents be written in? We here experiment with an assembler like language inspired by Ray's Tierra system [19]. We discuss this in Sect. 2.2.

Much of the strength of our Hayek2 was due to its ability to learn at the meta-level. We compared Hayek2 to a version with meta-learning turned off that we called NoML. Hayek2 succeeded in learning new ways of creating agents that enabled it to jump out of local maxima trapping NoML, or other approaches. This is evident not only in the superior results of Hayek2, but graphically. Performance of Hayek2 runs plateau repeatedly, then discover a new creation strategy and escape from the plateau, see Fig. 2 in Sect. 2.3. Performance of NoML runs simply plateau, see Fig. 3.

One of us has previously reported experiments with Hayek1, an economy of simple agents inspired in part by Holland's Classifier Systems [13] but correcting problems with the Classifier economy that cause a Tragedy of the Commons and a non-conservation of money, and with representational differences [5,4]. Hayek2 recurses the economic strategy on its own creation, generalizing the economy of Hayek1 to allow agents that can create other agents, and thus allowing meta-learning and much more sophisticated meta-computation.

The second approach we experimented with is hillclimbing, see Sect. 3. We maintain a Current Favorite Program(CFP)– a program which is the best solution we have yet found. We initialize this CFP with a random program. We iteratively generate a candidate by making small changes in the CFP and evaluate the candidate, replacing the CFP if the candidate is superior. This approach does explicit assignment of credit, since it makes small changes and evaluates their utility. Hillclimbing suffers, however, from two important problems. First, it is unclear how to evaluate whether a candidate is superior. Second, it is unclear how best to generate small changes. Our hillclimber implemented our best effort at addressing these problems, described in Sect. 3.

The third approach we tried was genetic programming (GP), see Sect. 4. We used strongly typed Genetic Programming [17], representing programs as S-expressions. GP does not attempt explicit credit assignment, but attempts what amounts to implicit credit assignment using crossover. There is considerable controversy over whether crossover is successful. [15] showed that crossover with random trees was substantially more effective than crossover within the population on a particular problem. Crossover with random trees was dubbed "headless chicken mutation" and considered as a macromutation operator. [18] and [1] compared macromutations to crossover on a suite of other problems and found no strong gains from crossover. To quote a recent text on GP ([3] p155)

> The empirical evidence lends little credence to the notion that traditional GP crossover is, somehow, a more efficient or better search operator than mutation-based techniques... On the state of the evidence as it exists today, one must conclude that traditional GP crossover acts primarily as a macromutation operator.

Accordingly we implemented as well a GP using the "headless chicken" macromutation. We found that crossover was in fact far superior to headless chicken mutation in our experiments, and our genetic programs exhibited what we regard as very promising results.

We tested these approaches on Blocks World problems. See Fig. 1. A series of instances are presented. Each instance consists of 4 stacks of blocks. The blocks come in three different colors. The learner controls a hand that if empty can pick up the top block on any but stack 0, and if full can place its block on top of any stack but 0. If it succeeds in copying stack 0 in stack 1 using no more than 100 actions, it receives a reward of n, where n is the number of blocks on stack 0. If it uses 100 actions without solving the instance, it gets 0 reward and goes on to the next instance.

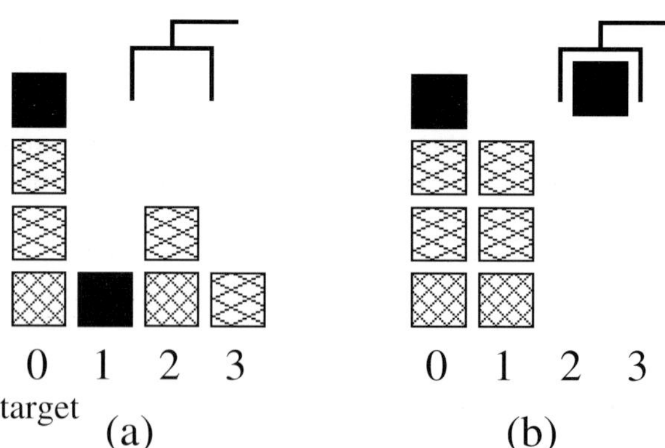

Fig. 1. A Blocks World Instance with 4 blocks. (a) shows the initial state. (b) shows the position just before solution. When the hand drops the white block on stack 1, the instance will be solved and Hayek will see another random instance.

All instances have $n \geq 1$ randomly colored blocks on stack 0, and n blocks randomly distributed elsewhere, with the same multiset of colors so that all instances are solvable. The instances are generated randomly from a distribution of instances. The distributions were tuned to promote training. For each method we used a distribution empirically optimized to that method. All learners were, however, tested on the same **test set**: 100 random instances each of size 1 through 10, and graded by total **score**, earning $i/10$ points for each instance of size i they solved.

Blocks World has been studied as a planning domain since the sixties [21]. Off the shelf, domain independent planners are able to solve related (but not identical) Blocks World problems involving about 5 blocks [2]. The literature applying learning algorithms to Blocks World has, however, addressed simplified "toy" versions of the problem which do not contain critical elements of having to discover an abstract goal (copying an arbitrary goal stack), and having to stack blocks on later need blocks. For example, [20] applied a temporal difference algorithm to learn to pick up a green block, initially under at most three other blocks. The schema mechanism of [9] moved a single block around an otherwise empty table. Genetic programming was applied to build the same fixed stack for every instance on an unbounded table, given built in sensors for such pertinent information as the next-needed block, and both the top block and the top correct block on the target stack [14]. An exception is the Hayek1 work of [4], which succeeded in solving arbitrary Blocks World problems given a single feature (top of stack) and intermediate reward. The work reported here uses neither intermediate reward nor handcoded features.

Results of our experiments are shown in table 1. We quote the average of the top half of the runs, and the best score, for each method. Hayek2's average score is seen to be better than the best score of any other method. GP's average is seen to be better than the best score of GPHC (the headless chicken version) or hillclimb.

All of our methods perform creditably in that they do much better than a search consisting of 100 random legal actions, and much better than an algorithm that simply stacks all the blocks on stack 1. This latter algorithm already involves considerable domain knowledge: it uses knowledge that stack 1 is important. Hayek2 has done substantially better than an algorithm that blindly stacks and unstacks on stack 1, using all 100 actions in a search. Planning programs are typically given a world model, allowing them to simulate the results of actions and then do an extensive search. Ultimately, though, in complex spaces it will be necessary to acquire domain knowledge and use it to eliminate or dramatically prune the search. Our learners have acquired such domain knowledge. They are given no world model and are allowed no search, but only 100 actions on the world. Initially, they are random, not even using many actions. All of them have learned extensive domain knowledge, guiding their actions so that, for example, they seem to "understand" the importance of stack 1, which they stack and unstack in an apparently "purposeful" way.

	Best	$\langle$Score$\rangle$	σ(Sc)
Hayek2	170	145	20
NoML	132	119	8
GP	136	130	3.7
GPHC	121	97	17
Hillclimb	125	102	15
100 Rndm.	51	47	1.7
Stack1	88	82	4
Search1	130	125	2

Table 1. Comparative rewards. Results are based on 8 4-day runs of each program on a 333 MHz PII (except hillclimb where 12 runs are reported and 100 rndm, stack1, and search1, which report 40 runs with no learning). 100 Rndm just did 100 random legal actions on each instance. Stack1 stacked all blocks on stack 1, in random order. Search1 repeatedly stacked and unstacked all blocks on stack 1, in random order. $\langle$Score$\rangle$ is the score averaged over the top half of the runs. σ(Sc) is the standard deviation in Score. Best is the best of the last four scores, for the best run in the sample. Since performance fluctuates a bit, this is not generally the best achieved overall.

A more extensive discussion of all the results in this paper, and other results, can be found in [7].

We have very recently obtained striking results with Hayek2-like economies with expanded powers. We call this system Hayek3. Hayek3 differs from Hayek2 in that the agents have been implemented in an S-expression-like language in which they can calculate numerical bids far more flexibly, and in that we have provided the agents access to a world model, so that they may simulate their actions on the world and use the results in calculating their bids. Hayek3 has been able to generate systems solving the vast majority of Blocks World instances with goal stacks up to 200 blocks high. We will report on this work elsewhere [8].

Sect. 2 discusses Hayek2. Sect. 3 discusses the hillclimber. Sect. 4 discusses our genetic program. Sect. 5 concludes the discussion.

2 Hayek2

In this section we discuss the Hayek2 model we tested. The section is broken into three subsections. Section 2.1 discusses the economic rules which motivated the agents to collaborate. Section 2.2 discusses the language the agents used. Section 2.3 discusses our experimental results.

2.1 Economic Framework

We first discuss Hayek1 [4], a simpler economy of agents, and then its generalization to Hayek2. Computation in Hayek1 proceeds in a series of steps.

In each step all the agents perform computations and can bid. The highest bidding agent buys the access to the world as its property. Its property rights are the same as any property owner's rights: it can take actions on the world, and it can sell the property. The agent pays its bid to the previous owner when it buys the world, and when it sells, the agent collects from the next buyer. The agent owning the world also collects rewards paid by the world. Because the agent owns the world, it is motivated to increase its value (immediate reward plus resale value). Note that this was our goal: a framework where each agent's individual motivation is to act so as to increase the value of our access to the world.

The winning bid converges to estimate the expected value of the world given that we use the winning agent. This happens as follows. If an agent bids lower than the value of the world, a new, similar agent can enter and bid higher. If the agent bids higher than the value of the world, it will go bankrupt and be removed. Thus choosing the agent with the highest bid is, roughly speaking, choosing the agent with highest expected world value– an eminently sensible choice. A new agent can outbid its competitors and enter the economy profitably if and only if it takes the world to states of higher value than its competitors' bids. Thus the entry and bankruptcy of agents stochastically hillclimbs in total rate of wealth extraction from the world.

Because one agent owns the world, there is an unambiguous assignment of credit. A sequence of agents may own the world and take actions between rewards from the world. Agents early in this chain are rewarded (if at all) by being able to sell the world for more than they bought it.

Hayek1 was inspired by Holland's classifier systems [13] in using a bucket-brigade-like assignment of credit, and in using condition, bid, and action agents but differs in key respects. Classifier systems have many agents active that split reward. This motivates agents to be active when reward is coming, even if their action harms performance, promoting a tragedy of the commons [11]. Classifiers enter with wealth, which breaks conservation of money, harming credit assignment. Classifier systems typically proceed in generations. For discussion of the differences see [4,5]. Hayek1 succeeded in achieving long chains of actions and solving hard Blocks World problems.

Hayek2, experimented with in the present paper, was organized identically to Hayek1 except in two regards. First, the computations the agents perform, in deciding whether to bid, or which action to take, are complex. Agents execute up to 400 lines of code in deciding whether to bid, and up to 8000 lines of code in acting. Second, at each auction the agents may choose to create a new agent (executing up to 8000 instructions in the process). What motivates an agent to create another agent? The child passes some of its profits to its parent. Thus the creation of a child is simply another way to attempt to earn money. The central idea of the Hayek Machine, that by auctioning off the access to the world we give the agents the motivation of improving performance of the system, remains intact. Wealth flows into the system in the form of payoffs from the world to agents who have won the auction and control the world at the time of the payoff. Wealth flows from them to other agents: agents who previously owned the world, or their creator, in a dynamic programming-like process designed to motivate every agent to improve the performance of the system.

We have explored several schemes for determining how much of the child's profits are passed to its parent. In all the Hayek runs reported, the child passes a portion of its profits to its parent that is specified in the child's code, which is written by the parent. All profits passed to a parent are regarded as profit to the parent as well, and thus a portion of them is recursively passed. Our most recent and best version is augmented by a prescription that all agents pay a small sum of money to their parent every single instance. This causes parents to eventually recover all money that is earned by any of their children, and seems to provide effective incentives for creation.

Every few auctions, we charged each agent a "rent" proportional to the computational time it has consumed. This implements the meta-computational notion of dynamic sensitivity to computational costs, since the population should evolve to perform only cost effective computation. Each agent's rent was set at

$$R = cyX$$

where c is a constant, y is the number of instructions executed by the agent per instance, and X is the total number of instructions executed by the system per instance (averaged over the last 100 instances). Having the "price" per instruction executed by an agent increase with the total number of instructions executed by the system per instance improved performance and was suggested to us by the market analogy: if there is more demand (from the system) for computer time, the cost of time should rise. It is not evident (to us) from this analogy that the cost of time should be linearly proportional to demand. Drexler and Miller have instead proposed sophisticated auction schemes [10].

In Hayek1, new agents were extended credit, so that they could initially bid. To maintain conservation of money, we implemented a "perfect credit check" by lookahead. In Hayek2, we have allowed no credit. Agents have creators who must endow them with initial capital else they cannot bid. This places a burden on the creation agents to learn not to produce new agents that will be fleeced. In principle the agents should have flexibility in determining how much they invested in each child, but for a first cut, we adopted a simple specification. We specified that the amount of capital an agent invested in each new child was at least l and as much as it had up to at most $\min(1.2h, h + 200)$, where l was the lowest *winning* bid and h the highest winning bid of the last 20 auctions. This seemed a sensible figure. The agent needs at least l to have a good chance of contributing, since it will either have to win an auction or create a child that will win an auction. To guarantee the agent will be able to win an auction, it needs in excess of h.

The system is initiated with a single root agent. This agent we wrote by hand, as will be described in Sect. 2.3. The root agent begins constructing children, who in turn construct more children. The root agent, which is initiated with 0 money, wins the initial auction with a bid of 0. It also creates a child. We imposed a rule that if two agents bid identically, the newer agent wins the bid. Thus the second auction is won by the child of the root, who also bids 0. His child in turn, or another child of the root, may win the third auction, also with a bid of 0. All bids are 0, until some agent succeeds in earning money in the world. Then it, and its children, can bid more. Eventually, the system is populated with a collection of agents with money.

We also experimented with various schemes for protecting intellectual property rights. We report here on the simplest (and in some experiments, the most effective). We required a brand new agent to bid more than a veteran agent was earning, on average, in order to beat the veteran in an auction. We calculated this bid by adding to the veteran agent's bid its wealth divided by the number of auctions it had won. On the one hand, this scheme promoted progress: a new agent could only enter if it could actually outperform the old agent, not merely exploit the old agent's underbid. Such exploitation of underbids gives rise (potentially) to the cherrypicking phenomenon discussed for Hayek1 in [4] where an agent can enter, even though it actually performs worse than a current agent, because the current agent is underbidding. On the other hand, this new scheme helped ensure that creators of useful agents would be compensated. When another agent immediately copies a creator's child, it deprives the creator of any profit. With this scheme, new agents enter paying out the value the old agent was attaining. They then can not be overbid till another agent actually adds value, and in the meantime they get to collect as profit the difference between the value they actually achieve (what they are paid) and the value the previous agent was achieving (which they pay, since this is where their bid is set).

Note that agents' capacity to create children is limited by the need for start-up capital; thus, they can create children in proportion to their wealth. Thus agents that are good at creating children are rewarded with opportunity to create more children directly in proportion to their profitability, and agents who create unprofitable children die off.

2.2   Representation Language

We next have to address the question of what language the agents use. We would like the language to be at least expressive enough that systems of agents can evolve capable of solving fairly general problems. The language must be able to talk naturally about agents, as well as the world, because agents have to be able to create other agents. Finally, the language must be a sufficiently good fit to the world that it is able to learn in an acceptable amount of time. We hypothesize that at least two-dimensional topology should be biased into the representation. The results reported here use languages that bias in a notion of strong typing as well. They will not compare addresses in a stack to small integers or to sensations in the world, for example, the color of a block at a particular region of the world. Our initial experiments utilized an untyped language, and did not learn as effectively as the language described here.

In these experiments we used an assembler like language inspired by Ray's Tierran [19]. Ray believed that Tierra's evolvability came from two features of his language: it has a small instruction set without numeric operands, and it has address by template. We have implemented languages with both these features, that like Tierra are stack and register based. These features give a language that is robust in the sense that random code will execute. (Some instructions, e.g. popping an empty stack, are illegal but simply ignored.) We have had to add many other commands so that Hayek's agents can sense the world, make bids, and act in the world, and our language is therefore rather more complex than Tierran.

Agents in the current implementation of Hayek consist of a program of up to 1000 instructions that reside in a "soup" stack S0. Each agent has 10 additional stacks– two integer stacks Sint and Sint2, two color stacks SC and SC2, three stacks of instruction addresses in the soup SIA, SIA2, and SRIP, and three stacks of instructions, Spatt1, Spatt2, and the child stack SIC. Each agent also has registers RIP, RIP2, and RIPR for instruction pointers, RX, RY, and RXY2 for integers, RC for colors. Each stack/register holds only values of the appropriate type. RIP points at the next instruction to execute, and is incremented when the instruction is executed. RIPR points to the next instruction to read, for copying into a child. We discuss instructions for creation of children below. The registers can be pushed onto or popped from the primary stack of the same type (e.g., Sint is the primary stack for RX, RY, and RY2). The secondary stacks (SIA2,SC2, and Sint2) and registers (RIP2 and RXY2) serve as swap space. The top of each stack can be duplicated and pushed, and in addition, the top two elements on each stack can be swapped, and in addition each stack can be swapped with the top of its dual stack. For SInt and SIAC there were addition and subtraction operations that add or subtract the top two positions in the stack, as well as increment and decrement operations that act on the topmost stack position.

As in Tierra, there are two pattern delimiters: noop0 and noop1. Hayek's jump instructions generalize Tierra's pattern match by allowing patterns to use arbitrary instructions instead of simply noops. When Hayek executes a push-patt-fwd (or push-patt-back) instruction, a set of following instructions demarked by a noop is is used as a template. If the set of following instructions ends in multiple noops, the last noop serves as the delimiter, and the template includes the other noops. This allows the template to include noops. Hayek searches forward (or backward) through the soup until it finds a matching template and pushes the address onto SIA. It can then be popped onto RIPR to be read from, or to RIP to be executed.

Of course, there are if statements allowing Hayek to branch. When it skips from an if statement, it jumps to the next occurrence of the immediately following instruction, which is thus used as a label.

All instructions that jump the instruction pointer, including skip instructions in conditionals and popping an instruction address stack onto RIP, also push the current location of RIP on the stack SRIP. The return instruction pops SRIP back onto RIP. This allows a call-return semantics.

Hayek also has look, move, and bid instructions that allow it to interact with the world. The world Hayek interacts with is a two-dimensional grid. World(i, j) is the value of the "pixel" at gridpoint i, j, and in the current application can take 4 possible values: one of 3 colors, or empty. The look instruction pushes the color at the gridpoint defined by RX and RY onto the color stack SC. The grab and drop instructions move the hand to stack RX and grab or drop, respectively. RX and RY can be controlled by instructions that push and pop onto Sint. Values can be prepared on Sint by incrementing, decrementing, adding, and subtracting, and then popped onto RX or RY.

We have experimented with a variety of special instructions for bidding. The version reported here has only a simple scheme: a bid-eps instruction. Each bid-eps instruction has an associated value. The first time that a bid-eps instruction is encountered during the bid phase, its associated value is fixed to be the minimal amount more than the highest competing bid. That particular bid-eps instruction then bids that amount every time the bid-eps instruction is encountered during the bid phase (including the first time).

Special instructions specify the fraction of profits paid to one's parent. Each agent has a variable f initiated at 0.5. Instructions in the agent can multiply or divide f by 1.1. When the agent earns profit, it pays a fraction $1 - f$ to its parent. By writing (multiple) increment or decrement instructions in the child, the parent can control this fraction. Profit is defined as wealth after the instance minus wealth before it if the result is positive, and zero otherwise.

A set of instructions was included for the purpose of creation of new agents. Children are created by copying instructions, drawn from other agents or the creator's own code. The copy process is pattern based, and is controlled by two stacks, Spatt1 and Spatt2. Instructions can be copied into the child using the copy_SIC_RIPR, copy_SIC_patt1, and copy_SIC_patt2 instructions. The instruction copy_SIC_RIPR copies the instruction at the address pointed to by RIPR to the next location in the child. The command copy_SIC_patt1 pops the stack Spatt1, appending the code there to the son's code. Likewise copy_SIC_patt2 does the same for the Spatt2 stack. The read pointer RIPR can be positioned in several ways: for example, there is a command to push the starting address of a random living agent onto the SIA stack, or one can alternatively push the address where a particular pattern is next found onto the SIA stack. The SIA stack can then be popped to RIPR. When the copy instruction is executed, the RIPR pointer is automatically incremented, so it is easy to copy long sequences of code. The Spattx stacks (for x=1,2) can be loaded using the command push_pattx, which pushes the following pattern (delimited as described above by a noop instruction) onto the pattern stack. The Spatt1 stack can also be loaded using the command read_p1_till_p2, which copies the code starting from the read pointer until the pattern in Spatt2 is encountered into the stack Spatt1. The child can also be edited using the substitute and rnd_substitute commands. The substitute command scans the child to find any occurrence of the set of instructions in stack Spatt1, substituting in place of these the set of instructions found in stack Spatt2. The rnd_substitute command does the same, but instead of substituting globally wherever pattern1 is found in the child, it substitutes in only one location in the child, chosen randomly if there are multiple matches. There are also explicit mutate instructions that write random lines of code in the child.

Note that in contrast with Tierra, the creation of new agents is not inherently noisy. Hayek can learn to copy old agents exactly. The only randomness comes in the form of some explicit instructions that Hayek can choose (not) to incorporate, such as explicit mutate instructions that place a random instruction in the child (at the current write location) or explicit calls for a random integer. This in principle allows the system to learn to use as much or as little randomness as it likes.

Evidently the language is complex, and we make no claim that it is optimal. A large number of different functions need to be allowed. Indeed there are many, many functionalities that Hayek might find useful but is not yet capable of. For example, Hayek's agents cannot currently sense their own wealth, the wealth of agents they might wish to copy, the identities of active agents, the identity of the agent they are copying from, or the ancestry or relationship of agents. They cannot communicate information to other agents, e.g., register values or strings, or arrange to pay other agents for information or services, except in limited ways. Exploring what makes a language effective for evolving good systems of agents is an open research topic, in which we are very actively engaged.

2.3   Experimental Results

Hayek2 is initiated with an initial agent called Root. The current version of Root we will call "Root1". Root1 bids 0. It takes no action on the world, but creates a child in one of three ways. With a probability of 3/4, it creates a child of the first type, with a probability of 1/8 it creates a child of the second type, and with a probability of 1/8 it creates a child of the third type. The first type of child is simply 256 lines of random code. The second type of child is a creation agent that bids 0, picks a random agent from the population, and copies it with mutations as follows. When this second type of child copies an agent, there is for each line in the agent being copied, a probability of 1/64 that it is instead deleted, else a probability of 1/64 that it is replaced with an instruction chosen randomly from the instruction set, else a probability of 1/64 that a randomly chosen additional instruction is inserted. The third type of child also bids 0 and creates a child. It creates by copying a random agent in the population without mutation, and then doing a global substitute of one pattern for another. When Root1 creates a child of type 3, it chooses two patterns randomly and writes them into the child, so they are fixed in any particular child. The patterns are of length i with probability $1/2^i$ for $i = 1, 2, \ldots$. Thus the children of Root1 may (depending on their randomized code) bid, take actions, or create more children.

We start the system, and Root begets children who beget children. After a while there is a collection of agents that are solving Blocks World problems. Hayek2 was trained on a distribution of instances initially weighted to exponentially favor small examples, and present larger ones as Hayek learned. The distribution was a mixture of two distributions as follows: 1/10 of the time the probability p_n of choosing task of size n was defined as $p_n = 1/2^n$ (except that $p(10) \equiv p(9)$), and the rest of the time the probability p_n of choosing the size n was proportional to: $p_n = 1/2^n(1 - \text{solved}(n))$, where $\text{solved}(n)$ was defined as the fraction of instances of size n that Hayek solved in the last 100 tasks presented.

We compared this system to several different types of controls. Type 1 control, called "NoML", controlled for meta-learning. NoML runs were identical to Hayek except that the agents created children differently. Whenever an agent in NoML attempted to create a child, instead of following the instructions in the agent, the child was created by picking a random agent from the population and copying it with exactly the probability of mutation, deletion, and insertion found in the children of the root in the Hayek runs (i.e. 1/64 each for Root1). Thus the creation of agents in NoML never evolves–all agents create offspring just as the (type 2) children of Root do originally. Thus NoML versions are identical to Hayek, except that meta-learning is turned off: NoML cannot learn better ways to create agents than by picking a random agent and making random changes.

The performance of Hayek2 and NoML, as well as GP, GPHC, and hill-climb (described below) are shown in Table 1.

The first conclusion is that Hayek2 is better than NoML– we are "succeeding in meta-learning". We have run Hayek in a large number of variations, as our research has progressed, using different economic rules and different instruction sets, and with and without intellectual property payments. Until recently we had a slightly different Root that created only type 1 and type 2 agents in both NoML and Hayek. In each different comparison we ran, metalearning always impacted performance in a quantitatively and qualitatively similar way. The cause of the difference emerges from examining graphs of individual runs. Payoff vs. time and bid for a Hayek2 run is shown in Fig. 2 and for a NoML run in Fig. 3. These runs are qualitatively representative of the shape of graphs for Hayek2 and NoML runs, respectively. The Hayek runs usually progress in jumps. The NoML runs never make these kinds of

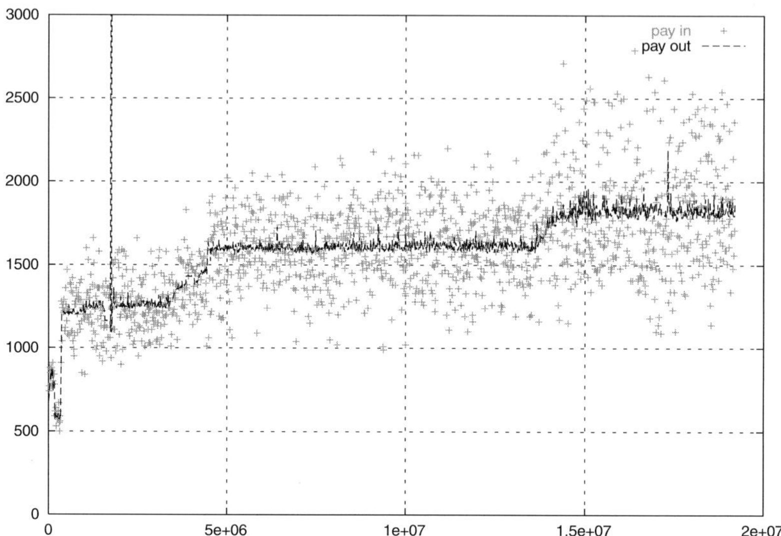

Fig. 2. The bid winning the last auction of each instance is graphed against instance number for a run of Hayek2. The reward paid by the world, averaged over the last 100 instances, is also shown as + signs.

jumps. The NoML runs rise up similarly to the Hayek runs at the beginning, which is not surprising since at the beginning the Hayek runs have not had time to discover better methods of creation. The NoML runs, like the Hayek runs, then get stuck roughly in a plateau. Many of the Hayek runs, however, later make jumps to perform better than the NoML runs. For most of the jumps we have examined, we have been able to identify one or several success-ful creation agents that are created immediately before or during the jump. It appears that Hayek2 discovers a fruitful method of creating new agents, causing a jump. These jumps are not vertical, but examined at a closer scale occur over tens of thousands of instances. They seem to be initiated by a discovery of a new creation method, allowing rapid subsequent incremental improvement for a while. Note, as seen in Fig. 2, Hayek occasionally made jumps late in the week. There is little reason to believe these runs are con-verged. Hayek2 runs allowed to go for 2 weeks or more have still continued to show improvement. NoML runs stall out.

We have attempted to examine the most successful creation agents to see what they are doing. We have been unable to decipher many of the most successful creation agents. The few we have been able to understand are not too different from type 2 children, but simply do more, or less, random mutation of one type or another. Of course, this doesn't say anything about the others – the reason we were able to decipher these is they had not mutated far yet.

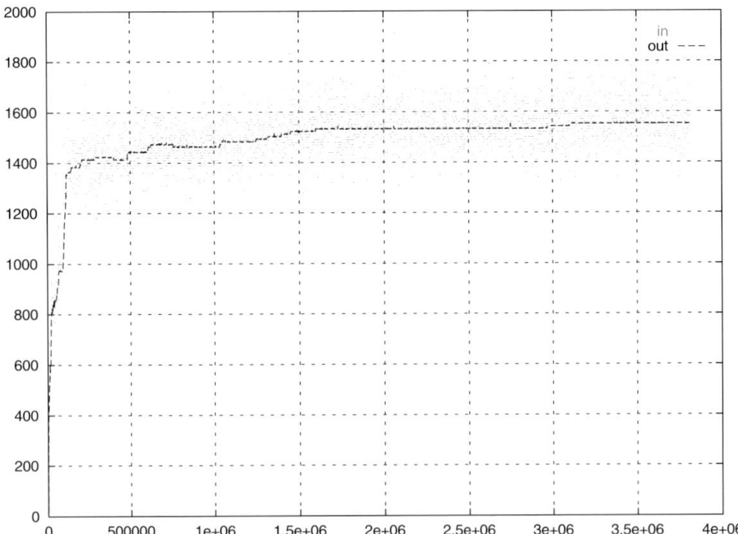

Fig. 3. The bid winning the last auction of each instance is graphed against instance number for a NoML run. Also shown as dots is reward paid by the world, averaged over the last 100 instances.

3   Hillclimbing

We wrote a program to hillclimb in code space. Hillclimb maintains a current favorite program (CFP) and iteratively attempts to replace it with a better favorite. The first CFP is generated in an initiation phase. In the initiation phase, hillclimb proposes programs consisting of 256 lines of random code (using the same instruction set as Hayek2 except that the creation instructions are omitted). These are tested on 1000 random instances. If one succeeds in solving 20% of the instances, it is declared the CFP and the initiation phase ends.

Once hillclimb has a CFP (which typically occurs in a matter of tens of minutes), it searches for a replacement. The CFP is scored on a test of 1000 instances. Potential replacements are created by copying the CFP, but inserting and deleting random lines. With a probability of $1/2^i$, $2i$ random lines are inserted. In addition, with a probability of $1/2^i$, $2i$ random lines are deleted. The replacement is tested on a similar test set of 1000 instances, but with cutoffs at 1, 10, and 100 instances. If the replacement either solves at least as high a fraction of instances as the CFP or earns at least as much money per instance, it proceeds through each cutoff; but if it fails at both tests, it is immediately discarded (without testing on the full 1000) and a new potential replacement generated. If the replacement earns more money than the CFP on its 1000-instance trial, it becomes the new CFP and we continue generating candidate replacements.

Hillclimb was found to work best with the following distribution of instances. In each test, the first instance is of size 1. Each time an instance of size n is solved, the next instance is of size n with a probability of $1/2$ and of size $n + 1$ with a probability of $1/2$. Each time it fails on an instance of size n, the next instance is of size $n - 1$ (or 1, if $n = 1$).

This hillclimbing heuristic is the result of an analytic and empirical effort at producing an effective hillclimber for this problem. It is necessary to test candidates on 1000 instances to minimize random variation. Using substantially less than 1000 instances results in getting rapidly stuck in a local minimum where one accepts a CFP simply because it got lucky on the test set, and it is then hard to generate a successful replacement. This flaw of hillclimbing algorithms in noisy environments was analyzed in [6] and is discussed in more detail in [7]. The cutoffs were introduced in order to speed up the testing of new replacements, minimizing the slowdown caused by this 1000 factor. The probabilities of insertion and deletion were arrived at after some empirical testing.

4   Genetic Program

We also ran a genetic program we called GP. GP was strongly typed [17], with types: VOID, BOOL, INT and COLOR as in the Hayek language. We grew S-expression trees. For a full description of functions and terminals, see [7]. After some experimentation, we used a population size of 100. The initial population was randomly grown, from the root down, respecting type constraints. Generated expression trees were dependent on two factors, p_f and p_m. At every node the probability p of choosing a function (probability $1 - p$ of choosing a literal) was defined as $p = p_f \times p_m^{\mathrm{depth}}$, where the root of the expression tree was defined to have depth 0. In our experiments we used:

$$p_f = 0.9, \quad p_m = e^{\frac{\log 0.5 - \log p_f}{\mathrm{avg.\ depth}}}, \quad \text{and} \quad \mathrm{avg.\ depth} = 3 \text{ or } 8.$$

No significant performance difference was observed between runs with avg. depth = 3 and runs with avg. depth = 8.

For each generation, we tested each S-expression in our population on a set of problems consisting of 20 random tasks of each size $i = 1, \ldots, k + 1$, where k was the largest size of problem that the previous generation had solved at least 50% of the time. As for Hayek and hillclimb, each S-expression was given only 1000 function calls to solve each example, and at most 100 actions. The S-expressions grew to be vast, a phenomenon known as genetic bloat [3], but only the first 1000 function calls encountered would be used in calculations. After each round of tests the next generation was produced as follows. The top 10 scoring members of the current generation were placed intact in the next generation. The additional 90 members were produced by breeding the top 50 scoring members of the present generation using a standard crossover. (We also experimented with drawing parents for the next generation proportional to fitness from the whole population. This performed slightly worse but the difference was not significant.) Finally the new generation was modified by having 10% of its new members further mutated by removing a random sub-expression and growing a new one.

The bloat phenomenon caused runs to grow to hundreds of megabytes. We improved the situation somewhat by instituting a "first money" rule to score S-expressions that were kept for several generations: we would not rescore such S-expressions for each generation; rather, they would simply keep their original score. We found that if we kept rescoring them, their score would fluctuate down (for random reasons) and they would be replaced by a version of themselves that might be longer. Such a version might often be functionally identical– recall that only the first 1000 function calls actually contribute to the behavior of an S-expression. Not rescoring them lessened such trivial expansion and kept our runs somewhat more manageable in size.

We also compared a genetic program without crossover we called GPHC for "genetic program headless chicken". This program was identical to GP except that no crossover between members of the population was used to produce new S-expressions. Whenever such a crossover would have been used in GP, the new S-expression was generated by crossing one of the top 50 scoring members of the present generation with a new randomly drawn S-expression.

As seen in Table 1, our results strongly favored crossover. Crossover achieved a best run of 136 and a top-half mean of 130, while headless chicken achieved, respectively, 121 and 97. (In fact, the one run of 121 was anomalous– no other runs were much above 100). Crossover was definitely able to find better modifications.

In some runs, crossover produced populations with no "looks" at all. The system just found optimized search algorithms that groped purposively, but without vision, never sensing the world, but searching through ways of stacking blocks on the first stack in hopes of achieving the correct stack. We were surprised to see how effective this could be. The system has discovered a search algorithm, optimized by significant learned domain knowledge, utilizing its 100 actions in systematically stacking and unstacking on column 1.

The fact that crossover was able to remove from the population all the "looks" and all the "if-statements", which depend on looks, shows that crossover was culling through the statements, achieving some degree of credit assignment.

5 The Bottom Line

We have begun exploring code evolution by artificial economies. Hayek2 succeeds in evolving code to solve Blocks World problems, and has been more effective at this than our hillclimbing program and our genetic program. All three algorithms perform creditably, succeeding in learning considerable domain knowledge to guide their search and vastly outperforming a search algorithm not utilizing any domain knowledge. We made some efforts to optimize our hillclimbing program and it has features that may be of some independent interest. Our genetic program exhibited strong gains from crossover compared to a version utilizing other macromutations. The relative strength of crossover and macromutations is a hotly debated issue within the GP community, and ours is the first unequivocal demonstration we are aware of where crossover is much better than the headless chicken mutation.

We have demonstrated meta-learning: Hayek succeeds in discovering new meta-level agents that improve its performance, getting it out of plateaus in which it has otherwise gotten stuck. Hayek's performance benefited from improvements in the algorithm deciding how creation agents give capital to their offspring, from improvements in how creation of intellectual property is rewarded, from improvements in how creation agents are paid by their offspring, from assessing a rent for computational time that was proportional to total demand, and from improvements in the language, including strong typing to bias the search for useful agents and expanding the representational power of the meta-instructions using pattern-based instructions. These are discussed in more detail in [7].

We have evidently not yet demonstrated that code evolution is a practical thing to do. The space we are searching is enormous– at minimum 100's of lines of code, each of which can be chosen from nearly 100 different possibilities. On the other hand, to the extent that an algorithm can learn in such a *tabula rasa* fashion, one might reasonably hope to be equally able to address other very hard environments. Note that we are by no means claiming that Hayek2 is the optimal artificial economy of agents. We have corrected several conceptual bugs in its constitution, and expect more may lurk. We are hopeful that further advances in constructing artificial economies might lead to useful systems.

Ongoing work, which we will report elsewhere [8], has studied Hayek2-like economies using an S-expression-like language and given the additional ability to simulate their actions on the world before bidding. These runs often develop extensive collaborations and have achieved results far stronger than any reported in this paper, solving problems with up to 200 blocks on the goal stack.

Acknowledgements. We are grateful to H. Stone and D. Waltz for comments regarding the presentation.

References

1. P. Angeline. Subtree Crossover: Building Block Engine or Macromutation. In Koza et al., editor, *Genetic Programming 1997, Proc 2nd ann.*, pages 9–17, 1997. Morgan Kaufmann, San Francisco, CA.

2. F. Bacchus and F. Kabanza. Using Temporal Logic to Control Search in Planning. In *European Workshop on Planning, 1995*. Unpublished document available from http://logos.uwaterloo.ca/tlplan/tlplan.html, 1995.

3. W. Banzhaf, P. Nordin, R. E. Keller, and F. D. Francone. *Genetic Programming, An Introduction*. Morgan Kaufmann, San Francisco, CA, 1998.

4. E. B. Baum. Toward a Model of Mind as a Laissez-faire Economy of Idiots, extended abstract. In L. Saitta, editor, *Proceedings of 13th International Conference on Machine Learning '96*, , pages 28–36, 1996. Morgan Kaufmann, San Francisco, CA.

5. E. B. Baum. Manifesto for an Evolutionary Economics of Intelligence. In C. M. Bishop, editor, *Neural Networks and Machine Learning*, pages 285–344, 1998. NATO ASI Series F, Computer and System Sciences, Vol 168, Springer-Verlag, Berlin.

6. E. B. Baum, D. Boneh, and C. Garrett. On Genetic Algorithms. In *COLT '95: Proceedings of the Eighth Annual Conference on Computational Learning Theory*, pages 230–239, 1995. Association for Computing Machinery, New York.

7. E. B. Baum and I. Durdanovic. Toward Code Evolution by Artificial Economies. Technical Report TR-98-065, NECI, 1998.

8. E. B. Baum and I. Durdanovic. Evolution of Cooperative Problem Solving in an Artificial Economy. *Neural Computation*, to appear, 2000.

9. A. Birk and W. J. Paul. Schemas and Genetic Programming. In *1994 Conference on Integration of Elementary Functions into Complex Behavior*, 1995. Bielefeld.

10. K. E. Drexler and M. S. Miller. Incentive Engineering for Computational Resource Management. In B. A. Huberman, editor, *The Ecology of Computation Studies in Computer Science and Artificial Intelligence 2*, pages 231–266, 1988. North Holland, New York.

11. G. Hardin. The Tragedy of the Commons. *Science*, 162:1243–1248, 1968.

12. J. H. Holland. . MIT Press, Cambridge,MA, 1975.

13. J. H. Holland. Escaping Brittleness: The Possibilities of General Purpose Learning Algorithms Applied to Parallel Rule-Based Systems. In T. M. Mitchell R. S. Michalski, J. G. Carbonell, editor, *Machine Learning II*, pages 593–623, 1986. Morgan Kauffman, Los Altos,CA.

14. J. R. Koza. *Genetic Programming*. MIT Press, Cambridge, MA, 1992.

15. K. Lang. Hill Climbing Beats Genetic Search on a Boolean Circuit Synthesis Task of Koza's. In *The Twelfth International Conference on Machine Learning*, pages 340–343, 1995.

16. M. S. Miller and K. E. Drexler. Comparative Ecology, a Computational Perspective. In B. A. Huberman, editor, *The Ecology of Computation, Studies in Computer Science and Artificial Intelligence 2*, pages 51–76, 1988. North Holland, New York.

17. D. J. Montana. Strongly Typed Genetic Programming. *Evolutionary Computation*, 3(2):199–230, 1994.

18. U. M. O'Reilly and F. Oppacher. Program Search with a Hierarchical Variable Representation: Genetic Programming, Simulated Annealing, and Hill Climbing. In H. P. Schwefel and R. Manner, editors, *Parallel Problem Solving from Nature -PPSN1, Lecture Notes in Computer Science Vol 866 pp 397-406*. Springer-Verlag, Berlin, 1994.

19. T.S. Ray. An Approach to the Synthesis of Life. In C. Langton and C. Taylor, editors, *Artificial Life II*, volume XI, pages 371–408, 1991. Addison-Wesley, Redwood City, CA.
20. S. D. Whitehead and D. H. Ballard. Learning to Perceive and Act. *Machine Learning*, 7(1):45–83, 1991.
21. T. Winograd. *Understanding Natural Language*. Academic Press, New York, 1972.

Natural Computing Series

W.M. Spears: Evolutionary Algorithms. The Role of Mutation and Recombination. XIV, 222 pages, 55 figs., 23 tables. 2000

H.-G. Beyer: The Theory of Evolution Strategies. XIX, 380 pages, 52 figs., 9 tables. 2001

L. Kallel, B. Naudts, A. Rogers (Eds.): Theoretical Aspects of Evolutionary Computing. X, 497 pages. 2001

M. Hirvensalo: Quantum Computing. XI, 190 pages. 2001

G. Păun: Membrane Computing. An Introduction. XI, 429 pages, 37 figs., 5 tables. 2002

A.A. Freitas: Data Mining and Knowledge Discovery with Evolutionary Algorithms. XIV, 264 pages, 74 figs., 10 tables. 2002

H.-P. Schwefel, I. Wegener, K. Weinert (Eds.): Advances in Computational Intelligence. VIII, 325 pages. 2003

A. Ghosh, S. Tsutsui (Eds.): Advances in Evolutionary Computing. XVI, 1006 pages. 2003

L.F. Landweber, E. Winfree (Eds.): Evolution as Computation. DIMACS Workshop, Princeton, January 1999. XV, 332 pages. 2002

M. Amos: Theoretical and Experimental DNA Computation. Approx. 200 pages. 2003

Printing (Computer to Plate): Saladruck Berlin
Binding: Stürtz AG, Würzburg